Effects of Temperature on Geotechnical Engineering

Effects of Temperature on Geotechnical Engineering

Guest Editors

Jiliang Pan
Jun Lu
Zhennan Zhu

MDPI

Basel • Beijing • Wuhan • Barcelona • Belgrade • Novi Sad • Cluj • Manchester

Guest Editors

Jiliang Pan
School of Civil and Resource Engineering
University of Science and Technology Beijing
Beijing
China

Jun Lu
Institute of Deep Earth Science and Green Energy Research
Shenzhen University
Shenzhen
China

Zhennan Zhu
School of Mechanics and Civil Engineering
China University of Mining and Technology
Xuzhou
China

Editorial Office
MDPI AG
Grosspeteranlage 5
4052 Basel, Switzerland

This is a reprint of the Special Issue, published open access by the journal *Applied Sciences* (ISSN 2076-3417), freely accessible at: https://www.mdpi.com/journal/applsci/special_issues/C6P3DGH5L4.

For citation purposes, cite each article independently as indicated on the article page online and as indicated below:

Lastname, A.A.; Lastname, B.B. Article Title. *Journal Name* **Year**, *Volume Number*, Page Range.

ISBN 978-3-7258-8495-7 (Hbk)
ISBN 978-3-7258-8496-4 (PDF)
https://doi.org/10.3390/books978-3-7258-8496-4

Contents

About the Editors . . . . **vii**

Jiliang Pan
Effects of Temperature on Geotechnical Engineering
Reprinted from: *Appl. Sci.* **2026**, *16*, 5911, https://doi.org/10.3390/app16125911 . . . . **1**

Xueqing Wu, Baoping Xi, Luhai Chen, Fengnian Wang, Jianing Chi and Yiyang Ge
Thermal Influence Zone Evolution Under THM Coupling in High-Geothermal Tunnels
Reprinted from: *Appl. Sci.* **2026**, *16*, 3952, https://doi.org/10.3390/app16083952 . . . . **6**

Xiaodi Xu, Qiunan Chen and Chen Zhang
Research on Creep Characteristics of Dredged Fill Soil in Humen Port Considering the Effect of Temperature
Reprinted from: *Appl. Sci.* **2026**, *16*, 2820, https://doi.org/10.3390/app16062820 . . . . **25**

Yuxi Guo, Yan Qin, Nengxiong Xu, Huayang Lei, Junhui Xu, Bin Zhang, et al.
The Influence of Periodic Temperature on Salt Rock Acoustic Emission, Strength, and Deformation Characteristics
Reprinted from: *Appl. Sci.* **2025**, *15*, 8848, https://doi.org/10.3390/app15168848 . . . . **44**

Jun Zhang, Jiguo Liu, Shuoren Fu and Zequn Hong
Theoretical Prediction and Safety Evaluation of Adjacent Pipeline Deformation Caused by Connecting Channel Excavation Reinforced with Freezing Method
Reprinted from: *Appl. Sci.* **2024**, *14*, 9274, https://doi.org/10.3390/app14209274 . . . . **65**

Yonglai Zheng, Zhengxie Zhang, Chenyu Hou, Tanbo Pan, Xin Lan and Yifan Ji
Experimental Study on Damage Constitutive Model of Rock under Thermo-Confining Pressure Coupling
Reprinted from: *Appl. Sci.* **2024**, *14*, 9122, https://doi.org/10.3390/app14199122 . . . . **81**

Haisheng Liu, Wan Sun, Jun Zheng and Bin Dou
Adaptive Kriging-Based Heat Production Performance Optimization for a Two-Horizontal-Well Geothermal System
Reprinted from: *Appl. Sci.* **2024**, *14*, 6415, https://doi.org/10.3390/app14156415 . . . . **110**

Yuegang Li, Mingyang Wu, Haoyong Huang, Yintong Guo, Yujie Wang, Junchuan Gui and Jun Lu
A Study on Three-Dimensional Multi-Cluster Fracturing Simulation under the Influence of Natural Fractures
Reprinted from: *Appl. Sci.* **2024**, *14*, 6342, https://doi.org/10.3390/app14146342 . . . . **125**

Yan Qin, Linqing Wu, Qiong Wu, Nengxiong Xu, Guanjun Cai, Yuxi Guo and Wenjing Zhou
An Experimental Study on the Physical and Mechanical Properties of Granite after High-Temperature Treatment Considering Anisotropy
Reprinted from: *Appl. Sci.* **2024**, *14*, 5585, https://doi.org/10.3390/app14135585 . . . . **145**

Zhongnian Yang, Zhenxing Sun, Xianzhang Ling, Guojun Cai, Rongchang Wang and Xiang Meng
Influence of Mixing Rubber Fibers on the Mechanical Properties of Expansive Clay under Freeze–Thaw Cycles
Reprinted from: *Appl. Sci.* **2024**, *14*, 5437, https://doi.org/10.3390/app14135437 . . . . **168**

Yaoxin Li, Zhibin Wang, Haiqing Cao and Tingyao Wu
Constitutive Characteristics of Rock Damage under Freeze–Thaw Cycles
Reprinted from: *Appl. Sci.* **2024**, *14*, 4627, https://doi.org/10.3390/app14114627 . . . . **183**

About the Editors

Jiliang Pan

Jiliang Pan is an Associate Professor and a Master's supervisor at the School of Resources and Safety Engineering, University of Science and Technology Beijing (USTB). He is also a council member of the Beijing Energy and Environmental Society. He received his Ph.D. in Civil Engineering from USTB in 2022 and conducted joint doctoral research at the University of British Columbia. His research focuses on mining rock mechanics and surrounding rock control, multiscale rock damage mechanics, and the efficient exploitation of geological energy and mineral resources. He has been selected for several competitive talent programs, including the Young Elite Scientist Sponsorship Program of the Beijing Association for Science and Technology and the National Postdoctoral Researcher Program of China. He has led more than ten research projects, including projects funded by the National Natural Science Foundation of China, the sub-project of the National Science and Technology Major Project, and the China Postdoctoral Science Foundation. He has published over 50 papers in peer-reviewed journals and has edited or co-authored three academic monographs.

Jun Lu

Jun Lu is an Assistant Professor and a Doctoral/Master's supervisor at the College of Civil and Transportation Engineering, Shenzhen University. He received his Ph.D. degree in Mining Engineering from Chongqing University in 2020. His research interests cover deep-mine rock mechanics, the prevention and control of mine dynamic disasters, and the high-efficiency exploitation of deep underground energy and mineral resources. He has been selected into the Young Talent Support Project of the China Association for Science and Technology, and recognized as a Class C talent under Shenzhen's Overseas High-Caliber Personnel Peacock Program. He has published more than 70 academic papers in peer-reviewed journals including *IJRMMS*, *RMRE*, and *TUST*. He has presided over more than ten research projects, including projects funded by the National Natural Science Foundation of China, sub-projects of the National Major Science and Technology Program for Deep Earth, and Special and General Grants from the China Postdoctoral Science Foundation.

Zhennan Zhu

Zhennan Zhu is an Assistant Professor and a Master's supervisor at the School of Mechanics and Civil Engineering, China University of Mining and Technology. He received his B.S. and Ph.D. degrees from Anhui University of Science and Technology and China University of Geosciences (Wuhan) in 2014 and 2021, respectively. His research focuses on high-temperature rock mechanics testing and simulation, rock constitutive theory and testing methods, multi-field coupled deformation and the failure mechanisms of rock mass, and geothermal energy extraction. He has led over eight projects funded by the National Natural Science Foundation of China, the China Postdoctoral Science Foundation, and key state laboratories. Zhu has published over 30 SCI/EI journal papers, including more than 19 as the first author. He has published extensively in prestigious geotechnical journals, including *Renewable Energy, Rock Mechanics and Rock Engineering, Journal of Cleaner Production, Geothermics*, etc.

applied sciences

MDPI

Editorial

Effects of Temperature on Geotechnical Engineering

Jiliang Pan

School of Resources and Safety Engineering, University of Science and Technology Beijing, Beijing 100083, China; panjiliang@ustb.edu.cn

1. Introduction

Temperature is a fundamental environmental and engineering variable in geotechnical problems. Its influence is reflected not only in the strength, stiffness, deformation, permeability, and durability of rocks and soils, but also in the long-term stability and serviceability of geotechnical engineering structures. With the expansion of human engineering activities into cold polar regions, high-altitude permafrost areas, deep hot rock formations, and even extreme environments associated with extraterrestrial construction, geotechnical engineering is increasingly exposed to complex and severe temperature conditions. In addition, emerging engineering fields, such as geological energy storage, geothermal resource development, natural gas hydrate extraction, and the geological disposal of high-level radioactive waste, have created new challenges related to temperature-induced material degradation, multi-field coupling, and long-term safety.

Temperature affects geotechnical engineering in two distinct but closely related ways. On the one hand, it acts as a passive environmental factor. Under low-temperature conditions, freeze–thaw cycles can cause frost heave, thaw settlement, pore-water phase transition, crack propagation, and progressive degradation of frozen soils and rocks [1,2]. Under high-temperature conditions, thermal expansion mismatch among mineral grains, thermal stress accumulation, and thermally induced cracking may significantly alter rock strength, deformation, and permeability [3,4]. In cyclic thermal environments, repeated temperature fluctuations may lead to fatigue damage, deterioration of mechanical properties, and changes in failure mode [5]. In deep underground engineering, temperature rarely acts alone; instead, it interacts with in situ stress, groundwater seepage, chemical environment, and excavation disturbance, forming complex thermo–hydro–mechanical (THM) or thermo–hydro–mechanical–chemical (THMC) coupling processes [6].

On the other hand, temperature can also serve as a controllable engineering factor for improving stratum stability, assisting mechanical rock breaking, optimizing thermal recovery performance, or assessing the influence range of thermal disturbance. Artificial ground freezing is a representative example, in which low temperature is deliberately used to convert water-bearing soft ground into a temporary frozen structure with improved strength, stiffness, and water-tightness [7,8]. This method is widely used in tunnel construction, shaft excavation, connecting-channel construction, and underground space development under difficult hydrogeological conditions. Similarly, high temperature or controlled temperature gradients may be used to assist rock breaking by inducing thermal stress, weakening mineral bonding, promoting crack initiation, and reducing mechanical cutting or drilling resistance [9,10]. Such temperature-assisted rock breaking has potential significance for deep mining, geothermal reservoir stimulation, hard-rock excavation, and planetary subsurface exploration.

This Special Issue was organized to present recent advances in understanding and exploring the influence of temperature on geological materials and geotechnical systems. The published papers cover freeze–thaw damage, high-temperature rock mechanics, temperature-confining pressure coupling, periodic thermal fatigue, temperature-dependent soil creep, artificial freezing construction, deep reservoir fracturing, geothermal system optimization, and THM coupling behavior in high-geothermal tunnels. From the perspectives of material testing, theoretical modeling, numerical simulation, and engineering application, these papers demonstrate the complexity and diversity of temperature effects in geotechnical engineering. Together, these studies provide valuable experimental data, theoretical models, numerical methods, and engineering insights for advancing this important field.

2. An Overview of Published Articles

2.1. Freeze–Thaw Deterioration

Li et al. (contribution 1) investigated the constitutive characteristics of rock damage under freeze–thaw cycles. By introducing a cumulative damage variable associated with the number of freeze–thaw cycles and assuming that rock strength follows a Weibull distribution, they established a cumulative damage constitutive model for rock shear strength attenuation based on the Mohr–Coulomb criterion. The proposed model can describe the mechanical behavior of rocks under different confining pressures and freeze–thaw cycles, including strain softening and residual strength characteristics. This work provides a useful theoretical framework for evaluating freeze–thaw-induced deterioration in cold-region rock engineering.

Yang et al. (contribution 2) examined the mechanical properties of rubber fiber-reinforced expansive clay subjected to freeze–thaw cycles. Their triaxial test results showed that adding rubber fibers from waste tires significantly reduces the adverse effects of freeze–thaw cycles on shear strength and elastic modulus. In contrast to unreinforced expansive clay, whose strength and stiffness decrease markedly with increasing freeze–thaw cycles, the reinforced clay exhibited only slight deterioration. This study not only contributes to the improvement of expansive soils in seasonal frozen regions but also provides a sustainable approach for recycling waste tires in geotechnical engineering.

2.2. High-Temperature Mechanical Behavior

Qin et al. (contribution 3) studied the physical and mechanical properties of granite after high-temperature treatment while considering anisotropy. Through wave velocity tests, microscopic observations, and uniaxial and triaxial compression tests on samples taken from three orthogonal directions, they showed that high temperature changes the density and orientation of thermal cracks in granite. Temperatures below 400 °C increased the anisotropy of granite, while more severe thermal damage occurred at higher temperatures. The study further showed that samples taken parallel to the dominant crack direction exhibited better mechanical properties under the same temperature condition. These findings are important for dry hot rock exploitation, nuclear waste disposal, and deep rock engineering where granite is often used as a key host or reservoir rock.

Zheng et al. (contribution 4) investigated the deformation and failure characteristics of four rock types—granite, red sandstone, gray sandstone, and shale—under coupled temperature and confining pressure. Their results indicated that high temperature causes internal structural damage and crack propagation, reducing compressive strength and elastic modulus, whereas high confining pressure can inhibit crack propagation and improve rock deformation capacity. They also found that different rock types exhibit distinct failure modes under thermo-confining pressure coupling. Based on the experimental results, they modified the Drucker–Prager criterion parameters and developed a damage constitutive

https://doi.org/10.3390/app16125911

model that can represent the stress–strain behavior of rocks under coupled temperature and pressure conditions.

2.3. Cyclic and Time-Dependent Thermal Effects

Guo et al. (contribution 5) investigated the influence of periodic temperature on acoustic emission, strength, and deformation characteristics of salt rock. Motivated by the cyclic injection and withdrawal processes in salt cavern gas storage, they performed uniaxial compression tests on salt rock samples after thermal fatigue treatment and combined acoustic emission monitoring with digital image correlation. Their results showed that increasing the number of cycles and the upper temperature limit promotes microcrack propagation, structural damage accumulation, radial deformation, and local displacement concentration. After 40 cycles, the compressive strength and elastic modulus decreased by 23.8% and 27.4%, respectively, and the failure mode gradually shifted from tension-dominated to tension–shear composite. This study provides important support for evaluating the long-term safety of underground salt cavern storage.

Xu et al. (contribution 6) analyzed the creep characteristics of dredged fill soil from Humen Port under different temperatures. Through triaxial creep tests under drained and undrained conditions, they showed that increasing temperature significantly accelerates creep deformation. The dredged fill soil was characterized by high water content, large void ratio, high compressibility, low shear strength, and low permeability, making its long-term settlement behavior highly sensitive to temperature. Based on the experimental results, they developed a Burgers-type creep constitutive model considering temperature effects, providing a useful tool for long-term stability assessment and settlement prediction of reclaimed foundations in coastal environments.

2.4. Numerical Methods and Engineering Insights

Zhang et al. (contribution 7) studied the deformation of adjacent pipelines caused by connecting-channel excavation reinforced with the freezing method. By combining field measurements, numerical simulation, and theoretical analysis based on Euler–Bernoulli beam theory, they evaluated the effects of freezing temperature and excavation parameters on pipeline settlement. Their results showed that excavation rate strongly affects pipeline deformation: when the excavation rate exceeds 1.0 m/d, settlement increases rapidly. Lower freezing temperatures can form thicker and stronger frozen soil walls, thereby enhancing support capacity and reducing ground disturbance. This paper highlights the active role of temperature control in underground construction safety.

Li et al. (contribution 8) developed a three-dimensional finite–discrete element model (FDEM) to simulate multi-cluster hydraulic fracturing under the influence of natural fractures. Their results demonstrated that increasing the number of fracturing clusters and natural fractures generally enlarges the total artificial fracture area and promotes more complex fracture networks. The presence of natural fractures can also alter artificial fracture propagation paths and failure modes, producing phenomena such as single-wing propagation, bifurcation, and fracture bending. Although the study focuses on deep shale reservoirs, its findings are also relevant to geothermal reservoir stimulation and deep energy engineering, where fracture-network geometry strongly controls fluid flow and heat extraction.

Liu et al. (contribution 9) proposed an adaptive Kriging-based optimization method for heat production performance in a two-horizontal-well geothermal system. To reduce the high computational cost of numerical simulations, they combined computer simulation with Kriging approximation and developed a parameterized lower confidence bounding sampling scheme to update the surrogate model adaptively. Their optimization framework

https://doi.org/10.3390/app16125911

considered fracture spacing, injection flow rate, and well spacing, with total power generation as the objective. The study provides an efficient computational strategy for geothermal system design and performance evaluation.

Wu et al. (contribution 10) studied the evolution of the thermal influence zone in high-geothermal tunnels under THM coupling. Unlike many conventional models that assume a uniform initial temperature field, their model incorporated an in situ geothermal gradient and quantitatively defined the thermal influence zone using an equivalent-radius method. The results showed that the thermal influence zone expands nonlinearly with increasing initial rock temperature and gradually stabilizes over time. Temperature and pore water pressure both promote plastic-zone development, but temperature plays the dominant role under the studied conditions by inducing thermal stress and degrading mechanical properties. This work provides quantitative guidance for support design and stability control in high-geothermal tunnels.

3. Conclusions

The papers in this Special Issue advance the field by deepening mechanistic understanding of temperature-induced damage, reinforcement, creep, anisotropy, and coupled deformation in geological materials; developing constitutive and numerical models for temperature-dependent behavior under realistic stress, drainage, and engineering conditions; and linking laboratory observations with field-scale problems such as frozen-ground construction, high-geothermal tunnels, geothermal energy extraction, salt cavern storage, reclaimed foundations, and deep reservoir stimulation.

Collectively, these studies show that temperature effects in geotechnical engineering are multi-scale, multi-process, and strongly coupled. At the material scale, freeze–thaw cycles, high-temperature treatment, cyclic thermal loading, and temperature-dependent creep can significantly alter the pore structure, crack development, wave velocity, strength, stiffness, deformation behavior, and failure mode of rocks and soils. At the modeling scale, temperature-dependent damage models, creep models, surrogate optimization methods, and THM coupling numerical models provide increasingly powerful tools for describing and predicting complex geotechnical responses. At the engineering scale, the published studies show that temperature is not only a source of deterioration and risk but also a controllable variable in artificial freezing, geothermal energy extraction, and tunnel thermal disturbance management. Overall, this Special Issue highlights the importance of temperature as a central variable in modern geotechnical engineering and provides a foundation for future studies on safer, more durable, and more sustainable geotechnical systems under complex environments.

Acknowledgments: Thanks to all the authors and peer reviewers for their valuable contributions to this Special Issue. We also appreciate the support of the Applied Sciences Editorial Office throughout the review and publication process.

Conflicts of Interest: The author declares no conflicts of interest.

List of Contributions:

1. Li, Y.; Wang, Z.; Cao, H.; Wu, T. Constitutive characteristics of rock damage under freeze–thaw cycles. *Appl. Sci.* **2024**, *14*, 4627.
2. Yang, Z.; Sun, Z.; Ling, X.; Cai, G.; Wang, R.; Meng, X. Influence of Mixing Rubber Fibers on the Mechanical Properties of Expansive Clay under Freeze–Thaw Cycles. *Appl. Sci.* **2024**, *14*, 5437.
3. Qin, Y.; Wu, L.; Wu, Q.; Xu, N.; Cai, G.; Guo, Y.; Zhou, W. An experimental study on the physical and mechanical properties of granite after high-temperature treatment considering anisotropy. *Appl. Sci.* **2024**, *14*, 5585.

https://doi.org/10.3390/app16125911

4. Zheng, Y.; Zhang, Z.; Hou, C.; Pan, T.; Lan, X.; Ji, Y. Experimental study on damage constitutive model of rock under thermo-confining pressure coupling. *Appl. Sci.* **2024**, *14*, 9122.
5. Guo, Y.; Qin, Y.; Xu, N.; Lei, H.; Xu, J.; Zhang, B.; Feng, S.; Chen, L. The Influence of Periodic Temperature on Salt Rock Acoustic Emission, Strength, and Deformation Characteristics. *Appl. Sci.* **2025**, *15*, 8848.
6. Xu, X.; Chen, Q.; Zhang, C. Research on Creep Characteristics of Dredged Fill Soil in Humen Port Considering the Effect of Temperature. *Appl. Sci.* **2026**, *16*, 2820.
7. Zhang, J.; Liu, J.; Fu, S.; Hong, Z. Theoretical Prediction and Safety Evaluation of Adjacent Pipeline Deformation Caused by Connecting Channel Excavation Reinforced with Freezing Method. *Appl. Sci.* **2024**, *14*, 9274.
8. Li, Y.; Wu, M.; Huang, H.; Guo, Y.; Wang, Y.; Gui, J.; Lu, J. A study on three-dimensional multi-cluster fracturing simulation under the influence of natural fractures. *Appl. Sci.* **2024**, *14*, 6342.
9. Liu, H.; Sun, W.; Zheng, J.; Dou, B. Adaptive Kriging-Based Heat Production Performance Optimization for a Two-Horizontal-Well Geothermal System. *Appl. Sci.* **2024**, *14*, 6415.
10. Wu, X.; Xi, B.; Chen, L.; Wang, F.; Chi, J.; Ge, Y. Thermal Influence Zone Evolution Under THM Coupling in High-Geothermal Tunnels. *Appl. Sci.* **2026**, *16*, 3952.

References

1. Qiu, P.; Tang, L.; Zheng, J.; Wang, W.; Li, Y.; Li, G.; Jin, L.; Yu, Y.; Duan, X. Experimental investigations on the shear strength and creep properties of soil-rock mixture under freeze-thaw cycles. *Cold Reg. Sci. Technol.* **2024**, *217*, 104037. [CrossRef]
2. Zhou, Z.; Xing, K.; Yang, H.; Wang, H. Damage mechanism of soil-rock mixture after freeze-thaw cycles. *J. Cent. South Univ.* **2019**, *26*, 13–24. [CrossRef]
3. Pan, J.; Feng, Z.; Zhang, Y.; Xi, X.; Miao, S.; Cai, M. Experimental study on evaluation of porosity, thermal conductivity, UCS, and elastic modulus of granite after thermal and chemical treatments by using P-wave velocity. *Geoenergy Sci. Eng.* **2023**, *230*, 212184. [CrossRef]
4. Pan, J.; Xi, X.; Wu, X.; Guo, Q.; Ren, F.; Cai, M. Physical properties evolution and microscopic mechanisms of granite modified by thermal and chemical stimulation. *Case Stud. Therm. Eng.* **2023**, *41*, 102633. [CrossRef]
5. Di Donna, A.; Laloui, L. Response of soil subjected to thermal cyclic loading: Experimental and constitutive study. *Eng. Geol.* **2015**, *190*, 65–76. [CrossRef]
6. Birkholzer, J.T.; Bond, A.E. DECOVALEX-2019: An international collaboration for advancing the understanding and modeling of coupled thermo-hydro-mechanical-chemical (THMC) processes in geological systems. *Int. J. Rock Mech. Min. Sci.* **2022**, *154*, 105097. [CrossRef]
7. Gong, Z.; Wang, G.; Dai, X. Field monitoring and numerical study of an artificial ground freezing reinforcement project for cross passage. *Appl. Sci.* **2025**, *15*, 9547. [CrossRef]
8. Alzoubi, M.A.; Xu, M.; Hassani, F.P.; Poncet, S.; Sasmito, A.P. Artificial ground freezing: A review of thermal and hydraulic aspects. *Tunn. Undergr. Space Technol.* **2020**, *104*, 103534. [CrossRef]
9. Li, J.; Xu, P.; Jiang, Y.; Fu, H. Response of hard rock to heating–cooling treatment and optimal temperature scheme for assisting rock-breaking. *Rock. Mech. Rock. Eng.* **2026**, *59*, 3959–3984. [CrossRef]
10. Cai, C.; Cao, W.Y.; Yang, X.P.; Xie, Q.G.; Li, B.R.; Tan, Z.B.; Zhang, C.L.; Peng, C.; Chen, H.; Zhao, Y.L. Rock breakage and temperature rising of rock cutting after high-temperature treatment. *Pet. Sci.* **2025**, *22*, 1289–1306. [CrossRef]

https://doi.org/10.3390/app16125911

applied sciences

MDPI

Article

Thermal Influence Zone Evolution Under THM Coupling in High-Geothermal Tunnels

Xueqing Wu [1], Baoping Xi [1,2,*], Luhai Chen [1], Fengnian Wang [3], Jianing Chi [1] and Yiyang Ge [1]

1 Department of Mining Engineering, Taiyuan University of Technology, Taiyuan 030024, China; 2023510709@link.tyut.edu.cn (X.W.)
2 Laboratory of In-Situ Property-Improving Mining of Ministry of Education, Taiyuan 030024, China
3 Transportation Industry Key Laboratory of Highway Construction and Maintenance Technology in Loess Areas, Shanxi Provincial Transportation Technology Research and Development Co., Ltd., Taiyuan 030024, China
* Correspondence: xibaoping@163.com; Tel.: +86-131-1106-1893

Abstract

High-geothermal tunnels are subjected to complex thermo–hydro–mechanical (THM) coupling effects, where the interaction of temperature, seepage, and stress significantly influences the stability of surrounding rock. To address the limitations of conventional models assuming uniform initial temperature, a THM-coupled numerical model incorporating an in situ temperature gradient is established based on the Sangzhuling Tunnel. The concept of the thermal influence zone is quantitatively defined by an equivalent-radius method, and its spatiotemporal evolution is systematically investigated. In addition, the distinct roles of temperature and pore water pressure in controlling deformation and plastic-zone evolution are comparatively clarified. The results show that the thermal influence zone expands nonlinearly with increasing initial rock temperature and gradually stabilizes over time. Temperature and pore water pressure both promote the development of the plastic zone, which predominantly propagates along directions approximately 45° to the horizontal. Under the geological and boundary conditions considered in this study, temperature plays a dominant role by inducing thermal stress and degrading mechanical properties, leading to significant expansion of the plastic zone and increased vault deformation. In contrast, pore water pressure mainly reduces effective stress, thereby influencing deformation distribution, especially at the tunnel invert. Overall, THM coupling significantly amplifies surrounding rock failure compared with single-field conditions. The findings provide quantitative insights into the evolution of the thermal influence zone and its coupled control on deformation and plasticity, offering a theoretical basis for support design and stability control in high-geothermal tunnels.

Keywords: high-temperature tunnel; thermal influence zone; THM coupling; surrounding rock mechanics; plastic zone

1. Introduction

With the continuous advancement of the Western Development Strategy in China, a large number of infrastructure projects have been launched, and numerous deep tunnels are confronted with severe high ground temperature challenges [1]. Such a high-temperature environment has become one of the key factors restricting the construction and long-term operational safety of deep tunnels. The complex interaction between the high-temperature characteristics of deep rock masses, the surrounding rock stress field, and the groundwater seepage field (i.e., the thermal–hydraulical–mechanical (THM) coupling problem) fundamentally changes the mechanical properties and stability of surrounding rock [2], bringing numerous problems and challenges to engineering construction.

To investigate the physical and mechanical properties of rocks under multi-physics coupling, extensive studies have been conducted by scholars worldwide. Studies by Dwivedi and Xi et al. [3,4] show that high temperature causes non-uniform expansion of mineral grains in hard rocks such as granite, reducing the elastic modulus and increasing the Poisson's ratio. The combination of high ground temperature and high stress significantly increases the risk of rockburst in brittle surrounding rock. Jia et al. [5] found that with the increase in the initial high temperature of granite, the energy dissipated during natural cooling increases while its uniaxial compressive strength decreases. Through triaxial compression tests, Long [6] discovered that within the temperature range of 25 °C to 90 °C, the peak strength and residual strength of sandstone first increase and then decrease with rising temperature. In the water pressure range of 2 to 8 MPa, the peak strength and residual strength of sandstone generally show a downward trend with increasing water pressure, while strength enhancement occurs under high confining pressure and high temperature. Tan et al. [7] described the deformation and failure of deep tunnel surrounding rock under thermal–mechanical coupling. Liu, Shu et al. [8,9] investigated the response law of thermal insulation layers in high-temperature tunnel linings and the deformation and failure of lining segment joints in fire tunnels based on thermal–mechanical coupling. Based on thermal–hydraulical coupling, Zhang et al. [10] revealed that fracture water flow exerts a considerable influence on the temperature field distribution of rock masses. Zhao et al. [11] found that the damage area induced by the hydraulical–mechanical coupling model during tunnel excavation in fractured rock masses is significantly larger than that of the solid model. Recent studies in underground coal gasification (UCG) have also highlighted the importance of coupled thermo–mechanical effects on rock stability. For example, numerical simulation results reported by Sakhno et al. [12] indicate that the thermo–mechanically affected zone around the UCG reactor is spatially limited, while thermal stresses mainly influence the local stability of the cavity roof and safety pillar. The findings confirm that thermal disturbance can play an important role in underground rock stability problems, although its extent and engineering significance depend strongly on the specific geological setting and engineering scenario.

The above review indicates that most existing studies focus on the mechanism of single or dual physical fields acting on rocks. Although some scholars have considered the THM coupling effect, current research still has limitations. First, most existing numerical simulation studies assume a uniform initial temperature inside the surrounding rock equal to the original rock temperature [13–15], which is inconsistent with the natural temperature gradient of surrounding rock in practical engineering. Such simplification may lead to deviations between analytical results and engineering practice, resulting in a lack of quantitative characterization for the spatial scope of surrounding rock disturbed by temperature (i.e., the thermal influence zone) in high-temperature tunnels under actual geological conditions, which is critical for determining the support scope and thermal damage control. Second, the independent and coupled influences of temperature and pore water pressure on plastic zone evolution and deformation field distribution, as well as their internal mechanisms and contribution rates, have not been systematically and thoroughly revealed. Clarifying these issues is a theoretical prerequisite for realizing the transition from empirical design to precise regulation in high-temperature tunnels.

In summary, the contribution of this study is not the development of a new THM governing framework itself, but the extension of THM analysis to a more realistic high-geothermal tunnel condition by incorporating an in situ geothermal gradient, propos-

https://doi.org/10.3390/app16083952

ing a quantitative definition of the thermal influence zone (TIZ), and systematically clarifying the respective roles of temperature and pore water pressure in surrounding-rock deformation and plastic-zone evolution. Using the Sangzhuling Tunnel of the Lhasa–Nyingchi Railway as the engineering background, a THM-coupled numerical model is established and validated against in situ stress measurements. On this basis, the spatiotemporal evolution of the TIZ is quantified, and its engineering significance for support design and stability control is discussed.

2. Methodology

2.1. Problem Description and Conceptual Framework

High-geothermal tunnels are characterized by the coupled interaction of temperature, seepage, and stress fields, which jointly control the deformation and failure of surrounding rock. After tunnel excavation, ventilation induces a significant thermal disturbance, forming a thermal influence zone within the surrounding rock. In this study, this region is defined as the thermal influence zone, referring to the area where the rock temperature decreases relative to the initial geothermal state. Unlike conventional studies that assume a uniform initial temperature field, the present work incorporates an in situ temperature gradient to better represent realistic geological conditions. The coupling processes among the thermal, hydraulic, and mechanical fields include: (1) temperature-dependent variation in rock mechanical properties, (2) pore pressure-induced effective stress changes, and (3) thermal stress generated by temperature gradients (Figure 1). Based on these considerations, a fully coupled THM framework is established to investigate the evolution of the thermal influence zone and the associated mechanical response of surrounding rock.

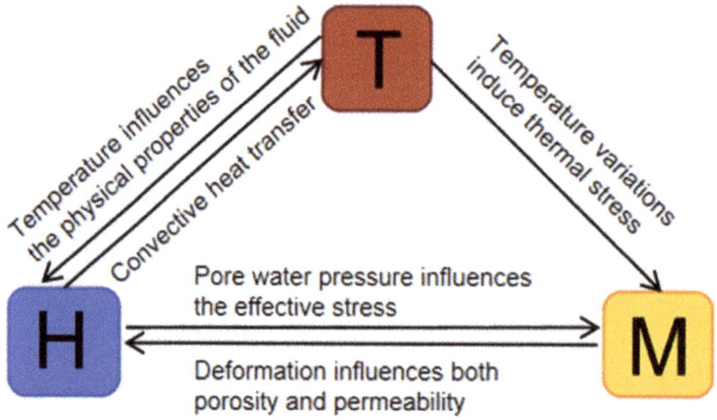

Figure 1. Schematic diagram of THM three-field coupling mechanism.

2.2. Governing Equations of THM Coupling

2.2.1. Heat Transfer Equation

Heat transfer in the surrounding rock is governed by conduction and convection induced by pore water flow, while thermal radiation is neglected. Based on Fourier's law and energy conservation, the governing equation can be expressed as:

$$(\rho C)_{eq}\frac{\partial T}{\partial t} + C_F\rho_F\mu_F\nabla T = \nabla\cdot(k_{eq}\nabla T) \tag{1}$$

where $(\rho C)_{eq}$ is the equivalent volumetric heat capacity of the porous medium at constant pressure, T is the temperature of the porous medium, t is time, C_F is the specific heat capacity, ρ_F is the fluid density, μ_F is the seepage velocity, ∇ is the divergence operator,

https://doi.org/10.3390/app16083952

and k_{eq} is the equivalent thermal conductivity of the porous medium. The temperature dependence of water density and dynamic viscosity is incorporated to capture the influence of thermal conditions on seepage behavior [16,17], as shown in the following equation:

$$\rho_F = \begin{cases} 972.757815 + 0.2084 \times T - 4.0 \times 10^{-4} \times T^2 & ,273.15 \le T < 283.0 \\ 345.28 + 5.749816 \times T - 1.57244 \times 10^{-2} \times T^2 + 1.264375 \times 10^{-5} \times T^3, & 283.0 \le T < 373.16 \end{cases} \quad (2)$$

$$\mu_F = e^{\frac{1.47297892\times10^{11}}{T^4} - \frac{1.56916127\times10^{9}}{T^3} + \frac{6.52798499\times10^{6}}{T^2} - \frac{1.10421451\times10^{4}}{T} - 2.86320560} \quad (3)$$

2.2.2. Seepage Equation

Fluid flow in the porous rock mass follows Darcy's law:

$$\rho_F \mu_F = -K \cdot \nabla h \quad (4)$$

where K is the hydraulic conductivity and h is the hydraulic head.

The continuity equation considering fluid compressibility and rock deformation is given as [18]:

$$S\frac{\partial h}{\partial t} = \nabla \cdot (K \cdot \nabla h) + \alpha_B \frac{\partial \varepsilon}{\partial t} \quad (5)$$

$$S = \rho_F g(\beta_S + \eta \beta_F) \quad (6)$$

$$\frac{\partial \varepsilon}{\partial t} \approx \frac{V_{t-1} - V_t}{V_{t-1}} \frac{1}{\nabla t} \quad (7)$$

where S is the storage coefficient, α_B is the Biot coefficient, $\frac{\partial \varepsilon}{\partial t}$ represents the volumetric change in the porous medium, ε is the volumetric strain, β_S and β_F are the compressibility of the solid matrix and that of the fluid, η is the porosity of the porous medium, V_{t-1} is the element volume at the previous time step $(t-1)$, V_t is the element volume at the current time step (t) after displacement, and Δt is the time step.

2.2.3. Mechanical Equilibrium Equation

The mechanical behavior of surrounding rock is governed by the equilibrium equation incorporating thermal stress and pore pressure [18]:

$$G\frac{\partial^2 d_i}{\partial x_i \partial y_j} + (G + \lambda)\frac{\partial^2 d_j}{\partial x_i \partial y_j} - \alpha_B \frac{\partial p}{\partial x_i} - \beta K_S \frac{\partial T}{\partial x_i} + F = 0 \quad (8)$$

$$\lambda = \frac{E\nu}{(1+\nu)(1-2\nu)} \quad (9)$$

$$K_S = \frac{E}{3(1-2\nu)} \quad (10)$$

where G is the shear modulus of the rock mass, d_i and d_j are the displacements along the *x*-axis and *y*-axis respectively, λ is the Lamé constant, p is the liquid pressure, β is the thermal expansion coefficient of the solid matrix, K_S is the bulk modulus of the solid matrix, F is the body force, E is Young's modulus, and ν is Poisson's ratio.

2.3. Temperature-Dependent Material Properties

The mechanical and hydraulic properties of rock are strongly influenced by temperature. In this study, temperature-dependent relationships for the elastic modulus, Poisson's ratio, and permeability are incorporated based on previous laboratory experimental data [4,19]. Specifically, the elastic modulus decreases with increasing temperature due to thermal damage, while permeability increases as microcracks develop under thermal stress.

https://doi.org/10.3390/app16083952

These relationships are introduced into the THM coupling model through fitted empirical functions, enabling a more realistic representation of rock behavior under high-temperature conditions, as shown in Figures 2 and 3.

$$E = 78.366e^{-0.0015T} \quad (11)$$

$$\mu = 0.0699 \ln T + 0.0028 \quad (12)$$

$$k = 6.42103 + 0.37897e^{\frac{T-20}{90.73292}} \quad (13)$$

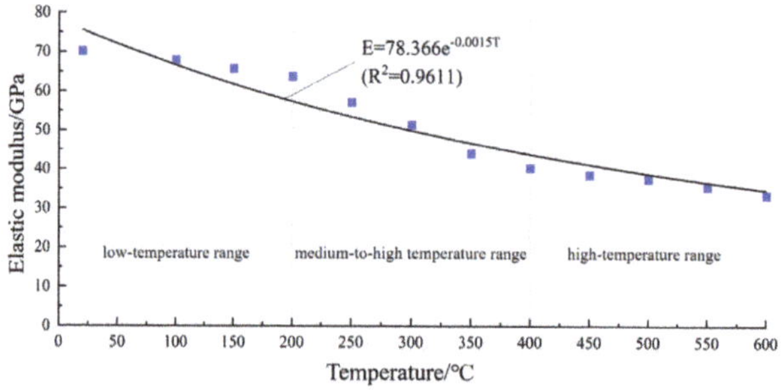

(a)

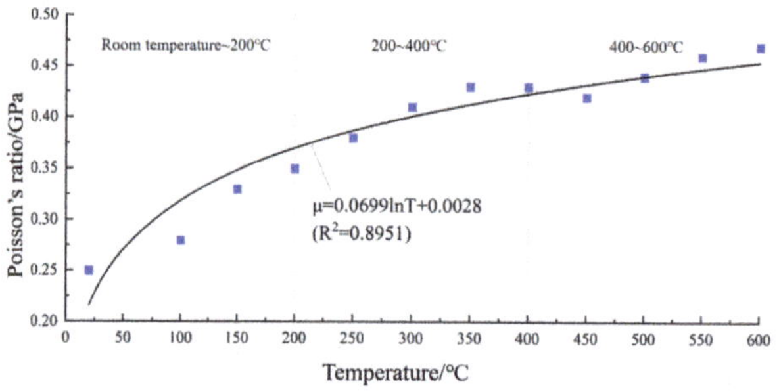

(b)

Figure 2. Thermophysical and mechanical properties of granite at different temperatures. (**a**) Calculated values of elastic modulus. (**b**) Calculated values of Poisson's ratio.

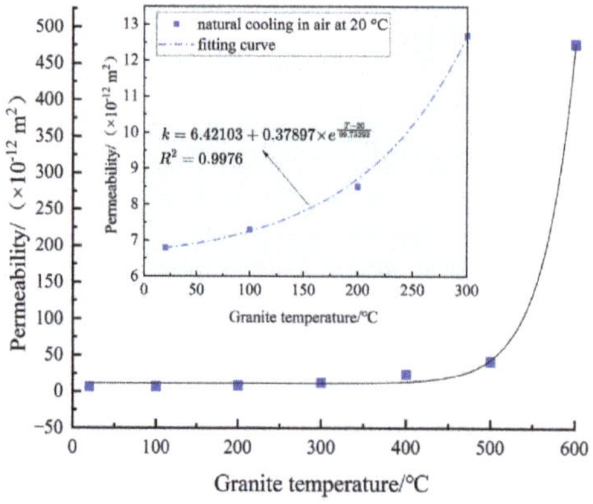

Figure 3. Permeability of granite at different temperatures.

https://doi.org/10.3390/app16083952

2.4. Numerical Model and Boundary Conditions

A two-dimensional plane strain model is established, with the tunnel simplified as a circular opening of radius 5 m. To minimize boundary effects, the model size is set to 100 m × 100 m, corresponding to five times the tunnel diameter. The boundary conditions are defined as follows:

(1) Thermal boundary: A geothermal gradient is applied to represent in situ temperature distribution, while the tunnel boundary is set to a constant temperature (28 °C) to simulate ventilation cooling.
(2) Hydraulic boundary: The model is assumed to be fully saturated. Lateral boundaries are defined as hydraulic boundaries, and pore water pressure is applied at the top boundary.
(3) Mechanical boundary: The bottom boundary is fixed, and the lateral boundaries are constrained with roller supports. Vertical stress is applied at the top boundary to simulate overburden load.

The above thermal, hydraulic and mechanical boundary settings and overall meshing of the numerical model are illustrated schematically in Figure 4. The present model is geometrically static after excavation, but the coupled THM response is solved in a transient manner. The numerical analysis was carried out sequentially. First, the initial geostress field, pore-pressure field, and in situ temperature gradient were established. Second, tunnel excavation was simulated by deactivating the tunnel elements (or equivalently removing the excavation domain) and converting the tunnel boundary into a free mechanical boundary and a no-flow hydraulic boundary. It represents a sealed condition similar to grouting in practice. Third, a constant tunnel-wall temperature of 28 °C was imposed to represent ventilation cooling, and the transient THM response was solved for different ventilation durations.

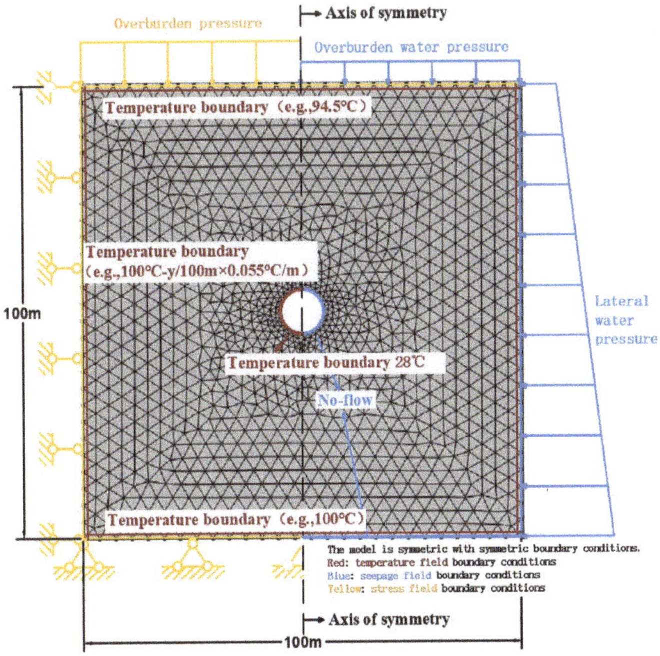

Figure 4. Schematic diagram of the physical model and grid division generation.

In addition, in the present model, the lining system was not explicitly considered. The tunnel boundary was simplified as an exposed excavation surface subjected to ventilation cooling. This treatment is suitable for investigating the intrinsic response of the surrounding rock, but it does not capture the thermal insulation and load-sharing effects of the lining. The influence of lining thickness, stiffness, and contact condition should therefore be addressed in future work.

2.5. Model Validation

To verify the reliability of the proposed THM model, numerical results are compared with in situ stress measurements from the Sangzhuling Tunnel. The maximum buried depth of the tunnel is 1500 m. The in situ stress data used for model validation were taken from hydraulic fracturing measurements reported for borehole D1K-SZLSD-2 near mileage D1K186+327 of the Sangzhuling Tunnel. The borehole had a designed depth of 475 m, an actual drilling depth of 604.45 m, and a static water level of about 65 m. Stress measurements were conducted at five depths, i.e., 205.6, 297.4, 391.8, 476.9, and 582.6 m, and the measured principal stresses were used for model validation [20]. According to the borehole location, the model is established based on the geological conditions at a buried depth of approximately 900 m, with a ground temperature of 56.6 °C and a geothermal gradient of 5.5 °C per 100 m. According to the cited geological investigation, the rock mass near borehole D1K-SZLSD-2 and the studied tunnel section is mainly granite or granodiorite, and the rock mass is relatively intact, with an RQD of about 95%. Relevant parameter values are listed in Table 1 [21].

Table 1. Model calculation parameters.

Young's Modulus E/GPa	**Poisson's Ratio μ**	**Density ρ/(kg·m^{-3})**	**Cohesion C/MPa**	**Internal Friction Angle ψ/°**
35	0.23	2.6×10^3	1.5	50
Coefficient of thermal expansion α/(1/K)	Thermal conductivity λ/(W·(m·K))	Isobaric specific heat capacity c/(J/(kg·K))	Porosity φ	Permeability K/m^2
8×10^{-6}	3.69	630	0.1987	5.05×10^{-18}

An independently established large-scale regional geostress field model within the depth range of 0–900 m shown was built to verify the rationality of the initial stress field and boundary condition settings of the numerical model. The comparison between simulated and measured in situ stress shows good agreement, as shown in Figure 5, indicating that the model can effectively capture the stress distribution characteristics of the surrounding rock under THM coupling conditions. Therefore, the developed model is considered reliable for subsequent analysis. It should be noted that the field comparison with measured in situ stress is used in this study mainly to verify the reasonableness of the initial stress field and boundary-condition setting of the numerical model. Due to the lack of long-term in situ monitoring data for temperature, pore water pressure, and deformation, the present model cannot be regarded as fully validated for all coupled THM responses over time. Rather, the subsequent analyses should be understood as predictions based on a mechanically validated and physically based THM framework under the geological and thermal conditions of the Sangzhuling Tunnel.

https://doi.org/10.3390/app16083952

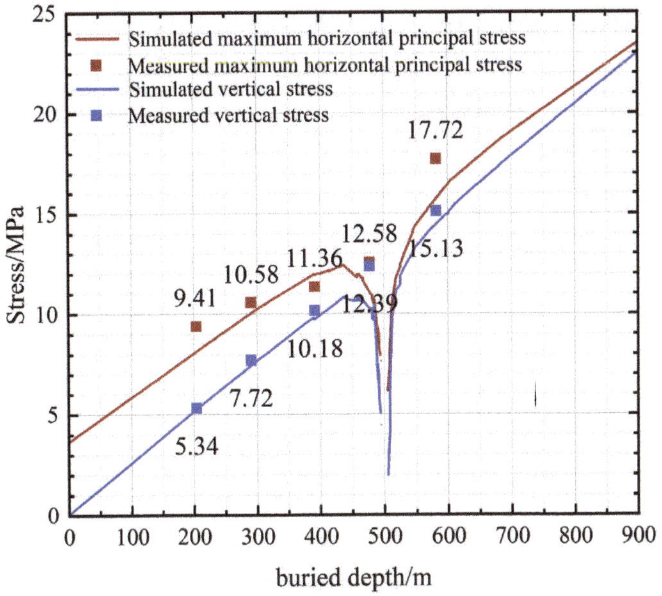

Figure 5. Comparison between measured and simulated in situ stress.

3. Evolution of the Thermal Influence Zone Under THM Coupling

3.1. Definition of the Thermal Influence Zone

After tunnel excavation, ventilation induces heat exchange between the tunnel air and the surrounding rock, leading to a gradual decrease in rock temperature near the tunnel boundary. As heat conduction continues, the temperature disturbance propagates outward into the surrounding rock, forming a thermal influence zone. In this study, the thermal influence zone (TIZ) is defined as the region where the rock temperature decreases relative to the initial geothermal state due to ventilation cooling [22]. Under coupled THM conditions, the evolution of the thermal influence zone is controlled by the combined effects of conductive and convective heat transfer under coupled THM conditions. When the permeability of the surrounding rock is relatively low, heat conduction plays the dominant role in thermal disturbance propagation. However, with tunnel excavation and the associated increase in rock permeability and seepage velocity, convective heat transfer becomes more significant and can further promote the outward expansion of the thermal influence zone. It should also be noted that, in the coupled process, conduction and seepage-enhanced convection are closely interconnected, and their individual contributions are difficult to distinguish strictly. Therefore, the TIZ evolution should be understood as the result of their joint action rather than two completely separable mechanisms. In addition, under THM coupling conditions, temperature variation further induces thermal stress and changes in rock mechanical properties, which may affect the deformation and permeability of the surrounding rock [3,4].

The spatial extent of the TIZ is determined by comparing the initial temperature field prior to excavation with the transient temperature field after excavation and ventilation. Based on the two-dimensional model, the coordinates and temperature values of all grid nodes are extracted to identify regions where the rock temperature is lower than the initial state. These regions, shown as gray areas in Figure 6, represent the actual distribution of the thermal influence zone. For quantitative characterization, an equivalent area method

https://doi.org/10.3390/app16083952

is adopted. The equivalent radius is defined by equating the area of a circle to the actual area of the thermal influence zone. It is introduced as a scalar descriptor of the disturbed thermal domain. It does not represent the actual geometric shape of the TIZ; instead, it converts the irregular influenced area into an equivalent circle with the same area, thereby facilitating comparison among different times and initial temperature conditions. The white solid line in Figure 6 indicates the corresponding equivalent circular boundary, and the equivalent radius is calculated as follows:

$$r_{eq} = \sqrt{\frac{A}{\pi}} \tag{14}$$

where r_{eq} is the equivalent radius of the thermal influence zone, and A is the actual area of the zone in which the rock temperature is lower than the initial geothermal state.

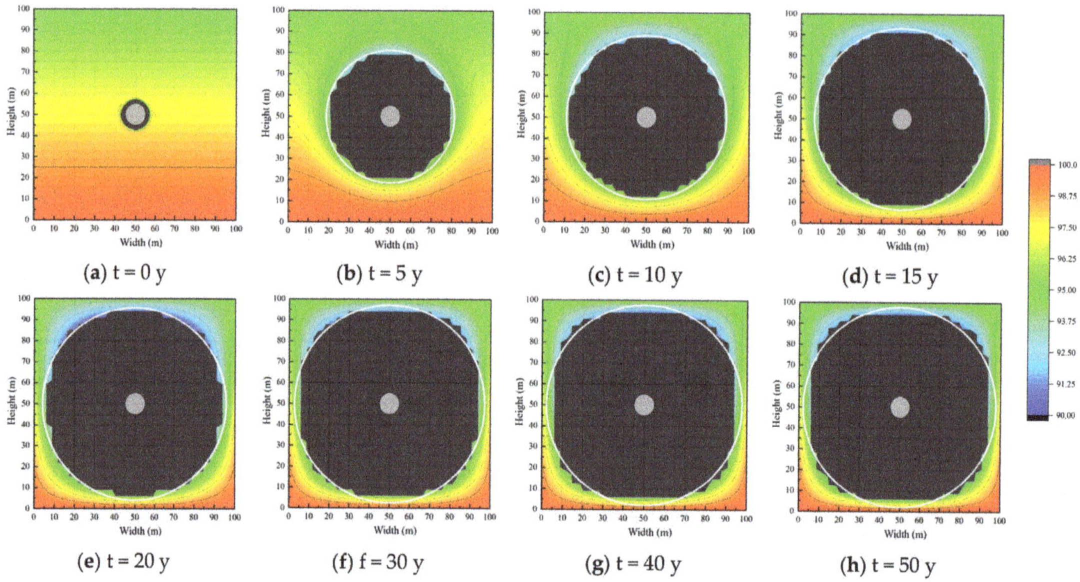

Figure 6. Temperature variation process of tunnel surrounding rock within 1~50 years (including the boundary of the thermal influence zone). The background lines are coordinate references rather than mesh lines.

3.2. Temporal Evolution of the Thermal Influence Zone

The temporal evolution of the thermal influence zone under THM coupling is illustrated in Figure 6, which presents the temperature distribution around the tunnel at different ventilation times. The results show that the temperature near the tunnel boundary decreases rapidly during the early stage of ventilation. As time progresses, the temperature disturbance gradually propagates into the surrounding rock, resulting in a continuous expansion of the thermal influence zone. However, the expansion rate of the TIZ decreases with time. In the early stage, from 0 to 5 y, the TIZ expands rapidly. As the temperature field gradually approaches equilibrium, the temperature gradient decreases, and the growth rate of the TIZ becomes progressively slower. After a sufficiently long ventilation period, from 1 year to 30 years, the temperature field tends to stabilize, and the expansion of the thermal influence zone becomes negligible.

As shown in Figure 7a, the TIZ radius expands rapidly during the initial cooling stage due to the large temperature difference between the surrounding rock and the tunnel wall.

After 15 years, its radius reaches 43.33 m. As the temperature gradient decreases, the heat flux density declines, leading to a reduced expansion rate. After 35 years, the radius increases to 47.78 m, and eventually stabilizes at approximately 48.07 m as the surrounding rock approaches thermal equilibrium. This trend is consistent with previously reported temperature field evolution patterns [15].

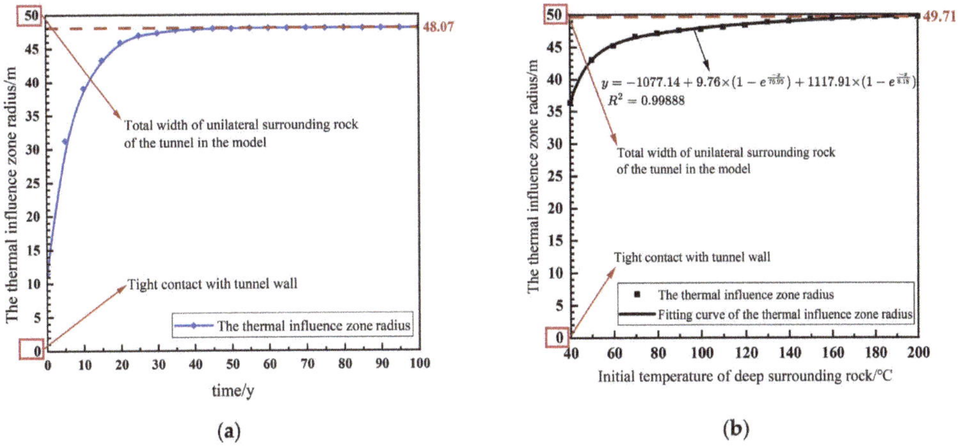

Figure 7. Variation in the radius of the thermal influence zone with time and temperature. (**a**) Relationship curve between TIZ radius and time. (**b**) Relationship curve between the TIZ radius and the initial temperature of deep surrounding rock.

The initial rock temperature is one of the key factors controlling the development of the thermal influence zone. Figure 7b presents the TIZ range under different initial geothermal conditions. The results indicate that the extent of the TIZ increases nonlinearly with increasing initial rock temperature. When the initial temperature is relatively low, the temperature disturbance is mainly confined to the near-field region around the tunnel. In contrast, under high geothermal conditions, the temperature disturbance propagates further into the surrounding rock, resulting in a significantly larger TIZ. This finding highlights the importance of considering realistic geothermal gradients in the analysis of high-geothermal tunnels. This finding differs from results obtained under the assumption of a uniform initial temperature field [13–15].

4. Mechanical Response of Surrounding Rock Under THM Coupling

4.1. Influence of Temperature on Mechanical Response

Temperature plays a dominant role in controlling the mechanical response of surrounding rock, including the tunnel deformation and plastic zone evolution. As shown in Figure 8a, the deformation response of different tunnel positions varies significantly with increasing initial temperature of the deep surrounding rock. The vault displacement increases continuously and exhibits the largest variation, indicating the highest sensitivity to temperature. In contrast, the haunch displacement first decreases slightly and then increases, while the invert displacement gradually decreases, with a smaller variation amplitude than that of the vault. At the same temperature, the deformation pattern also differs among tunnel positions. In the low-temperature range (40–80 °C), the displacement follows the order of invert < vault < haunch, with the maximum deformation occurring at the haunch and averaging 0.0418 m. In the high-temperature range (80–200 °C), the order changes to invert < haunch < vault, and the maximum deformation shifts to the vault, reaching 0.0678 m at 200 °C. As illustrated in Figure 8b, the overall tunnel deformation presents a gourd-shaped pattern with inward contraction in the middle. With increasing temperature, this deformation pattern gradually becomes flatter. These results indicate that the tunnel crown is the most temperature-sensitive region under high-temperature conditions, while deformation at the haunch remains significant and should be closely monitored in engineering practice.

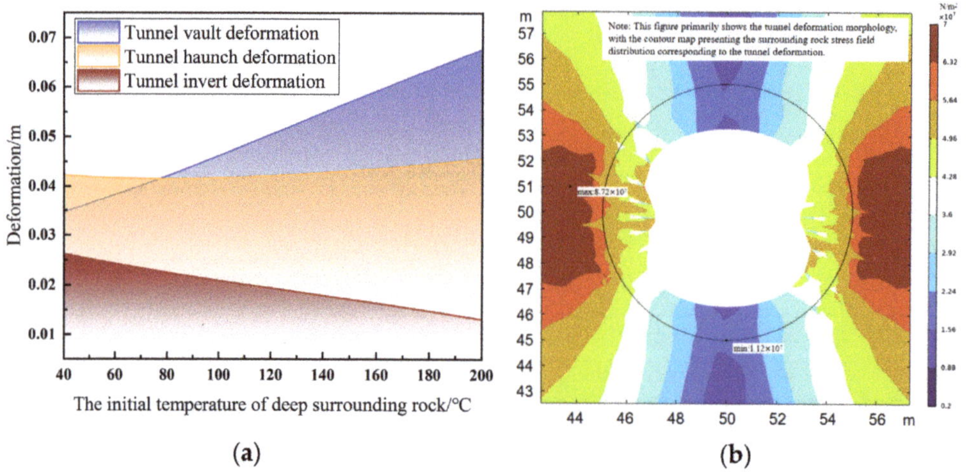

Figure 8. Tunnel deformation under temperature influence. (**a**) Variations in displacements of tunnel vault, haunch and bottom with temperature. (**b**) Schematic diagram of tunnel deformation.

Figure 9 shows the distribution of plastic strain around the tunnel. The color scale represents the magnitude of plastic strain, where red indicates a larger value and blue indicates a smaller value. As shown in the figure, the plastic strain is concentrated near the excavation boundary, where the surrounding rock appears mainly red, while the plastic strain decreases gradually away from the tunnel, as indicated by the transition from red to blue. It shows that the plastic zone expands predominantly along directions approximately 45° to the horizontal as the initial temperature of the deep surrounding rock increases, which is generally consistent with the failure pattern reported by Liu et al. [23]. At low temperatures (40–60 °C), the plastic zone is localized around the tunnel and exhibits a relatively regular shape, with an area of 159.92–219.62 m^2 and an equivalent radius of 7.67–8.82 m. When the temperature exceeds 80 °C, the plastic zone expands rapidly and gradually evolves into a butterfly-shaped pattern. At 200 °C, the plastic zone area reaches 1856.8 m^2, corresponding to an equivalent radius of 24.47 m. This transition from localized concentration to outward expansion indicates that elevated temperature significantly aggravates the plastic deterioration of surrounding rock. This is mainly due to two mechanisms. First, elevated temperature induces additional thermal stress, which alters the stress distribution around the tunnel. Second, temperature-dependent degradation of mechanical properties, such as the elastic modulus and strength, weakens the rock mass. As a result, high-temperature conditions significantly reduce the stability of the surrounding rock and increase the risk of large deformation and failure.

https://doi.org/10.3390/app16083952

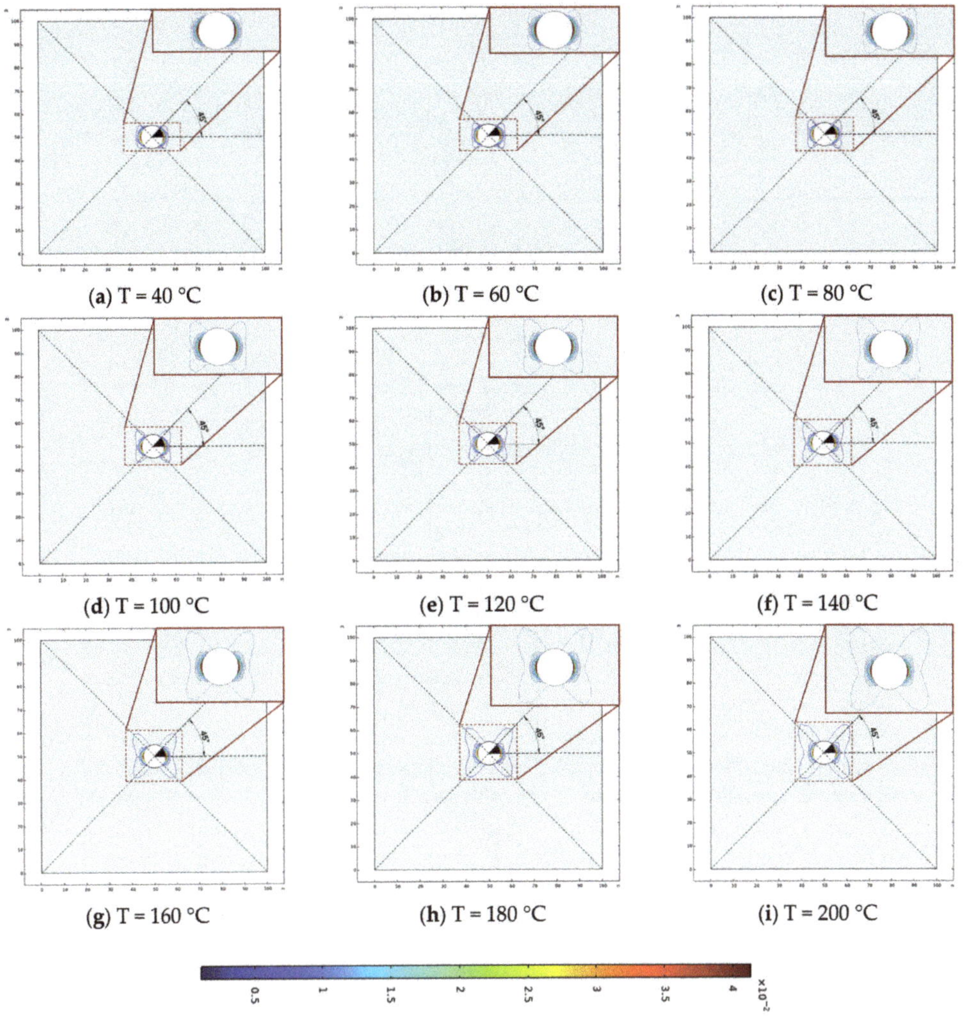

Figure 9. Distribution of the plastic zone under temperature changes.

As shown in Figure 10, both the plastic zone area and its equivalent radius increase with the initial temperature of the deep surrounding rock, indicating that elevated temperature promotes plastic zone expansion and aggravates surrounding rock failure. Within the temperature range of 40–100 °C, the growth of the plastic zone is relatively gradual. Above 120 °C, however, the expansion rate increases markedly. On average, the plastic zone area increases by 10.6055 m^2 and the equivalent radius by 0.1050 m for every 1 °C rise in temperature. The synchronous increase in these two parameters reflects the continuous outward propagation of the plastic zone and the increasing depth of thermal damage around the tunnel. This trend suggests that high temperature may weaken the radial confinement provided by the support system, and the fitted relationship of the equivalent radius can therefore serve as a useful reference for determining bolt support depth.

https://doi.org/10.3390/app16083952

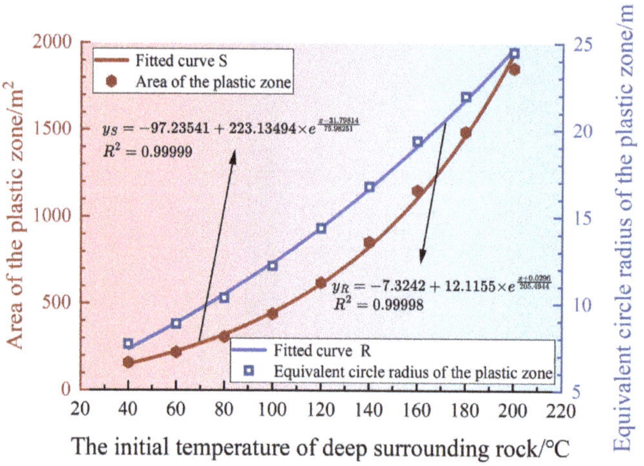

Figure 10. Curves of plastic zone area and equivalent circle radius of the plastic zone varying with temperature.

4.2. *Influence of Pore Water Pressure on Mechanical Response*

Here, the different pore water pressure was considered. The pore water pressure is applied at the top boundary with an increment of 2.5 MPa, ranging from 0 to 15 MPa. This range was selected because a detailed measured distribution of groundwater pressure was not available for the tunnel section considered. Therefore, the upper bound was taken with reference to the maximum hydrostatic pressure corresponding to the deepest tunnel section (approximately 1500 m burial depth). This range was selected to cover a representative interval from dry/undrained local conditions to high-pressure groundwater environments that may occur in deep tunnel sections. The applied hydraulic boundary conditions should thus be understood as a simplified engineering representation for comparative analysis of pore-pressure effects under THM coupling, rather than a direct one-to-one reconstruction of the full in situ groundwater system.

As shown in Figure 11a, the deformation of different tunnel positions exhibits distinct responses to increasing pore water pressure. The vault displacement decreases continuously, reaching a maximum of 0.04271 m at 0 MPa. In contrast, the invert displacement increases significantly with pressure, reaching 0.06413 m at 15 MPa, indicating the highest sensitivity to pore pressure. The haunch displacement shows a slight decrease at low pressure and then increases gradually beyond 2.5 MPa. Overall, pore water pressure significantly affects the deformation distribution, with the invert being the most sensitive region.

As illustrated in Figure 11b, the tunnel deformation exhibits a gourd-shaped pattern, which gradually transitions from inward contraction to outward bulging with increasing pore pressure. These results suggest that deformation at the tunnel invert should be carefully monitored under high pore pressure conditions, while haunch stability remains critical at lower pressures.

Figure 12 shows that the plastic zone is initially concentrated on both sides of the tunnel and attenuates outward, indicating that excavation-induced stress concentration dominates its initial distribution. With increasing pore water pressure, the plastic zone becomes more pronounced near the tunnel floor and tends to penetrate downward, increasing the risk of instability. As shown in Figure 13, both the plastic zone area and its equivalent radius increase with pore water pressure. On average, the plastic zone area increases by 3.0307 m^2 and the equivalent radius by 0.0405 m for every 1 MPa rise in pressure. This behavior is primarily controlled by effective stress reduction. According to the Mohr–Coulomb criterion:

$$\tau = \sigma' \tan \psi + c \tag{15}$$

where τ is the shear strength, σ' is the normal effective stress, ψ is the internal friction angle, and c is the cohesion. Increasing pore pressure reduces effective stress, thereby lowering shear strength and promoting plastic deformation. However, the expansion of the plastic zone induced by pore pressure is significantly smaller than that caused by temperature, indicating its secondary role in controlling overall rock failure.

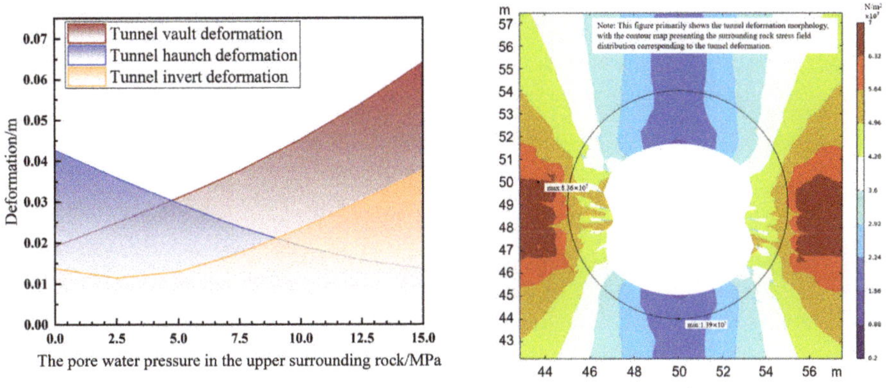

(a) (b)

Figure 11. Tunnel deformation under water pressure influence. (**a**) Diagram of displacement of tunnel vault, haunch and bottom with water pressure. (**b**) Schematic diagram of tunnel deformation.

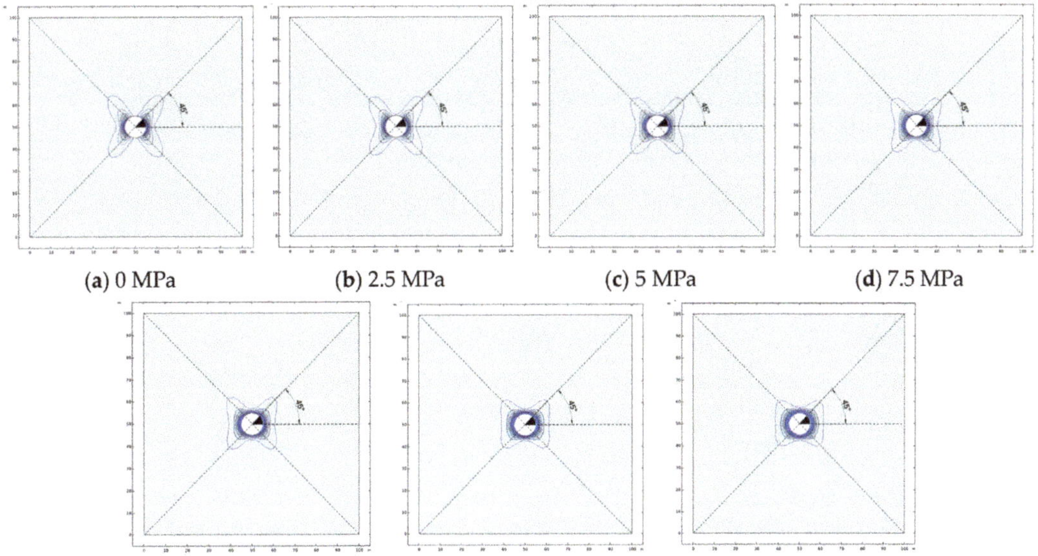

(**a**) 0 MPa (**b**) 2.5 MPa (**c**) 5 MPa (**d**) 7.5 MPa

(**e**) 10 MPa (**f**) 12.5 MPa (**g**) 15 MPa

Figure 12. Distribution of the plastic zone under water pressure changes.

https://doi.org/10.3390/app16083952

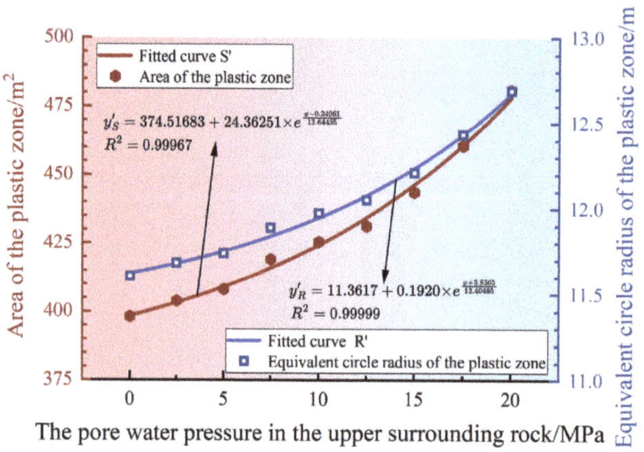

Figure 13. Curves of plastic zone area and equivalent circle radius of plastic zone varying with water.

5. Discussion

5.1. Mechanism of Thermal Influence Zone Evolution Under THM Coupling

The evolution of the thermal influence zone (TIZ) is governed by the coupled processes of heat transfer, fluid flow, and mechanical response, as shown in Figure 14. Unlike conventional analyses that consider heat conduction alone, the present results demonstrate that the coupled THM model indicates that seepage within the saturated surrounding rock can modify the temperature field through advective–convective transport. However, because the tunnel boundary is treated as a no-flow boundary in the present model, seepage does not enter the tunnel but may redistribute heat transfer within the surrounding rock mass under coupled THM conditions. More importantly, temperature variation induces thermal stress due to differential thermal expansion, which modifies the stress field around the tunnel. This thermally induced stress redistribution not only contributes directly to deformation but also promotes the initiation and propagation of microcracks, leading to permeability enhancement. The increased permeability, in turn, facilitates fluid flow and further intensifies convective heat transfer. This feedback loop among temperature, permeability, and seepage represents a typical THM coupling mechanism. As a result, the evolution of the TIZ is not a purely thermal diffusion process but a nonlinear, multi-field coupled process characterized by mutual reinforcement among the involved physical fields.

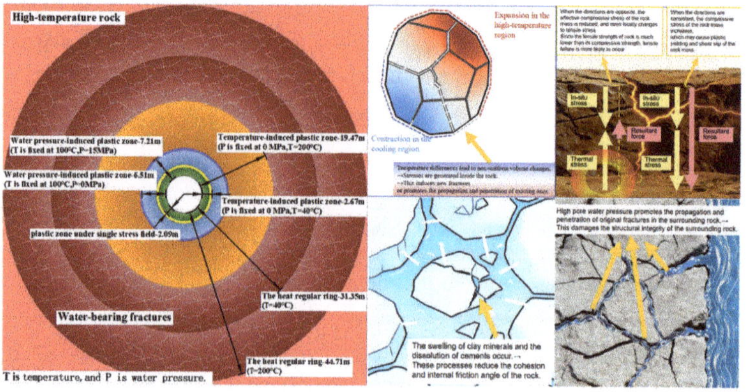

Figure 14. Schematic diagram of the thermal influence zone, and the plastic zone and failure mechanisms in the surrounding rock.

https://doi.org/10.3390/app16083952

It should be noted that the relative contributions of temperature and pore water pressure depend on the geological, hydraulic, and boundary conditions, and therefore should not be generalized as a universal rule. Under the conditions considered in this study, the temperature plays a dominant role in controlling the mechanical behavior of surrounding rock under high-geothermal conditions. Compared with pore water pressure, temperature affects both the stress state and the intrinsic mechanical properties of the rock. On the one hand, temperature gradients generate thermal stress, which alters the original stress distribution around the tunnel. On the other hand, elevated temperature leads to the degradation of mechanical properties, such as the elastic modulus and strength, due to thermally induced microstructural damage. In contrast, pore water pressure mainly influences the effective stress without directly altering the material properties. Therefore, although pore pressure contributes to deformation and plastic zone development, its effect is secondary compared with that of temperature. This distinction highlights the necessity of explicitly considering temperature-dependent material behavior in the analysis of high-geothermal tunnels, as neglecting thermal effects may lead to a significant underestimation of deformation and failure risk.

5.2. Engineering Implications for High-Geothermal Tunnels

The engineering implications of the present results are reflected mainly in three aspects. First, the quantified evolution of the thermal influence zone provides a reference for evaluating the long-term spatial extent of thermal disturbance around high-geothermal tunnels, which is relevant to the layout of thermal insulation and disturbance-control measures. Second, the expansion characteristics of the plastic zone under different temperature and pore-pressure conditions provide guidance for support design, especially for determining the required reinforcement range and support depth in thermally and hydraulically sensitive zones. Third, the different deformation responses of the vault, haunch, and invert indicate that monitoring priority and stability-control measures should be condition-dependent: under high-temperature conditions, more attention should be paid to the vault and adjacent plastic-expansion zone, whereas under high pore-pressure conditions, the tunnel invert should be given higher monitoring priority and stronger local control. These results provide a more direct basis for support optimization and long-term stability management in high-geothermal tunnels.

5.3. Limitations and Future Work

Despite the insights provided by this study, several limitations should be acknowledged. First, it should be noted that the present numerical analysis is based on a 2D plane-strain model with a circular tunnel idealization and full-face excavation assumption. These simplifications were adopted to focus on the dominant thermo–hydro–mechanical coupling mechanisms and to facilitate quantitative analysis of the thermal influence zone and surrounding-rock response. However, they may not fully capture three-dimensional effects, the actual geometric characteristics of the tunnel section, or the sequential influence of staged excavation and support installation. Therefore, the results should be interpreted as a simplified THM evaluation under idealized conditions, and further studies using more realistic three-dimensional and construction-stage-dependent models are needed. Second, the temperature-dependent material parameters adopted in this study were derived from published laboratory tests on granite and are used here as engineering-scale constitutive approximations. They are intended to capture the first-order trends of thermal damage, stiffness degradation, and permeability enhancement. It should be noted that the in situ rock mass is generally more heterogeneous and discontinuous than intact laboratory samples; therefore, these parameters do not represent a strict one-to-one characterization of

https://doi.org/10.3390/app16083952

field rock-mass properties, but rather an engineering idealization for coupled THM analysis under high-geothermal conditions. Further calibration with site-specific rock-mass tests or long-term field monitoring data would improve the reliability of the parameterization. Third, the model does not explicitly consider fracture network evolution or large deformation effects, which may play a significant role under extreme conditions. In addition, a limitation of this study is that model validation is presently restricted to the mechanical field. Further validation of the transient thermal, hydraulic, and deformation responses will require long-term field monitoring or laboratory/field-coupled datasets, which should be addressed in future work. Future work should focus on developing three-dimensional THM models, incorporating more advanced constitutive relationships that account for damage evolution, and validating the results through laboratory experiments and field monitoring data.

6. Conclusions

This study investigates the evolution of the thermal influence zone and the mechanical response of surrounding rock in high-geothermal tunnels under thermo–hydro–mechanical (THM) coupling conditions. The main conclusions are as follows:

(1) A THM-coupled numerical model incorporating an in situ temperature gradient is established, and the thermal influence zone (TIZ) is quantitatively characterized. The results show that the TIZ radius increases rapidly at the early stage and gradually stabilizes. The steady-state radius increases nonlinearly with the initial rock temperature, indicating that geothermal gradient plays a critical role in determining the spatial extent of thermal disturbance.

(2) Temperature has a dominant effect on the mechanical response of surrounding rock under the present model conditions. As the initial temperature increases from 40 °C to 200 °C, the tunnel crown displacement increases significantly, reaching a maximum of 0.0678 m, while the plastic zone expands from 159.92–219.62 m^2 to 1856.8 m^2, with the equivalent radius increasing from 7.67–8.82 m to 24.47 m. On average, the plastic zone area and equivalent radius increase by 10.6055 m^2/°C and 0.1050 m/°C, respectively.

(3) Pore water pressure mainly influences deformation distribution through effective stress reduction. As pore pressure increases from 0 MPa to 15 MPa, the invert displacement increases from 0.04271 m to 0.06413 m, showing the highest sensitivity. Meanwhile, the plastic zone area and equivalent radius increase at rates of approximately 3.0307 m^2/MPa and 0.0405 m/MPa, respectively, which are significantly lower than those induced by temperature.

(4) The plastic zone develops preferentially along directions approximately 45° to the horizontal and evolves from a localized distribution to a butterfly-shaped pattern with increasing temperature and pore pressure. This evolution reflects stress redistribution from the tunnel boundary to deeper surrounding rock and indicates an increasing depth of damage under high-temperature conditions.

(5) THM coupling significantly amplifies surrounding rock deformation and failure compared with single-field conditions. Temperature controls the intensity of damage through thermal stress and material degradation, while pore pressure regulates the spatial distribution of deformation via effective stress reduction. These findings highlight the necessity of considering multi-field coupling in the design of support systems, particularly for determining support depth based on the expanded plastic zone.

https://doi.org/10.3390/app16083952

Author Contributions: X.W. performed the numerical simulation and wrote the manuscript. B.X. developed the idea of the research and supervised the research. L.C. participated in manuscript reviewing, editing and project supervision. F.W. contributed to conceptualization, methodology development, as well as manuscript revision. J.C. and Y.G. provided overall supervision for this research. All authors have read and agreed to the published version of the manuscript.

Funding: This research was sponsored by the Joint Key Program of the National Natural Science Foundation of China (No. U24A2089), the National Natural Science Foundation of China (No. 51874207), and the Natural Science Foundation of Shanxi Province (Nos. 202303021211042 and 202303011222006).

Institutional Review Board Statement: Not applicable.

Informed Consent Statement: Not applicable.

Data Availability Statement: The data used to support the findings of this study are available from the corresponding author upon request.

Conflicts of Interest: Author Fengnian Wang was employed by the company Shanxi Provincial Transportation Technology Research and Development Co., Ltd. The remaining authors declare that the research was conducted in the absence of any commercial or financial relationships that could be construed as a potential conflict of interest.

References

1. Guo, P.Y.; Bu, M.H.; Zhang, P.; Li, Q.B.; He, M.C. Review on catastrophe mechanism and disaster countermeasure of high geotemperature tunnels. *Chin. J. Rock Mech. Eng.* **2023**, *42*, 1561–1581. [CrossRef]
2. Zhao, Y.S. Retrospection on the development of rock mass mechanics and the summary of some unsolved centennial prob. *Chin. J. Rock Mech. Eng.* **2021**, *40*, 1297–1336. [CrossRef]
3. Dwivedi, R.; Goel, R.; Prasad, V.; Sinha, A. Thermo-mechanical properties of Indian and other granites. *Int. J. Rock Mech. Min. Sci.* **2008**, *45*, 303–315. [CrossRef]
4. Xi, B.P.; Zhao, Y.S. Experimental study of thermophysico-mechanical property of drilling surrounding rock in granite under high temperature and high pressure. *Chin. J. Rock Mech. Eng.* **2010**, *29*, 1245–1253.
5. Jia, P.; Xi, B.P.; Li, X.K.; Xie, J.; Cai, J.H.; Gao, P.L. Energy evolution pattern in destructive process of unsteady state heat transfer i. *Rock Soil Mech.* **2024**, *45*, 3013–3023+3036. [CrossRef]
6. Long, L.J. Study on Mechanical and Seepage Properties of Sandstone Under the Coupling of Temperature-Seepage-Stress. Doctoral Dissertation, Chongqing University, Chongqing, China, 2021. [CrossRef]
7. Tan, Y.L.; Zhang, Q. Numerical analysis of surrounding rock deformation and failure in deep roadway under the condition of thermo-mechanical coupling. *J. Shandong Univ. Sci. Technol. (Nat. Sci. Ed.)* **2016**, *35*, 29–37. [CrossRef]
8. Liu, X.C.; Huang, F.; Zhou, Y.; Yang, D.; Hu, Z. Optimization of thermal insulation layer thickness and thermal coupling response analysis of lining in high-temperature tunnels. *Mod. Tunn. Technol.* **2022**, *59*, 108–117. [CrossRef]
9. Shu, L.S.; Xue, Y.D. Numerical simulation of high-temperature mechanical properties of lined segment joints based on thermal-mechanical coupling. *Mod. Tunn. Technol.* **2021**, *58*, 64–72. [CrossRef]
10. Zhang, S.G.; Li, Z.J.; Xu, Y.H.; Zhang, C.C. Three-dimensional numerical simulation and analysis of fluid-heatcoupling heat-transfer in fractured. *Rock Soil Mech.* **2011**, *32*, 2507–2511. [CrossRef]
11. Zhao, C.X.; Zhang, Z.X.; Wang, S.F.; Lei, Q.H. Hydro-mechanical coupling model for tunnel excavation in fractured rocks and analysis on excavation-induced damage. *J. Harbin Inst. Technol.* **2024**, *56*, 28–36.
12. Sakhno, I.; Sakhno, S.; Vovna, O. Surface subsidence response to safety pillar width between reactor cavities in the underground gasification of thin coal seams. *Sustainability* **2025**, *17*, 2533. [CrossRef]
13. Chen, L.; Han, F. Distribution laws of temperature field in mine surrounding rock. *Saf. Coal Mines* **2017**, *48*, 56–59+64. [CrossRef]
14. Zhang, J.; Zhang, P.Y.; Wei, J.P.; Si, G.; Xu, B.; Xu, Z.G. Study on distribution character of thermal influence zone in roadway surrounding rock: Taking Yangchangwan coal mine as an example. *China Saf. Sci. J.* **2023**, *33*, 125–133. [CrossRef]
15. Zhang, S.G.; Chang, J.; Wang, H.W. Characteristics of heat-adjusting ring and the influence of thermal insulation support structure in high-temperature roadway. *Coal Geol. Explor.* **2019**, *47*, 179–185.
16. Kestin, J.; Sokolov, M.; Wakeham, W.A. Viscosity of liquid water in the range 8 °C to 150 °C. *J. Phys. Chem. Ref. Data* **1978**, *7*, 941–948. [CrossRef]
17. Lide, D.R. *CRC Handbook of Chemistry and Physics*, 79th ed.; National Institute of Standards and Technology: Gaithersburg, MD, USA, 1998; pp. 3–6.

https://doi.org/10.3390/app16083952

18. Wu, D.; Deng, T.F.; Zhao, R.K.; Wang, Y.H. THM modeling of ground subsidence induced by excavation of subway tunnel. *Comput. Geotech.* **2018**, *94*, 1–11. [CrossRef]
19. Cheng, Z.P.; Xi, B.P.; Yang, X.X.; He, S.X.; Li, X.X. Experimental study on the evolution of granite permeability under thermal shock. *J. Taiyuan Univ. Technol.* **2021**, *52*, 198–203. [CrossRef]
20. Wang, Q.W.; Ju, N.P.; Huang, J.; Du, L.L. Regression analysis of initial geostress field of Sangzhuling Super-long tunnel. *Sci. Technol. Eng.* **2016**, *16*, 137–143.
21. Yan, J.; He, C.; Wang, B.; Meng, W.; Yang, J.F. Prediction of rock bursts for Sangzhuling Tunnel located on Lhasa-Nyingchi railway under coupled thermo-mechanical effects. *J. Southwest Jiaotong Univ.* **2018**, *53*, 434–441.
22. Zhang, Y.G.; Huang, X.J. Using the tunneling work surface cannon to determine the ground temperature. *Coal Geol. Explor.* **1981**, *9*, 19–25.
23. Liu, H.T.; Jin, Y.W.; Guo, X.F.; Han, Z.J. Evolution of butterfly leaf angle in the butterfly plastic zone of the surrounding rock of a circular roadway. *J. China Coal Soc.* **2025**, *50*, 3388–3399. [CrossRef]

https://doi.org/10.3390/app16083952

applied sciences

Article

Research on Creep Characteristics of Dredged Fill Soil in Humen Port Considering the Effect of Temperature

Xiaodi Xu, Qiunan Chen * and Chen Zhang

College of Civil Engineering, Hunan University of Science and Technology, Xiangtan 411201, China; 22300201004@mail.hnust.edu.cn (X.X.); zhan2022chen@163.com (C.Z.)
* Correspondence: cqndoc@163.com

Abstract

Dredged Fill Soil, as a primary foundation material in reclamation projects, exhibits complex physical and mechanical properties, characterized by a high plasticity index, high water content, low density, high compressibility, large void ratio, and low bearing capacity. Its creep behavior is highly sensitive to temperature changes. This study systematically investigates the temperature-dependent creep behavior of reclaimed soil from Humen Port through laboratory experiments, theoretical modeling, and experimental validation. Triaxial creep tests conducted at different temperatures (5 °C, 15 °C, 25 °C, 35 °C) show that increasing temperature significantly exacerbates creep deformation: under undrained conditions, creep strain at 35 °C is nearly 300% higher than at 5 °C, while drainage reduces the strain by approximately 29.3%. Based on these results, a Burgers-type creep constitutive model considering temperature effects is developed, incorporating the impact of temperature on viscosity and elastic modulus. The model's predictions show good agreement with the experimental results (15 °C: R^2 = 0.9788; 35 °C: R^2 = 0.9890), confirming the model's validity. The research findings provide theoretical and practical references for the long-term stability evaluation and engineering design of reclaimed foundations in complex marine environments.

Keywords: dredged fill soil; creep characteristics; temperature effect; creep test; constitutive model

1. Introduction

As a major maritime country, China possesses approximately 3 million km^2 of maritime territory and a coastline extending approximately 14,500 km [1]. With the rapid advancement of coastal infrastructure construction under the "Maritime Silk Road" initiative, land reclamation has become a widely adopted engineering method. In reclamation areas such as Humen Port, dredged marine sediments are widely used as primary foundation fill materials. These dredged fills are typically characterized by high water content, large void ratio, high compressibility, low shear strength, and pronounced time-dependent deformation behavior. Under long-term service loading, creep deformation becomes a governing factor controlling post-construction settlement and foundation stability, posing significant challenges to the long-term durability and safety.

Extensive research has been conducted on the creep behavior of marine soft soils. One-dimensional consolidation and triaxial creep tests have demonstrated that soft clays generally exhibit attenuation and steady-state creep stages, with rheological behavior strongly dependent on stress level and drainage condition [2–6]. Lei et al. [6] reported distinct nonlinear transitions under varying stress ratios, while Yang et al. [7] revealed that particle rearrangement and pore structure evolution play critical roles in long-term deformation. Combined experimental and numerical investigations further clarified the coupling mechanism between consolidation pressure and creep settlement [8,9], and cyclic loading tests confirmed progressive creep accumulation and stiffness degradation [10]. More recent triaxial and direct shear creep tests on Hangzhou Bay and Qianhai soft soils indicated that creep modulus is significantly influenced by confining pressure and deviator stress, and that Burgers and modified Singh–Mitchell models can effectively capture multi-stage creep characteristics [11]. Collectively, these studies indicate that creep behavior is governed by stress state, soil structure, and loading path, with microstructural evolution serving as the intrinsic mechanism.

Constitutive modeling has gradually evolved from empirical formulations toward viscoelastic–viscoplastic and fractional derivative frameworks. Classical models such as Singh–Mitchell, Merchant, and Burgers provide simplified rheological descriptions but often lack long-term predictive robustness under complex stress paths [3,11,12]. Fractional derivative and fractal-based creep models have shown improved capability in characterizing long-term attenuation behavior of silty clays [13]. Experimental research on consolidation creep and microstructure evolution further established the linkage between macroscopic creep deformation and pore collapse mechanisms [14]. In addition, one-dimensional compression creep tests considering temperature and suction effects revealed significant hydro-thermal coupling influences on creep evolution [15], and analytical solutions for three-dimensional consolidation and creep processes of layered fractional viscoelastic soils incorporating temperature effects have provided theoretical support for multi-field coupling problems [16].

Temperature is a critical environmental factor affecting soil mechanical behavior, particularly in coastal reclamation areas subject to seasonal variations and artificial thermal disturbances [17]. Experimental studies have shown that increasing temperature accelerates creep strain accumulation and reduces creep modulus [18], while investigations on Hong Kong marine deposits demonstrated enhanced secondary compression and long-term settlement under elevated temperatures [19]. Thermodynamic-based constitutive theories and thermo-elastic viscoplastic (TEVP) models have incorporated temperature variables into consolidation–creep formulations [20–22]. Although significant progress has been achieved in understanding thermo-mechanical coupling in conventional soft clays and frozen soils [23,24], systematic investigations on temperature-dependent triaxial creep behavior of dredged reclamation fills under realistic stress conditions remain limited. Therefore, based on the Humen Port reclamation project, this study conducts temperature-controlled triaxial creep tests on dredged fill materials and develops a thermo-mechanical coupled constitutive model. The aim is to clarify the temperature–stress–time interaction mechanism and to provide theoretical support for long-term stability evaluation and design optimization of coastal reclamation foundations.

2. Experimental Overview

2.1. Preparation of Soil Samples

The soil samples used in this study were collected from a land reclamation project at a wharf in Humen Port, Dongguan City, Guangdong Province. The sampling location and surrounding conditions are shown in Figures 1 and 2. The physical and mechanical properties of the soils were determined through laboratory geotechnical testing, with the results presented in Table 1.

The soil is classified as dredged fill, a typical marine deposit commonly encountered in coastal reclamation areas. It exhibits the following engineering characteristics: high liquid and plastic limits, indicating strong water affinity and plasticity; high natural water content, often exceeding the liquid limit; high compressibility, making it susceptible to significant consolidation settlement; low shear strength in its undisturbed state, which poses challenges for immediate load-bearing applications; and low permeability, resulting in slow drainage and prolonged consolidation periods.

Figure 1. Photograph of the soil sampling area.

Figure 2. Collected Soil Samples.

Table 1. Basic Physical Properties of Dredged Fill Soil at Humen Port Area.

Soil Sample	Moisture Content W /%	Wet Density ρ_0 /g·cm^{-3}	Pore Ratio e	Dry Density ρ_d /g·cm^{-3}	Specific Gravity ρ_s	Liquid Limit W_L /%	Plastic Limit W_P /%	Plasticity Index I_P	Liquidity Index I_L
Dredged Fill Soil	78.16	1.84	1.52	1.34	2.71	57.75	29.87	27.8	1.21

In engineering practice, this type of soil is primarily intended for use as wharf foundation ground after appropriate ground improvement and consolidation treatments. The long-term performance of such foundations is critically dependent on the creep behavior of the underlying dredged fill, as excessive time-dependent deformation can compromise the serviceability and stability of wharf structures. Understanding the creep characteristics of this material under various thermal and drainage conditions is therefore essential for predicting post-construction settlements and ensuring the safe operation of port infrastructure.

https://doi.org/10.3390/app16062820

2.2. Experimental Apparatus

The primary equipment used in the experiments included a low-temperature, high-pressure soil creep triaxial testing machine manufactured by Xi'an Kangtuoli Instrument Equipment Co., Ltd (Xi'an, China). This soil creep triaxial testing machine is equipped with an automatic data acquisition system, enabling real-time acquisition and storage of experimental data. Figure 3 shows the configuration of the low-temperature high-pressure triaxial creep testing system. The main technical parameters of this equipment are as follows: maximum triaxial load of 256 kN, maximum confining pressure of 30 MPa, maximum back pressure of 20 MPa, axial displacement range of 100 mm, and temperature control range from −30 °C to 100 °C.

Figure 3. Low-Temperature High-Pressure Triaxial Creep Testing System: (**a**) Experimental Apparatus; (**b**) Soil Specimen Installed in the Triaxial Cell.

2.3. Experimental Research Plan

This study investigates the fill soil from Humen Port in Dongguan, conducting triaxial consolidated undrained creep tests and triaxial consolidated drained creep tests. Cylindrical remolded soil specimens (38 mm diameter, 76 mm height) were used. Test conditions were determined based on the original depth of the dredged fill material (0–15 m), with consolidation conducted at a confining pressure of 200 kPa followed by triaxial creep testing. This study aimed to investigate the influence of temperature on the long-term creep behavior of the dredged material. Based on test results obtained under ambient temperature conditions, test plan incorporating temperature gradients (5 °C, 15 °C, 25 °C (ambient), 35 °C) was designed. Triaxial drained (CD) and undrained (CU) creep tests were subsequently conducted at different temperatures. Throughout the tests, confining pressure was maintained at 200 kPa, and deviatoric stress was incrementally applied in 50 kPa steps. The cumulative variation in creep strain was recorded. Detailed test procedures are presented in Table 2. A total of twelve cylindrical specimens were prepared following a standard test method [25]: the soil was air-dried, oven-dried at 105 °C for 24 h, ground, sieved through a 0.025 mm mesh, mixed with distilled water to target water content, sealed for 18 h, compacted in five layers with interlayer scarification, and vacuum-saturated at –0.1 MPa for 12 h. Eight specimens were used for creep tests under various drainage and temperature conditions, and four were kept as backups.

https://doi.org/10.3390/app16062820

Table 2. Design of Triaxial Creep Tests for Reclaimed Soil.

Test No.	σ_1 /kPa	σ_2 /kPa	σ_3 /kPa	Temperature	Incremental Shear Stress Δq_s /kPa	Drainage Condition
CU01	200	200	200	5 °C	50	Undrained
CU02				15 °C		Undrained
CU03				25 °C (Ambient)		Undrained
CU04				35 °C		Undrained
CD01	200	200	200	5 °C	50	Drained
CD02				15 °C		Drained
CD03				25 °C (Ambient)		Drained
CD04				35 °C		Drained

3. Results and Discussion

Representative photographs of selected test specimens and experimental observations are shown in Figure 4. The corresponding strain–time relationship curves of the Humen Port dredged fill material processed using the "Chen Method" under different conditions are presented in Figure 5.

Figure 4. Representative Post-Test Morphology of Remolded Dredged Fill Soil Specimens After Triaxial Creep Tests.

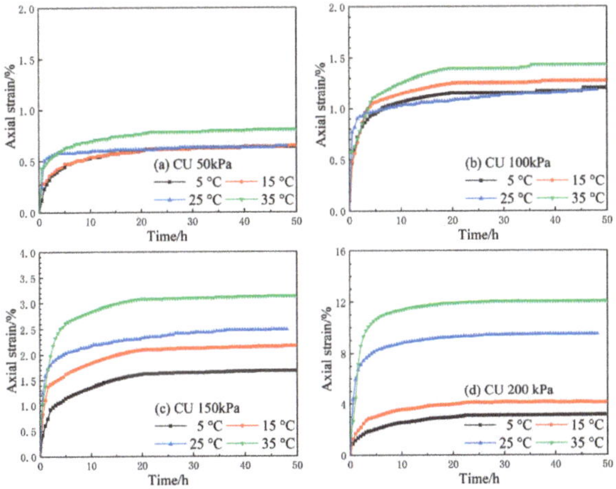

Figure 5. *Cont.*

https://doi.org/10.3390/app16062820

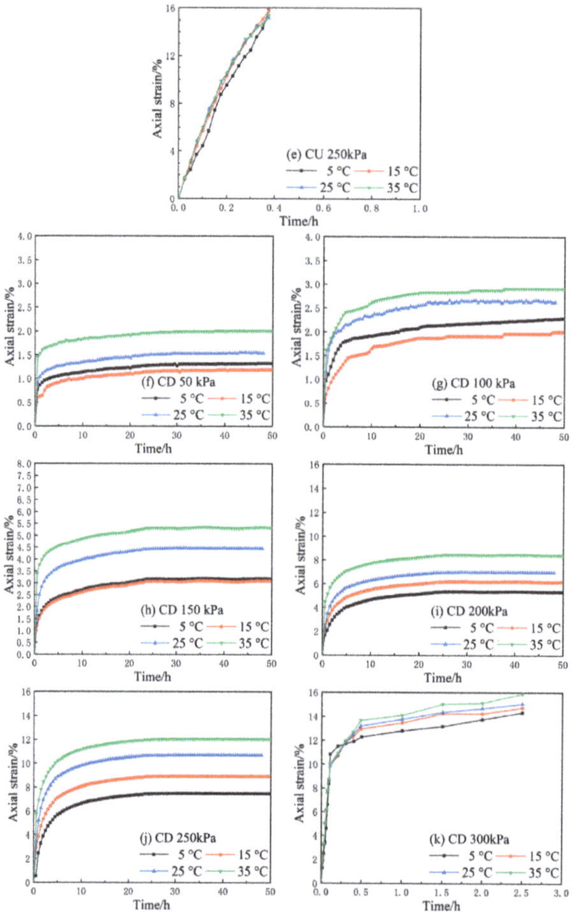

Figure 5. Stress–Strain Curves of Remolded Soil at Different Temperatures Obtained from Triaxial Tests: (**a**–**e**) CU tests with axial stresses of 50, 100, 150, 200, and 250 kPa; (**f**–**k**) CD tests with axial stresses of 50, 100, 150, 200, 250, and 300 kPa.

To comprehensively present the long-term creep behavior characteristics of dredged fill under varying temperature conditions, this study details the creep properties of dredged fill under both undrained and drained conditions in Tables 3 and 4, respectively.

Table 3. Creep Characteristics of Consolidated Undrained Dredged Fill Soil at Different Temperatures Under a Confining Pressure of 200 kPa.

Temperature /°C	Shear Stress /kPa	Instantaneous Strain/%	Creep Strain /%
5	50	0.20	0.41
	100	0.45	0.85
	150	0.65	1.50
	200	1.20	3.02
15	50	0.25	0.50
	100	0.53	1.01
	150	0.82	2.04
	200	1.53	4.07

https://doi.org/10.3390/app16062820

Table 3. *Cont.*

Temperature /°C	Shear Stress /kPa	Instantaneous Strain/%	Creep Strain /%
25 (ordinary temperatures)	50	0.34	0.61
	100	0.54	1.13
	150	0.95	2.43
	200	2.86	9.55
35	50	0.44	0.78
	100	0.73	1.55
	150	1.16	3.11
	200	3.52	12.02

Table 4. Creep Characteristics of Consolidated Drained Dredged Fill Soil at Different Temperatures Under a Confining Pressure of 200 kPa.

Temperature /°C	Shear Stress /kPa	Instantaneous Strain/%	Creep Strain /%
5	50	0.53	1.22
	100	0.60	2.03
	150	0.75	3.01
	200	1.03	5.04
	250	1.33	7.52
15	50	0.55	1.50
	100	0.65	2.31
	150	0.84	3.50
	200	1.12	6.02
	250	1.52	9.02
25 (ordinary temperatures)	50	0.65	1.51
	100	0.71	2.63
	150	0.95	4.41
	200	1.28	7.01
	250	1.74	10.70
35	50	0.81	2.00
	100	0.93	3.02
	150	1.15	5.51
	200	1.60	8.50
	250	2.10	12.03

1. Mechanism by which temperature influences creep behavior

Temperature exerts a pronounced regulatory effect on the creep behavior of dredged fill under both undrained and drained conditions, a phenomenon clearly observed in the experimental results. Taking the undrained condition as an example, when the uniaxial stress was maintained at 200 kPa, an increase in temperature from 5 °C to 35 °C resulted in an increase in instantaneous strain from 1.20% to 3.52%, representing a rise of 193.3% (Table 3). This change stems from the reduction in inter-particle friction resistance within the soil caused by elevated temperatures, coupled with enhanced activity of clay minerals, which accelerates the initial deformation response. The sensitivity of long-term creep strain to temperature is even more pronounced: under identical unconfined stress, the creep strain at 35 °C reached 12.02%, representing an increase of nearly 300% compared to 3.02% at 5 °C. This indicates that elevated temperatures significantly enhance the soil's viscous flow characteristics. By contrast, drainage conditions markedly suppressed creep development at elevated temperatures. For instance, under 200 kPa uniaxial stress at

35 °C, the creep strain during drainage amounted to 8.50% (Table 4), representing a reduction of approximately 29.3% compared to undrained conditions at the same temperature. This indicates that introducing drainage substantially mitigates the temperature-induced intensification of creep behavior. This phenomenon is mainly attributed to the dissipation of excess pore water pressure, which partially counteracts the viscoplastic deformation induced by temperature.

2. Nonlinear driving of deformation evolution by shear stress

Building upon the temperature effect, the level of uniaxial stress plays a decisive role in the accumulation of creep strain, a phenomenon further validated through data analysis. Taking the undrained conditions at 25 °C at an example, as the uniaxial stress progressively increased from 50 kPa to 200 kPa (i.e., 50, 100, 150, and 200 kPa), the creep strain surged sharply from 0.61% to 9.55% (Table 3), exhibiting a pronounced exponential growth trend. This pattern indicates a critical stress level likely occurs between 150 kPa and 200 kPa, beyond which a distinct shift in behavior is observed: the damage rate within the soil structure accelerates significantly. This is accompanied by the progressive failure of inter-particle connections, with the development of shear bands emerging as the primary controlling mechanism. Although similar trends were observed under drained conditions, the strain increase was somewhat suppressed. For instance, at 25 °C and a deviatoric stress of 200 kPa, the creep strain was 7.01% (Table 4), approximately 26.6% lower than under undrained conditions. This indicates that the drainage process, through adjustments to effective stress, mitigates the rapid progression of plastic deformation.

3. Deformation Suppression Effect of Drainage Conditions

Comparative analysis shows that drainage mitigates the combined influence of temperature and stress. At 35 °C and 200 kPa, creep strain decreased from 12.02% (undrained) to 8.50% (drained), a 29.3% reduction. Temperature sensitivity further confirms this trend: creep strain increased by approximately 0.35%/°C under undrained conditions but only 0.25%/°C under drained conditions. Thus, drainage effectively delays temperature-sensitive deformation by controlling pore pressure evolution, supporting long-term stability design of coastal reclamation foundations.

4. Establishment of Creep Constitutive Model

4.1. Selection and Construction of Creep Model

The Burgers model comprises a series connection of a Maxwell body (spring and dashpot in series) and a Kelvin body (spring and dashpot in parallel). Its advantage lies in simultaneously describing both transient creep (initial rapid deformation) and steady-state creep (deformation at a constant rate) in soil materials, rendering it suitable for characterizing soil behavior under complex stress histories.

Maxwell's strain equations:

$$\varepsilon_1 = \frac{\sigma}{E_1} + \frac{\sigma}{\eta_1}t \tag{1}$$

In the equation: σ denotes the applied stress; E_1 denotes the elastic modulus; η_1 denotes the viscosity.

Kelvin strain equation:

$$\varepsilon_2 = \frac{\sigma}{E_2}\left(1 - e^{-\frac{E_2}{\eta_2}t}\right) \tag{2}$$

In the equation: E_2 denotes the elastic modulus; η_2 denotes the viscosity.

The total strain expression is given by the superposition of Maxwell and Kelvin body strains:

$$\varepsilon = \varepsilon_1 + \varepsilon_2 \tag{3}$$

https://doi.org/10.3390/app16062820

From (3), the strain-time relationship for the Burgers model (under constant stress σ_0) is obtained:

$$\varepsilon(t) = \frac{\sigma_0}{E_1} + \frac{\sigma_0}{\eta_1}t + \frac{\sigma_0}{E_2}\left(1 - e^{-\frac{E_2}{\eta_2}t}\right) \tag{4}$$

4.2. Three-Dimensional Stress Decomposition and Extension of the Model

Stress tensor decomposition: Triaxial stresses (σ_1 being the axial stress, σ_2 being the confining pressure) may be decomposed as follows:

Mean stress (hydrostatic pressure):

$$p = \frac{1}{3}(\sigma_1 + 2\sigma_3) \tag{5}$$

Shear stress:

$$q = \sigma_1 - \sigma_3 \tag{6}$$

Strain decomposition and assumptions: The total strain is divided into volumetric strain (ε_v caused by spherical stress) and shear strain ε_s (caused by oblique stress):

$$\varepsilon = \varepsilon_v + \varepsilon_s \tag{7}$$

Volume strain hypothesis:

$$\varepsilon_v = \frac{p}{K} \tag{8}$$

Shear strain: Under constant shear stress q, the time response of shear strain is:

$$\varepsilon_q(t) = \frac{q}{G_1} + \frac{q}{G_2}\left(1 - e^{-\frac{G_2}{\eta_2}t}\right) + \frac{q}{\eta_1}t \tag{9}$$

Under confining pressure conditions, the axial strain ε is jointly determined by the volumetric strain ε_v and the shear strain ε_s:

$$\varepsilon = \varepsilon_v + \frac{2}{3}\varepsilon_s \tag{10}$$

Substitute the expressions for volumetric strain and shear strain:

$$\varepsilon(t) = \underbrace{\frac{p}{K}}_{\text{Volume strain}} + \frac{2}{3}\left[\underbrace{\frac{q}{G_1}}_{\text{Instantaneous shear}} + \underbrace{\frac{q}{G_2}\left(1 - e^{-\frac{G_2 t}{\eta_2}}\right)}_{\text{Delayed shear}} + \underbrace{\frac{q}{\eta_1}t}_{\text{Viscous flow}}\right] \tag{11}$$

Relationship between shear modulus G_1, G_2 and Young's modulus E_1, E_2:

$$G = \frac{E}{2(1+v)} \tag{12}$$

Relationship between the bulk modulus K and E_1:

$$K = \frac{E_1}{3(1-2v)} \tag{13}$$

4.3. The Effect of Temperature on Model Parameters

The elastic moduli (E_1, E_2) and viscosity coefficients (η_1, η_2) in the Burgers model are both temperature-dependent. The effect of temperature on the viscosity coefficient is as

https://doi.org/10.3390/app16062820

follows: the viscosity coefficient η decreases with increasing temperature, following the Arrhenius equation:

$$\eta(T) = \eta_0 \cdot e^{\frac{Q}{R}\left(\frac{1}{T} - \frac{1}{T_0}\right)} \tag{14}$$

4.4. Derivation of Temperature-Dependent Creep Equation

Equation (15) comprehensively describes the evolution of total strain in materials subjected to complex thermal cycling environments by coupling multistage deformation mechanisms with temperature effects. The total strain expression is:

$$\varepsilon_1(t,T) = \frac{p}{K(T)} + \frac{2}{3}\left[\frac{q}{G_1(T)} + \frac{q}{G_2(T)}\left(1 - e^{-\frac{G_2(T)}{\eta_2(T)}t}\right) + \frac{q}{\eta_1(T)}t\right] + \alpha(T - T_0) \tag{15}$$

Equation (15) comprehensively describes the total strain evolution of materials under complex thermal cycling conditions by coupling multistage deformation mechanisms with temperature effects. Its mathematical expression decomposes into four distinct components, each corresponding to the physical essence of the creep process: (1) Instantaneous elastic strain: This term characterizes the reversible elastic deformation occurring at the instant of loading, whose magnitude is jointly determined by the deviatoric stress level q and the temperature-dependent bulk modulus $K(T)$; (2) Delayed elastic strain: This term describes the time-dependent elastic response under constant stress. As $E_2(T)$ decreases and $\eta_2(T)$ diminishes, delayed deformation progresses more rapidly, with the steady-state value σ_0/E_2 increasing; (3) Steady-state viscous flow: This term quantifies the irreversible viscoplastic deformation after the material enters the steady-state creep phase; (4) Thermal expansion strain: This term reflects purely thermally induced deformation due to temperature variations.

In the creep model for offshore reclamation soils, the thermal expansion effect is negligible in magnitude. Therefore, the thermal expansion strain term is omitted to simplify the model and avoid excessive parameters. The total strain expression is:

$$\varepsilon_1(t,T) = \frac{p}{K(T)} + \frac{2}{3}\left[\frac{q}{G_1(T)} + \frac{q}{G_2(T)}\left(1 - e^{-\frac{G_2(T)}{\eta_2(T)}t}\right) + \frac{q}{\eta_1(T)}t\right] \tag{16}$$

4.5. Validation of Creep Constitutive Model Considering Temperature Effects

Based on Table 3, the instantaneous strain and creep strain data for CU at different temperatures (5 °C, 15 °C, 25 °C, 35 °C) and partial stresses (50 kPa, 100 kPa, 150 kPa, 200 kPa) are organized. We derived and calculated the parameters for the fill soil creep test model under varying temperatures and uniaxial stress conditions based on the triaxial creep constitutive model under temperature effects, as shown in Table 5.

Analysis results indicate that temperature and uniaxial stress exert significant and opposing effects on the rheological parameters of dredged fill material. With increasing temperature, both the transient and steady-state viscosity coefficients (η_1, η_2) and elastic moduli (G_1, G_2) exhibit a gradual decrease. This demonstrates that under elevated temperature conditions, the elastic and viscous constraints within the soil diminish, enhancing its deformation capacity and accentuating its creep characteristics. In contrast, increasing uniaxial stress causes η_1, η_2, G_1 and G_2 to progressively increase. This indicates that at high stress levels, the soil requires greater intrinsic resistance to withstand sustained strain, thereby exhibiting enhanced viscous and elastic properties. Overall, elevated temperatures promote soil rheological behavior, while uniaxial stress enhances its deformation resistance. The coupled interaction of these factors jointly governs the creep evolution patterns of reclaimed fill soils.

https://doi.org/10.3390/app16062820

Table 5. Model Parameters of Creep Tests for Dredged Fill Soil under Different Temperatures and Axial Stress Conditions at a Confining Pressure of 200 kPa.

Temperature /°C	Shear Stress /kPa	Model Parameters				R^2
		η_1 /kPa·s	η_2 /kPa·s	G_1 /kPa	G_2 /kPa	
5	50	250.12	121.95	250.43	121.95	0.951
	100	222.22	117.65	222.22	117.65	0.977
	150	230.77	100	230.77	100	0.963
	200	166.67	66.23	166.67	66.23	0.938
15	50	200	100	200	100	0.965
	100	188.68	99.01	188.68	99.01	0.972
	150	182.93	73.53	182.93	73.53	0.959
	200	130.72	49.14	130.72	49.14	0.936
25 (ambient temperature)	50	147.06	81.97	147.06	81.97	0.932
	100	185.19	88.5	185.19	88.5	0.941
	150	157.89	61.73	157.89	61.73	0.954
	200	69.93	20.94	69.93	20.94	0.967
35	50	113.64	64.1	113.64	64.1	0.979
	100	136.99	64.52	136.99	64.52	0.968
	150	129.31	48.23	129.31	48.23	0.986
	200	56.82	16.64	56.82	16.64	0.978

The variation in model parameters (η_1, η_2, G_1, G_2) under different temperatures and different shear stress conditions is fitted to a surface as shown in Figure 6.

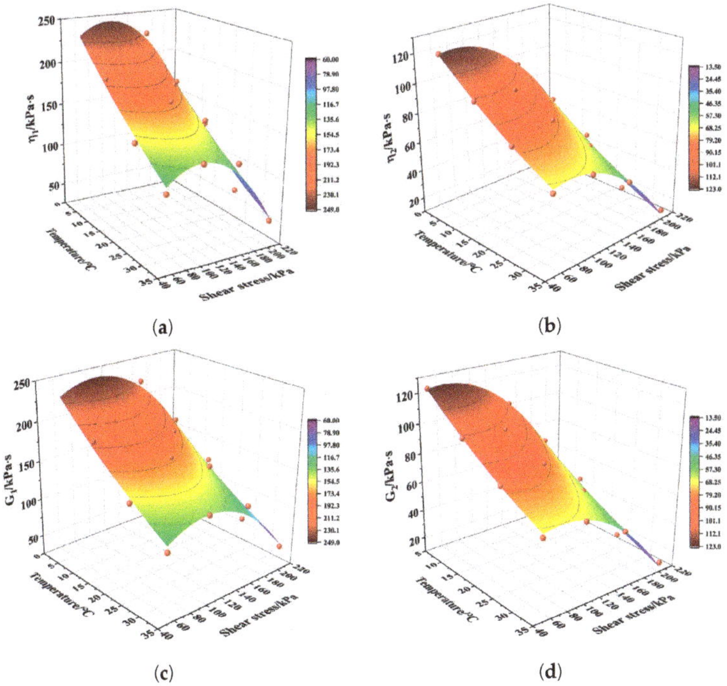

Figure 6. Fitted surface of model parameters for temperature versus in-plane stress: (**a**) η_1; (**b**) η_2; (**c**) G_1; (**d**) G_2.

https://doi.org/10.3390/app16062820

Figure 6 demonstrates that temperature and deviatoric stress exert a significant influence on the long-term creep behavior of dredged fill under drained conditions. The variation trends of parameters η_1, η_2, G_1 and G_2 under different environmental conditions reveal the rheological behavior of the soil when subjected to changing external conditions. Here, η_1 and η_2 represent the viscous resistance of the soil, while G_1 and G_2 reflect the shear stiffness. Changes in these parameters directly influence the soil's deformation characteristics and bearing capacity. Results indicate that η_1 and η_2 decrease with increasing deviatoric stress, suggesting that heightened external loading diminishes the soil's resistance to deformation, rendering it more susceptible to long-term creep. Furthermore, elevated temperatures similarly cause η_1 and η_2 to decline. In high-temperature environments, this intensifies relative slip between soil particles, thereby accelerating rheological effects and compromising long-term stability.

The variation trends of shear moduli G_1 and G_2 are analogous to those of viscous resistance parameters, both decreasing with increasing deviatoric stress. Higher deviatoric stresses cause the internal structure of the soil to loosen, weakening the inter-particle constraints and consequently reducing shear resistance. Furthermore, elevated temperatures further diminish the soil's stiffness, rendering it more susceptible to deformation under prolonged loading. Under elevated temperatures, the structural stability of soil diminishes, increasing susceptibility to settlement or shear failure.

By comparing the predicted curves from the creep theoretical model with experimental data at both 15 °C and 35 °C operating conditions (Figure 7), it was found that both exhibited high consistency during both the steady-state and accelerated creep phases (15 °C: R^2 = 0.9788; 35 °C: R^2 = 0.9890), validating the temperature effect-blowfill soil creep constitutive model.

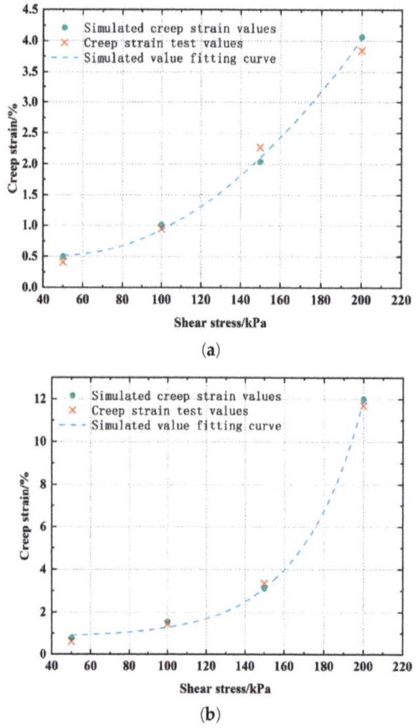

Figure 7. Comparison of Creep Fitting Curves with Experimental Results under Consolidated Undrained (CU) Conditions: (**a**) 15 °C; (**b**) 35 °C.

https://doi.org/10.3390/app16062820

5. Numerical Simulation and Engineering Applications

5.1. Numerical Simulation of Triaxial Creep Testing

To perform the engineering numerical simulation, a numerical reproduction of the triaxial creep tests was first conducted. In this section, numerical simulations of consolidated drained (CD) and consolidated undrained (CU) triaxial creep tests were carried out under a confining pressure of 0.2 MPa. In order to investigate the influence of temperature on creep behavior, two temperature conditions were considered, namely, 25 °C and 35 °C. Accordingly, the simulations were divided into four groups: 25 °C–CU, 25 °C–CD, 35 °C–CU, and 35 °C–CD.

In the model design, a cylindrical specimen with a diameter of 38 mm and a height of 76 mm was adopted as the geometric model for the consolidation creep simulation. The mesh discretization of the model is illustrated in Figure 8. The bottom surface of the cylinder was constrained in displacement, while the lateral surface was subjected to a confining pressure identical to that applied in the laboratory triaxial compression creep tests, set to 0.2 MPa.

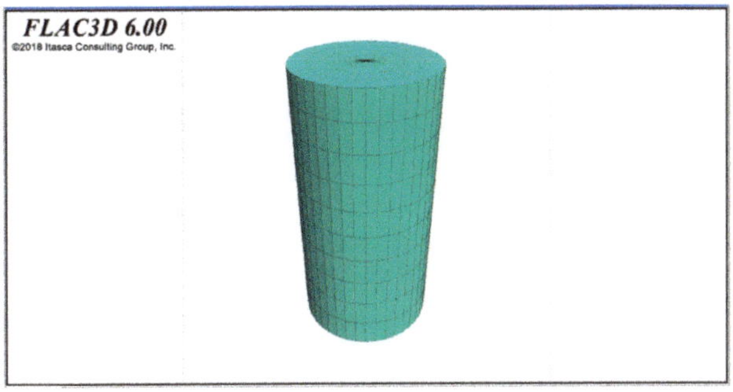

Figure 8. Finite Element Mesh of the Triaxial Creep Specimen Model.

At the top surface, a uniform axial load was applied in a stepwise loading manner along the negative y-direction, ensuring that the magnitude of the applied load was consistent with the axial loading conditions used in the laboratory tests. The numerical loading scheme implemented in the model is presented in Figure 9.

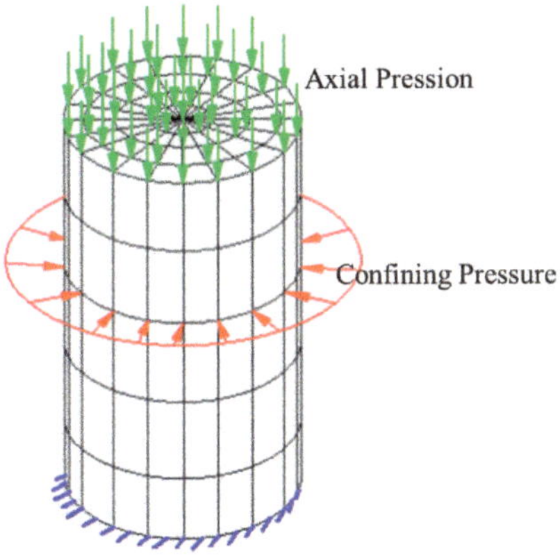

Figure 9. Stepwise Axial Loading Scheme Applied in the Numerical Simulation.

5.2. Numerical Simulation Results of Triaxial Creep Tests

Figure 10 presents the numerical simulation results of the triaxial step-loading creep tests for the dredged fill soil. Based on the numerical model established according to the laboratory testing conditions, the time-dependent deformation behavior of the soil specimen under staged loading was simulated. The numerical results successfully reproduce the main characteristics of the creep process, including the instantaneous deformation occurring immediately after loading and the subsequent gradual development of creep strain with time.

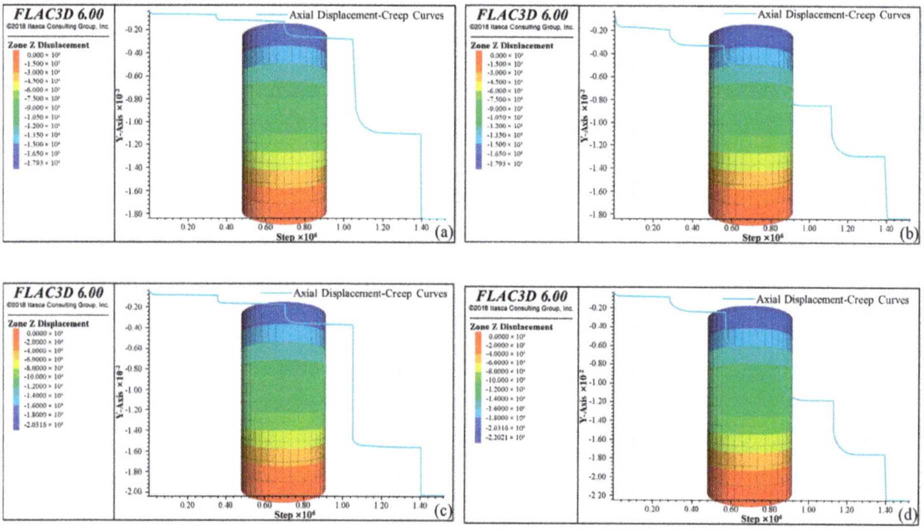

Figure 10. Axial Displacement-Creep Curves of Dredged Fill Soil under a Confining Pressure of 0.2 MPa at Different Temperatures and Drainage Conditions: (**a**) CU, 25 °C, (**b**) CD, 25 °C, (**c**) CU, 35 °C, (**d**) CD, 35 °C. (Unit: m).

To further evaluate the reliability of the numerical model, the simulation results were compared with the experimental data obtained from laboratory triaxial creep tests, as illustrated in Figure 11. The comparison indicates that the numerical curves generally follow the same evolution trend as the experimental curves, and the strain development under different loading stages shows a good level of agreement. Although minor deviations can be observed at certain time intervals, the overall consistency between the numerical simulation and experimental results is satisfactory.

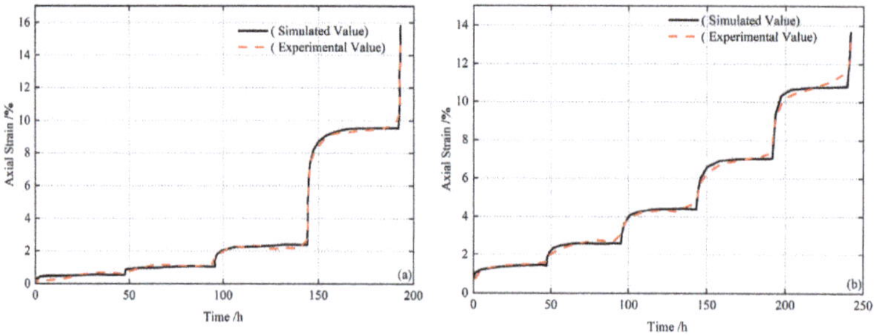

Figure 11. *Cont.*

https://doi.org/10.3390/app16062820

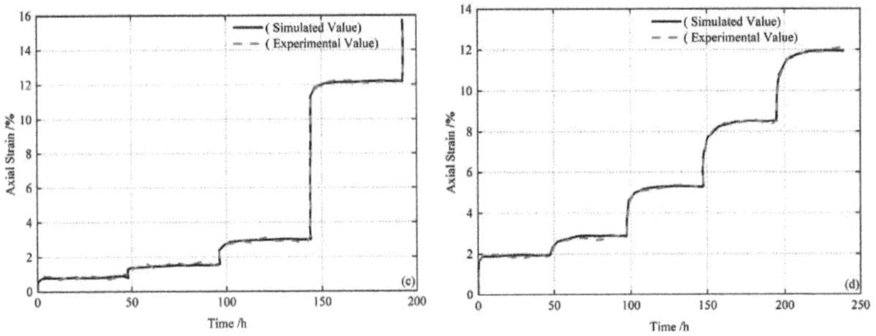

Figure 11. Comparison Between Numerical Simulation Results and Triaxial Creep Test Curves: (**a**) CU, 25 °C, (**b**) CD, 25 °C, (**c**) CU, 35 °C, (**d**) CD, 35 °C.

These results suggest that the adopted model parameters are capable of reasonably representing the creep behavior of dredged fill soil under triaxial loading conditions. Therefore, the proposed numerical model provides a reliable basis for further engineering simulations and settlement prediction of dredged fill foundations.

5.3. Computational Models

Taking the reclamation foundation project at Humen Port in Dongguan City, Guangdong Province as the subject of study, and incorporating previous research on reclaimed soil, finite difference simulation settlement analysis was conducted using soil layer calculation parameters as per Table 6. The simulation process was divided into two stages. In the first stage, conducted at ambient temperature (25 °C), the focus was on the initial phase—specifically, soil settlement under preloading without considering soil rheological effects. This stage simulation lasted 100 days. The accuracy of the model was validated by comparing simulation results with measured in situ settlement values.

Table 6. Soil Calculation Parameters.

No.	Soil Layer	Thickness H /m	Saturated Unit Weight γ_{sat} /kN·m^{-3}	Poisson's Ratio ν	Cohesion c/kPa	Friction Angle Φ /°	Initial Porosity e_o	Modulus of Deformation Es /MPa	Horizontal Permeability k_h /m·s^{-1}	Vertical Permeability k_v /m·s^{-1}
1	Dredged fill soil	5.6	18.4	0.33	10.4	9.40	1.438	2.53	1.82×10^{-9}	5.12×10^{-9}
2	Silt	6.8	17.8	0.40	9.8	12.99	1.220	1.95	1.41×10^{-8}	3.28×10^{-9}
3	Silty silt	10.3	19.3	0.35	11.75	8.57	1.712	3.04	2.76×10^{-8}	1.04×10^{-10}
4	Silty clay	12.5	18.7	0.38	9.77	7.78	1.775	2.15	9.64×10^{-10}	2.12×10^{-10}

In the second stage, the preloading effect was removed and soil rheological properties were incorporated. This stage simulated conditions at 35 °C during summer, aiming to predict the influence of creep parameters on long-term settlement. To more accurately model soil creep behavior, the constitutive creep model incorporating temperature effects from Section 3 was employed. The numerical calculations considered the effects of removing preloading after its conclusion and analyzed the long-term impact of soil rheological properties on settlement. Through this stage of simulation, axial settlement was predicted for the six months following construction, and corresponding creep curves were plotted to further investigate the creep characteristics during the settlement process. Additionally, preloading consolidation tests were conducted on the reclaimed land foundation in the Humen Port area, with the test section width reaching 30 m. In model construction, the foundation's computational width was set at 100 m to ensure simulation results align with actual test conditions. The schematic diagram of the model is shown in Figure 12, and the model mesh division is shown in Figure 13.

https://doi.org/10.3390/app16062820

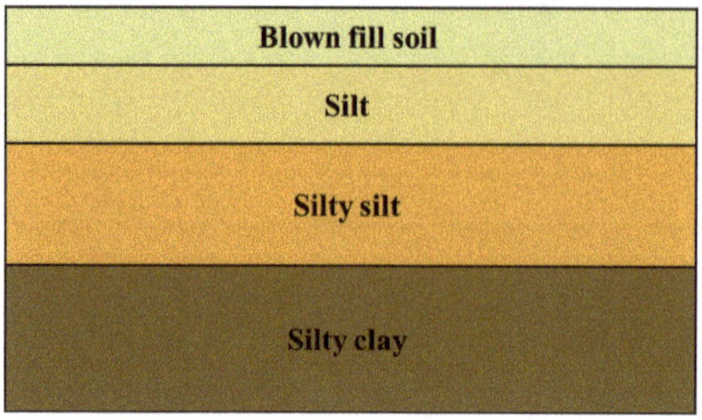

Figure 12. Schematic Diagram of the Model.

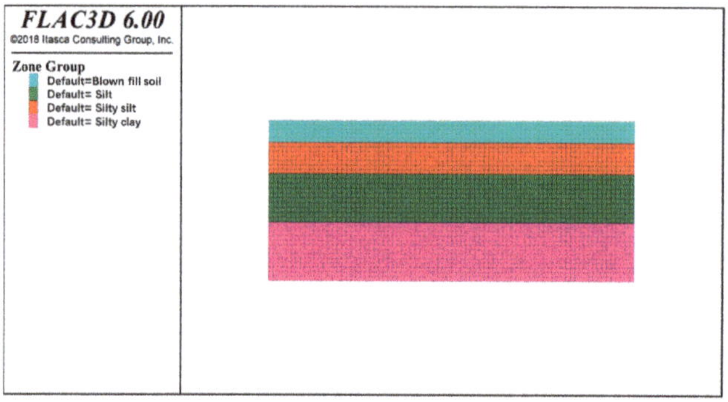

Figure 13. Model Mesh Partitioning.

5.4. Soil Settlement Under Preloading

At ambient temperature of 25 °C, the numerical calculation for preloading at 100 kPa includes monitoring soil settlement during the initial stage under preloading conditions, sustained for 100 days. This is compared with actual measured settlement values as shown in Figure 14.

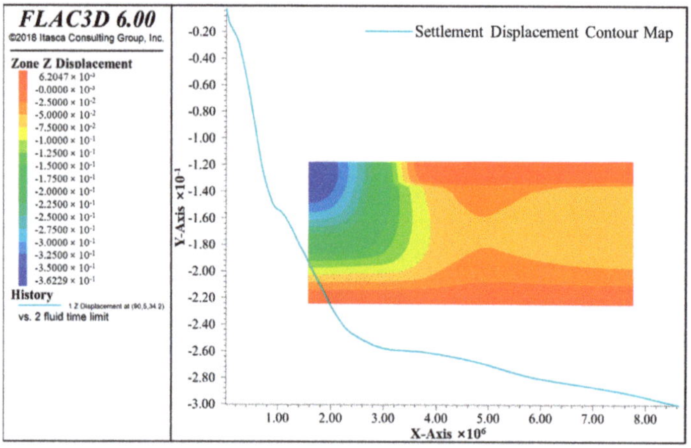

Figure 14. Settlement Displacement Contour Map for the Preloading Stage (Unit: m).

Calculated settlement curve for the center surface of the reinforced zone:

https://doi.org/10.3390/app16062820

As shown in Figure 15, although the computational model employed in this study exhibits certain discrepancies from actual conditions, the general patterns remain consistent. Furthermore, the simulated results are largely comparable to the measured values. The simulated ground settlement in this instance amounts to approximately 36.2 cm, with the model's calculated value differing only marginally from the actual measured value.

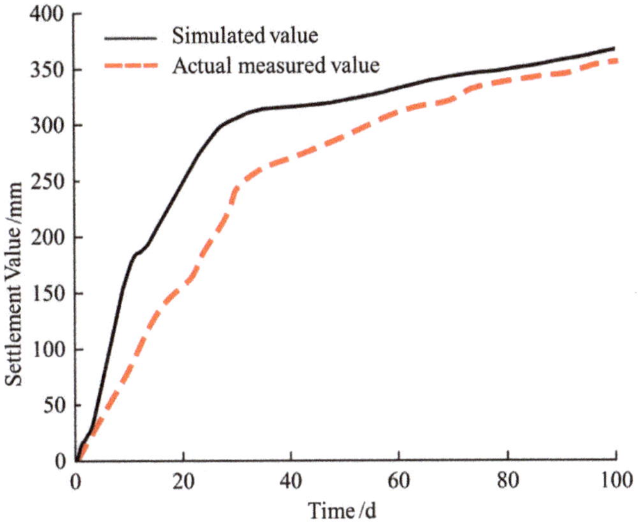

Figure 15. Surface Settlement Diagram for the Preloading Stage.

5.5. Prediction of Ground Subsidence Depth

The summer temperature value of 35 °C was calculated to predict the impact of soil rheological properties on long-term settlement after lifting preloading. To account for creep parameters, axial settlement was forecasted for the six months post-construction using a derived constitutive creep model incorporating temperature effects. Figure 16 presents the predicted axial settlement and creep curve for the six-month post-construction period. The figure indicates a settlement of approximately 39.5 cm at this stage, which is approximately 3.3 cm greater than the settlement value during the preloading phase.

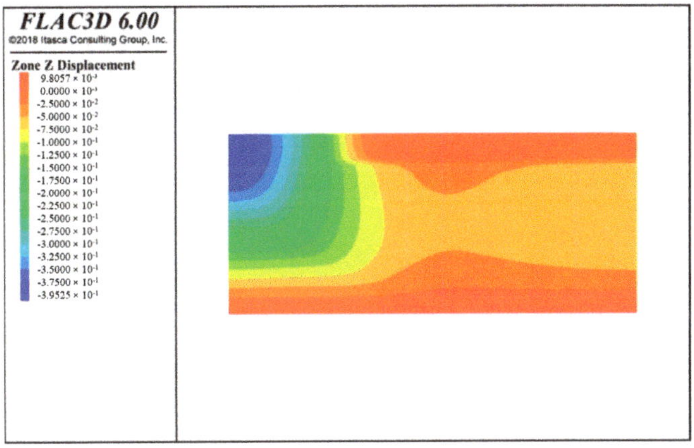

Figure 16. Axial Settlement Prediction at 35 °C Six Months Post-Construction (Unit: m).

https://doi.org/10.3390/app16062820

6. Conclusions

This study systematically investigates the influence of temperature effects on the creep behavior of dredged fill material from Humen Port, Dongguan City, Guangdong Province, through a series of triaxial creep tests. A creep constitutive model incorporating temperature effects was developed, yielding the following principal conclusions:

1. Temperature elevation markedly intensifies creep deformation in the fill material, with heightened temperature sensitivity under undrained conditions; drainage effectively suppresses creep progression. Shear stress exerts a nonlinear driving effect on creep strain, accelerating structural damage and approaching instability beyond a critical threshold (approximately 150 kPa).
2. Temperature and deviatoric stress exert opposing effects on rheological parameters: elevated temperatures reduce η_1, η_2, G_1, and G_2, enhancing soil fluidity; increased deviatoric stress, conversely, elevates these parameters, improving soil resistance.
3. The temperature-dependent creep constitutive equation derived from the Burgers model accurately describes the creep behavior of reclaimed fill soil under varying temperature and stress conditions ($R^2 > 0.97$). This provides a theoretical foundation for long-term foundation stability analysis and engineering design.

Author Contributions: Conceptualization, Q.C. and X.X.; methodology, Q.C.; software, C.Z.; validation, Q.C., X.X. and C.Z.; formal analysis, X.X.; investigation, X.X.; resources, C.Z.; data curation, C.Z.; writing—original draft preparation, X.X.; writing—review and editing, X.X.; visualization, C.Z.; supervision, Q.C.; project administration, Q.C.; funding acquisition, Q.C. All authors have read and agreed to the published version of the manuscript.

Funding: Qiunan Chen discloses support for the research of this work from National Natural Science Foundation of China [grant number 52478341] and Key Project of the Hunan Provincial Natural Science Foundation [grant number 2025JJ30022]. Xiaodi Xu discloses support for publication of this work from Hunan Province Postgraduate Research Innovation Programme [grant number CX20231018].

Data Availability Statement: The data utilized in this study were collected and processed by the authors themselves and have not been publicly released. However, the authors are willing to share these data with other researchers within reasonable limits to facilitate academic exchange and collaboration. Interested readers are invited to contact [Xiaodi Xu and 22300201004@mail.hnust.edu.cn].

Conflicts of Interest: The authors declare no conflicts of interest.

References

1. Yang, S.; Yang, A.; Lang, R. Research on Creep Characteristics of Blow-filled Cured Lightweight Soils under the Action of Wet-dry Cycles. *J. Disaster Prev. Mitig. Eng.* **2024**, *44*, 478–487. [CrossRef]
2. Yu, X.; Liu, S.; Miao, L. Creep Properties of Lianyungang Soft Clay and Its Engineering Application. *Rock Soil Mech.* **2003**, *6*, 1001–1006. [CrossRef]
3. Zhu, H.; Chen, X.; Cheng, X.; Zhang, B. Study on creep characteristics and model of soft soil considering drainage condition. *Rock Soil Mech.* **2006**, *5*, 694–698. [CrossRef]
4. Liu, W.; Cao, D.; Li, H.; Zhao, C. Experimental Study on Creep Behavior of Marine Clay in Shenzhen. *Ind. Constr.* **2011**, *41*, 74–77+81. [CrossRef]
5. Xie, X.; Li, J.; Wang, W.; Liu, K.; Zhu, X. Rheological Test and Empirical Model of Ningbo Soft Soil. *J. Zhejiang Univ. Eng. Sci.* **2012**, *46*, 64–71. [CrossRef]
6. Lei, H.; Jia, Y.; Li, X. Experimental Study of Nonlinear Creep Property of Soft Soil in Littoral Area. *Chin. J. Rock Mech. Eng.* **2013**, *32*, 2806–2816.
7. Yang, A.; Kong, L.; Zhang, X. Analysis of evolution of particles and pores in creep process of dredger fill soft soil. *Rock Soil Mech.* **2014**, *35*, 1634–1640. [CrossRef]
8. Luo, Q.; Chen, X. Experimental Research on Creep Characteristics of Nansha Soft Soil. *Sci. World J.* **2014**, *2014*, 968738. [CrossRef]
9. Tan, R.; Jiao, Y.; Xu, W. Tests on the Creep Parameters of the Soft Soil in Tianjin Coastal Region and Analysis on the Effect of Subgrade Settlement Deformation. *Hydrogeol. Eng. Geol.* **2015**, *42*, 67–73. [CrossRef]

10. Liu, P.; Jia, J.; Du, Y. Experimental Study on Creep Behavior of Marine Soft Clay Under Low Frequency Cyclic Loading. *Chin. J. Undergr. Space Eng.* **2015**, *11*, 1193–1198. [CrossRef]
11. Deng, H.; Dai, G.; Qiu, G.; Chen, Z.; Lin, X. Drained creep test and component creep model of soft silty clay in Hangzhou Bay. *J. Southeast Univ. Nat. Sci. Ed.* **2021**, *51*, 318–324. [CrossRef]
12. Liang, R.; Zhang, Q.; Zhang, C.; Li, D.; Su, D.; Li, Q. Creep characteristics and model parameters of Qianhai soft soil. *Chin. J. Geotech. Eng.* **2021**, *43*, 133–136. [CrossRef]
13. Yin, Q.; Dai, J.; Dai, G.; Gong, W.; Zhang, F.; Zhu, M. Study on Creep Behavior of Silty Clay Based on Fractal Derivative. *Appl. Sci.* **2022**, *12*, 8327. [CrossRef]
14. Yuan, J.; Gan, Y.; Chen, J.; Tan, S.; Zhao, J. Experimental Research on Consolidation Creep Characteristics and Microstructure Evolution of Soft Soil. *Front. Mater.* **2023**, *10*, 1137324. [CrossRef]
15. Dahhaoui, H.; Belayachi, N.; Zadjaoui, A.; Nishimura, T. One-Dimensional Compression Creep Change under Temperature and Suction Effects. *Int. J. Geotech. Eng.* **2022**, *16*, 670–681. [CrossRef]
16. Wang, L. An Analytical Model for 3D Consolidation and Creep Process of Layered Fractional Viscoelastic Soils Considering Temperature Effect. *Soils Found.* **2022**, *62*, 101124. [CrossRef]
17. Campanella, R.G.; Mitchell, J.K. Influence of Temperature Variations on Soil Behavior. *J. Soil Mech. Found. Div.* **1968**, *94*, 709–734. [CrossRef]
18. Chen, Z.; Zhao, R.; Chen, W.; Wu, P.; Yin, J.; Feng, W. Effects of Temperature on the Time-Dependent Compression and Shear Behavior of a Soft Marine Clayey Soil. *Eng. Geol.* **2023**, *314*, 107005. [CrossRef]
19. Li, A.; Chen, Z.-J.; Feng, W.-Q.; Yin, J.-H. The Effects of Temperature on One-Dimensional Consolidation and Creep Behaviors of Hong Kong Marine Deposits. *Int. J. Geomech.* **2023**, *23*, 04023215. [CrossRef]
20. Chen, Z.; Cheng, X. Thermodynamic constitutive theory and analysis of consolidation compression and creep of saturated soils. *Chin. J. Geotech. Eng.* **2014**, *36*, 489–498. [CrossRef]
21. Deng, Y.; Mao, W.; Yu, L.; Zhu, Y.; Xie, K. Consolidation and Creep Coupling Model for Soft Soil Considering Temperature Effect. *Eng. Mech.* **2022**, *39*, 103–113. [CrossRef]
22. Chen, Z.; Yin, J. A New One-Dimensional Thermal Elastic-Viscoplastic Model for the Thermal Creep of Saturated Clayey Soils. *J. Geotech. Geoenviron. Eng.* **2023**, *149*, 04023010. [CrossRef]
23. Yao, Z.; Zhang, W.; Guo, M. An Internal-variable Creep Model for Frozen Clay in Consideration of Temperature Effect. *J. Yangtze River Sci. Res. Inst.* **2020**, *37*, 81–85+91. [CrossRef]
24. Luo, F.; Zhang, Y.; Zhu, Z.; Zhang, D.; He, J. A Creep constitutive model for frozen sand of Qinghai-Tibet Plateau. *J. Harbin Inst. Technol.* **2020**, *52*, 26–32. [CrossRef]
25. *JTG 3430-2020*; Standard Test Methods of Soils for Highway Engineering. Ministry of Transport of the People's Republic of China, China Communications Press: Beijing, China, 2020.

https://doi.org/10.3390/app16062820

applied sciences

MDPI

Article

The Influence of Periodic Temperature on Salt Rock Acoustic Emission, Strength, and Deformation Characteristics

Yuxi Guo [1,2], Yan Qin [1,2,*], Nengxiong Xu [1,2,*], Huayang Lei [3], Junhui Xu [4], Bin Zhang [1,2], Shuangxi Feng [3] and Liuping Chen [4]

1 School of Engineering and Technology, China University of Geosciences (Beijing), Xueyuan Road 29, Beijing 100083, China; 3002230011@email.cugb.edu.cn (Y.G.); sc_zhb@cugb.edu.cn (B.Z.)
2 Engineering and Technology Innovation Center for Risk Prevention and Control of Major Project Geosafety, MNR, Beijing 100083, China
3 Department of Civil Engineering, Tianjin University, Tianjin 300354, China; leihuayang74@163.com (H.L.); shuangxi@tju.edu.cn (S.F.)
4 China National Salt Industry Group Co., Ltd., Beijing 100055, China; jhxu@chinasalt-jt.com (J.X.); jsb@chinasalt-jt.com (L.C.)
* Correspondence: yanqin@cugb.edu.cn (Y.Q.); xunengxiong@cugb.edu.cn (N.X.)

Abstract

During the long-term operation of salt cavern gas storage, multiple injections and extractions of gas will cause periodic temperature changes in the storage, resulting in thermal fatigue damage to the surrounding rock of the salt cavern and seriously affecting the stability of the storage. This article takes the salt rock samples after thermal fatigue treatment as the research object, adopts a uniaxial compression test, and combines DIC and Acoustic Emission (AE) technology to study the influence of different temperatures and cycle times on the mechanical properties of salt rock. The results indicate that as the number of cycles and upper limit temperature increase, thermal stress induces continuous propagation of microcracks, leading to continuous accumulation of structural damage, enhanced radial deformation, and intensified local displacement concentration, causing salt rock to enter the failure stage earlier. The initial stress for expansion and the volume expansion at the time of failure both show a decreasing trend. After 40 cycles, the compressive strength and elastic modulus decreased by 23.8% and 27.4%, respectively, and the crack failure mode gradually shifted from tension-dominated to tension-shear composite. At the same time, salt rock exhibits typical "elastic-plastic creep" behavior under uniaxial compression, but the uneven expansion and thermal fatigue effects caused by periodic temperature changes suppress plastic slip, resulting in an overall decrease in peak strain energy. The proportion of elastic strain energy increases from 21% to 38%, and the deformation process shows a trend of enhanced elastic dominant characteristics. The changes in the physical and mechanical properties of salt rock under periodic temperature effects revealed by this study can provide an important theoretical basis for the long-term safe operation of underground salt cavern storage facilities.

Keywords: salt rock; periodic temperature; mechanical properties; failure characteristics; acoustic emission characteristics

1. Introduction

During the long-term geological evolution, saturated multi-source brines undergo processes such as concentration, crystallization, sealing, and compaction to ultimately form salt rocks [1]. Salt rock, due to its low permeability, plasticity, and self-healing properties [2,3], has become an ideal medium for underground strategic energy storage, particularly suitable for constructing large-scale underground gas storage facilities [4,5] to meet the requirements of long-term stability and sealing of storage facilities [6]. During the long-term operation of salt cavern gas storage, multiple gas injections and extractions will cause periodic changes in the temperature of the salt rock inside the chamber [7,8]. Unlike constant high-temperature conditions, periodic temperature changes cause salt rock to repeatedly undergo thermal expansion and contraction, resulting in continuous adjustment of internal structural stress, which in turn affects its mechanical properties and seriously threatens the stability of salt cavern gas storage.

A large number of studies have found that the compressive strength of salt rock is lower than that of general limestone, sandstone, dolomite, etc., and the deformation modulus is 1–2 orders of magnitude lower, belonging to the category of soft rock [9–11]. In the stage division of stress-strain curves, salt rock shares common characteristics with other rocks and can be divided into four stages: compaction, elastic, inelastic, and failure [12,13]. However, compared with common rocks, salt rock, which is mainly composed of NaCl and has a polycrystalline structure, still exhibits distinct and unique mechanical properties [14,15]. Figure 1a–f shows the stress-strain curves prior to the peak under uniaxial compression for rocks, including the linear elastic type, elastic-plastic type, plastic-elastic type, plastic-elastic-plastic type I, plastic-elastic-plastic type II, and elastic-plastic-creep type. At low stress levels, the strain recovery ability is limited. As the load is continuously applied, the stress-strain curve rises gently until failure [11,12,16], exhibiting typical "elastic-plastic-creep" characteristics [12,17,18], as shown in Figure 1f. Especially in the storage and release cycle of salt cavern gas storage, the plastic strengthening stage of salt rock is inevitable [1,12], which has a decisive impact on the long-term stability of the gas storage. Given the unique "elastic-plastic-creep" stress–strain behavior of salt rock, its polycrystalline structure tends to evolve under the influence of cyclic temperature variations, thereby leading to corresponding changes in its mechanical properties.

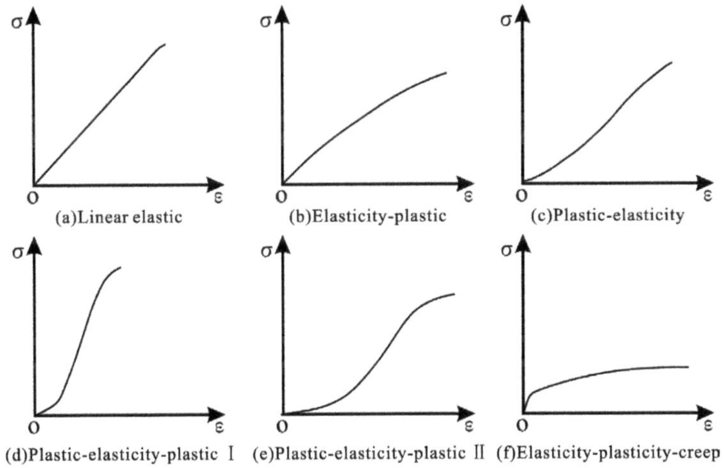

Figure 1. Types of rock stress-strain curves before peak [17]. Note: (**a**) Linear elastic; (**b**) Elasticity-plastic; (**c**) Plastic-elasticity; (**d**) Plastic-elasticity-plastic I; (**e**) Plastic-elasticity-plastic II; (**f**) Elasticity-plasticity-creep.

Temperature changes have a significant impact on the mechanical properties of salt rocks [11]. Numerous studies have shown that the mechanical properties of salt rock exhibit different trends of change in different real-time high-temperature ranges. When the temperature is below 120 °C, the compressive strength and elastic modulus of salt rock gradually decrease with increasing temperature [19–21]. At 120–200 °C, the compressive strength and elastic modulus of salt rock increase with temperature, especially at 170 °C, where the stress-strain characteristics of salt rock undergo a sudden change. This temperature is considered the threshold for the sudden change in the mechanical properties of salt rock [20]. As the temperature exceeds 500 °C, the internal structure of salt rock undergoes significant changes, leading to a significant decrease in peak intensity [20,22]. In addition, the changes in the mechanical properties of salt rock after cooling under high temperature are consistent. As the temperature increases, the mechanical properties of salt rock gradually deteriorate, manifested by a decrease in elastic modulus and peak stress, as well as an increase in peak strain [11,23,24]. Through analysis of acoustic emission energy and counting, it was found that microcracks in salt rock are mainly concentrated during the process of temperature rise. Further analysis of the acoustic emission spectrum characteristics indicates that the cracks formed under low temperature conditions are mainly large-sized intergranular cracks, while under high temperature conditions, small-sized intragranular cracks are mainly generated [25,26]. On this basis, experts and scholars further studied the relationship between the elastic modulus of salt rock and temperature and derived a salt rock damage constitutive model based on temperature stress coupling conditions [27,28]. At present, research on the effect of temperature on the mechanical properties of salt rocks mainly focuses on the deformation laws and strength characteristics of salt rocks after real-time constant temperature or single heating and cooling. However, during the long-term operation of salt cavern gas storage, the internal temperature undergoes periodic fluctuations, and the resulting thermal fatigue damage in the salt rock cannot be ignored [7,24,29]. Previous studies on rocks such as granite, marble, and sandstone have shown that periodic temperature changes have cumulative destructive effects on the macroscopic mechanical properties and microstructure of rocks [30,31]. The cyclic temperature effect can promote the propagation of internal cracks and the increase in grain boundary spacing in rocks such as marble and sandstone [31], resulting in a gradual decrease in strength, wave velocity, and deformation ability [30,32,33], while the volume and permeability show an increasing trend [34,35]. In contrast, there are few reports on the correlation between periodic temperature changes and the mechanical properties of salt rocks. With the advancement of salt cavern gas storage construction, it is urgent to conduct in-depth research on the mechanical properties of salt rock under periodic temperature effects.

This article takes salt rock samples after thermal fatigue treatment as the research object and conducts physical and mechanical tests on the samples to study the influence of different upper limit temperatures and cycle times on the physical and mechanical properties of salt rock. Combined with DIC and acoustic emission technology, qualitative and quantitative analysis is conducted on the evolution process and expansion behavior of salt rock damage, revealing the mechanical behavior of salt rock under thermal fatigue damage. The research results contribute to a deeper understanding of the impact of temperature changes on the mechanical properties of salt rock during the storage and release cycle of salt cavern gas storage and provide a theoretical basis and technical reference for ensuring the long-term stability and sealing of salt cavern gas storage projects.

2. Experimental Materials and Methods

2.1. Specimen Preparation

The salt rock used in this experiment appears reddish-brown and semi-transparent in appearance, mainly composed of NaCl, with a natural density of about 2.16 g/cm^3. Due to the easy dissolution of salt rock in contact with water, dry sawing and grinding methods were used to process the specimens, resulting in a standard cylindrical specimen with a height of 100 mm and a diameter of 50 mm. The basic physical and mechanical properties of the salt rock were determined in accordance with the International Standard for Rock Mechanics Testing (ISO 22282, 2012 [36]). The testing procedures and results are shown in Figure 2. The wave velocity and hardness of salt rock are mainly determined by RSM-SY6 (C) wave velocity meter and Leeb hardness tester (Produced in Beijing, China), which are 4.3–4.4 km/s and 256–269 HL, respectively (Figure 2a). According to the stress-strain curve of salt rock, the compressive strength is 37.35 MPa and the elastic modulus is 2.08 GPa (Figure 2b).

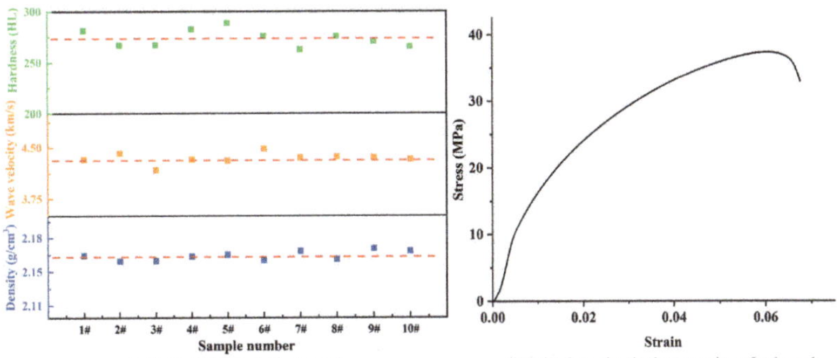

(a)Initial physical of salt rock (b)Initial mechanical properties of salt rock

Figure 2. Basic physical and mechanical properties testing of salt rock samples.

2.2. Experiment Scheme and System

Due to current limitations in experimental techniques, it is not yet possible to ensure that the heating-cooling process of the rock sample occurs simultaneously with the loading process, making it difficult to conduct truly coupled temperature-stress tests. Therefore, in order to achieve the research objectives, a simplified method was adopted, which first conducted temperature cycling tests on salt rock specimens, followed by physical property tests [37] and uniaxial compression tests. A total of 9 sets of experiments were designed for this experiment (Table 1), each using 3 parallel specimens. If there is significant variability in the test results, abnormal data should be removed and additional tests should be conducted to ensure data reliability. The specific experimental plan is as follows.

Table 1. Temperature cycle setting scheme.

Number	T_{max}/°C	T_{min}/°C	Number of Cycles	Temperature Cycle Diagram
T1	Normal temperature		0	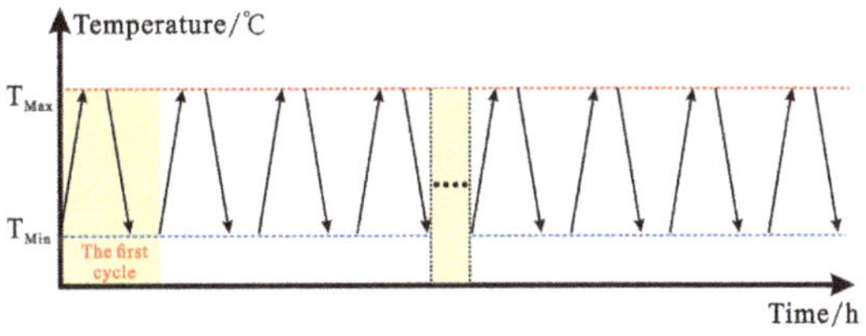
T2	60	40	5	
T3	60	40	10	
T4	60	40	20	
T5	60	40	40	
T6	50	40	10	
T7	60	40	10	
T8	70	40	10	
T9	80	40	10	

First, temperature cycling tests were conducted on salt rock specimens under different numbers of temperature cycles and various upper limit temperatures (Figure 3).

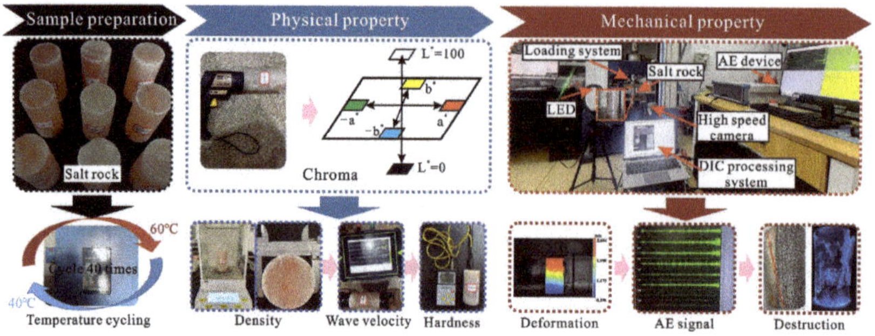

Figure 3. Experimental equipment and process.

(1) Considering that salt cavern gas storage facilities are typically buried at depths of 600–800 m, and based on a geothermal gradient of approximately 3 °C per 100 m with a surface temperature of about 20 °C [15,20], as well as referring to actual temperature variations during the storage–withdrawal cycles of salt caverns [7,38], the upper and lower temperature limits were finally set as T_{max} = 60 °C and T_{min} = 40 °C, respectively. In low-frequency and high-amplitude operation mode, salt cavern gas storage facilities usually undergo 1–2 injection production cycles per year, and their designed service life is generally 40–50 years [39,40]. Based on this, this study set four types of cycle times, namely 5, 10, 20, and 40, representing the initial, middle, and long-term service stages of the storage operation, respectively, to ensure that the experimental plan has good representativeness and engineering relevance (Table 1).

(2) Considering the extreme changes in salt cavern temperature under different storage conditions, different upper limit temperatures T_{max} are set at 50 °C, 60 °C, 70 °C, and 80 °C, respectively. The lower limit temperature and cycle times are the same, set at 40 °C and 10 times, respectively (Table 1).

In addition, each temperature cycle is set to 4 h, with a heating and cooling rate of 3–5 °C/min. After heating or cooling to the target temperature, the temperature is maintained for 2 h. Then, the salt rock samples that have completed the temperature cycle are taken out of the heating test chamber and subjected to physical property testing using a TES-135A colorimeter (Produced in Beijing, China), RSM-SY6 (C) wave velocity meter, and Leeb hardness tester [41]. Finally, mechanical property tests were conducted on the specimens, during which acoustic emission (AE) sensors were installed and full-field deformation was monitored using DIC. The test procedures and equipment are shown in Figure 3. The Acoustic Emission test was conducted using a DS5 multi-channel monitoring system (Produced in Beijing, China), with a threshold set at 10 dB. During the test, six acoustic emission sensors were evenly arranged and fixed at positions 15 mm from both the top and bottom end faces of the specimen to collect acoustic emission signals. To compare and analyze the changes in the physical and mechanical properties of salt rock specimens before and after temperature cycling tests, corresponding tests were conducted on specimens that had not undergone temperature cycling tests.

The color measurement of salt rock was primarily conducted using the TES-135A color analyzer to extract color parameters, with the CIE-Lab color space employed for color representation. This coordinate system quantifies a color using three primary values: L^*, a^*, and b^*. Specifically, L^* represents the dimensionless lightness, ranging from 0 to 100, where 0 corresponds to black and 100 to white; a^* denotes the green-red component, ranging from −100 (pure green) to 100 (pure red); and b^* indicates the yellow-blue component, ranging from −100 (pure blue) to 100 (pure yellow).

3. Results

3.1. Changes in Physical Properties

Considering that salt cavern gas storage may experience temperature cycles of different amplitudes and frequencies during service, the temperature range mainly affects the strength of thermal fatigue, while the number of cycles controls its cumulative evolution process, thereby affecting the physical properties of salt rock to varying degrees. Figures 4 and 5, respectively, show the experimental results of the influence of different temperature cycles and temperature ranges on the physical properties of salt rock.

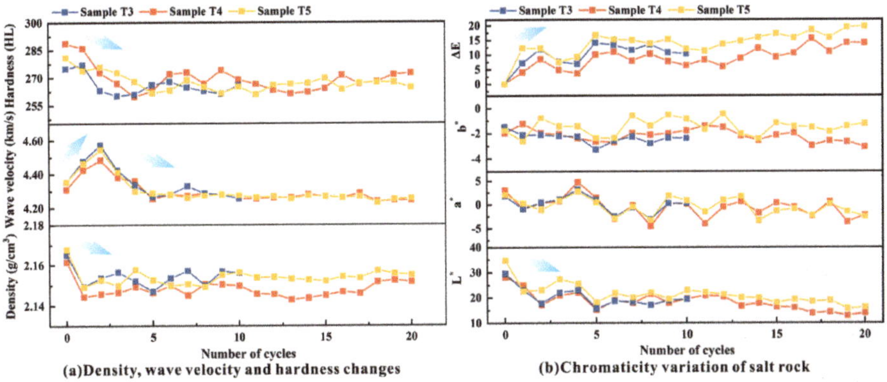

Figure 4. Changes in physical properties of salt rock under different cycle times.

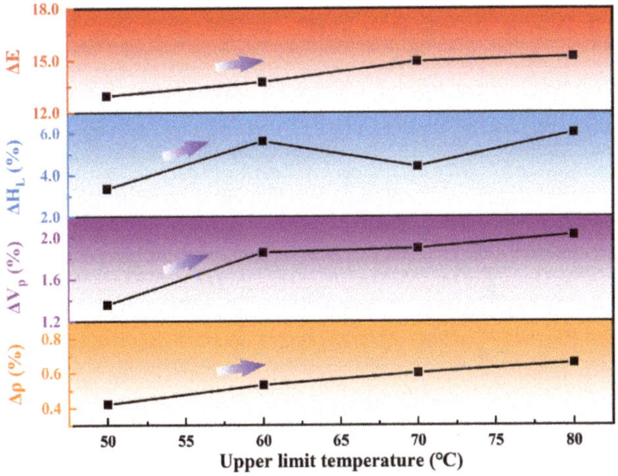

Figure 5. Changes in physical properties of salt rock under different upper temperature limits. Note: (The samples T6, T7, T8, and T9 in Table 1).

To compare and analyze the effect of temperature cycling on the physical properties of salt rock, salt rock samples that have undergone 10, 20, and 40 thermal cycles, respectively, within the same temperature range (40–60 °C) were selected for analysis. As shown in Figure 4a, the variation of salt rock density, wave velocity, and hardness with the number of temperature cycles under periodic temperature changes. The main component of the salt rock used in this experiment is NaCl, which contains a small amount of impurity minerals such as gypsum, calcite, and dolomite. Under the temperature cycling of 40–60 °C, a large amount of free water in the salt rock sample evaporates, and gypsum partially dehydrates to form hemihydrate gypsum [15,42], resulting in a decreasing trend in salt rock density. And as the number of cycles increases, the water loss in salt rock gradually decreases, and the density decreases and eventually stabilizes. In addition, as the number of temperature cycles increases, the overall hardness of salt rock shows a decreasing trend, while the wave velocity briefly increases at the beginning of the cycle and then also shows a decreasing trend. This is because in the initial stage of temperature cycling (0–3 times), the natural cracks inside the rock sample partially close under the action of thermal expansion, but the bonding force between grain boundaries decreases, and local new cracks emerge. This leads to a short-term increase in wave velocity, while the hardness, which reflects local resistance to deformation, begins to decrease. Entering the midterm stage of temperature cycling (3–8 times), thermal stress accumulates and crack propagation dominates [43]. The wave velocity and hardness of salt rock samples show a decreasing trend. In the later stage of temperature cycling (8–20 times), the development of internal fractures in the rock sample tends to stabilize, and at this time, the changes in salt rock wave velocity and hardness tend to stabilize.

The chromaticity changes of salt rock under different temperature cycles are shown in Figure 4b. The parameter L^*, which characterizes the brightness of salt rock, gradually decreases with the increase in temperature cycles, while the color difference ΔE shows a gradually increasing trend. In contrast, the values of parameters a^* and b^*, which reflect the changes in the red-green and yellow-blue chromaticity of salt rock, fluctuate overall with the increase in temperature cycles, without showing a significant upward or downward trend. This is because the salt rock used in the experiment is mainly NaCl, which has stable chemical properties and is not prone to redox reactions within the temperature cycling range of 40–60 °C. In addition, salt rocks are formed in terrestrial or shallow marine sedimentary environments [40], where a large amount of iron elements have been oxidized to Fe^{3+}, further reducing the impact of temperature changes on their chromaticity. In summary, under periodic temperature changes, the physical properties of salt rocks undergo a dynamic adjustment process and eventually tend towards a new stable state.

As the upper limit temperature of the thermal cycle increases, the physical properties of salt rock exhibit a certain degree of temperature sensitivity. As shown in Figure 5, the decrease in salt rock density ($\Delta\rho$), wave velocity (ΔV_p), hardness (ΔH_L), and color difference (ΔE) all gradually increase with the increase in the upper limit temperature, but their trends are not linear. Especially at low temperatures (50–60 °C), the changes in various parameters are significant, while at high temperatures (70–80 °C), the changes tend to stabilize. This indicates that as the upper limit temperature increases, the internal structure of the salt rock gradually weakens, but in the higher temperature stage, the structural damage tends to saturate, and further heating stabilizes its impact.

3.2. Stress-Strain Curve

Figure 6a–i shows the stress-strain curves and energy evolution characteristics of salt rock specimens under different temperature cycling cycles and upper temperature limits, respectively. From the figure, it can be seen that under uniaxial compression conditions, the axial stress-strain curve of the salt rock specimen exhibits a typical "elastic-plastic-creep" mechanical behavior [12,17], and its deformation process can be roughly divided into four stages, each corresponding to specific energy evolution characteristics:

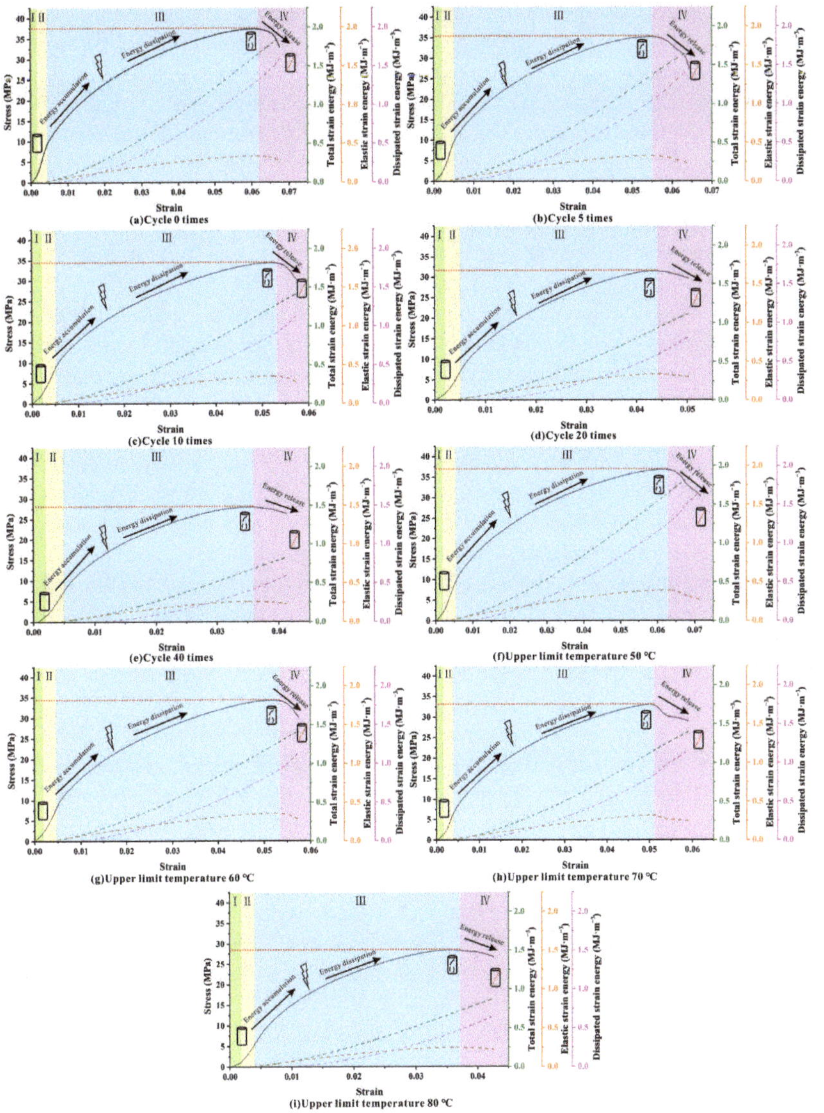

Figure 6. Stress-strain curve and energy evolution characteristics of salt rock under periodic temperature action. Note: (The samples T1~T9 in Table 1).

① In the compaction stage (Stage I), under the action of axial stress, the original defects (microcracks and micropores, etc.) inside the salt rock quickly close, and the mineral particles inside the salt rock become more compact. The axial stress-strain curve shows a concave shape.

② In the elastic deformation stage (stage II), salt rock transitions from the compaction stage to the deformation stage dominated by elastic interaction between grains under continuous axial stress. The axial strain increases approximately linearly with the increase in axial stress, and the amount of axial deformation is relatively small. This stage mainly involves energy accumulation. Due to the low initial stress, the strain energy curves change smoothly, and the total strain energy absorbed by the sample is mainly elastic strain energy.

The growth rate of the dissipated strain energy curve is lower than that of the elastic strain energy curve.

③ In the plastic deformation stage (stage III), when the axial stress exceeds the yield point, the stress-strain curve of salt rock shows an upward convex shape until it reaches the ultimate compressive strength. During this process, dislocations slide along the grain boundaries, exacerbating the micro rupture of pores and cracks, promoting stable propagation of micro cracks, and leading to a significant increase in axial deformation. In terms of energy evolution, this stage is mainly characterized by energy dissipation. In the initial stage, the energy absorbed by the sample is still mainly stored in the form of elastic strain energy, with less energy consumed for compaction and microcrack propagation. In the later stage, as the axial strain continues to increase, although the elastic strain energy is still increasing, the growth rate decreases, while the growth rate of dissipated strain energy gradually accelerates. This indicates that during the plastic deformation stage of salt rock, the energy dissipation process plays a dominant role and becomes a key factor affecting its mechanical behavior.

④ In the stage of failure (Stage IV), when the axial stress exceeds the compressive strength, the salt rock specimen undergoes deformation and failure, and the axial stress-strain curve rapidly drops. A large number of microcracks generated inside the salt rock specimen eventually penetrate to form a macroscopic fracture surface, and the bearing capacity rapidly decreases. At this stage, the elastic strain energy is rapidly released, the dissipated strain energy increases rapidly, and most of the input total strain energy is converted into dissipated strain energy. In addition, due to the presence of residual stress, although the structural integrity of the specimen is compromised, the total strain energy continues to rise.

As shown in Figure 7, the changes in compressive strength and elastic modulus of salt rock under different temperature cycle times and different upper limit temperature conditions are presented. From Figure 7a, it can be seen that the compressive strength and elastic modulus of the salt rock sample without periodic temperature action are 37.35 MPa and 2.1 GPa, respectively. After 10, 20, and 40 temperature cycles, the compressive strength of the salt rock decreased by 8.2%, 16.8%, and 23.8%, respectively, and the elastic modulus decreased by 20.4%, 29.7%, and 27.4%, respectively. It can be seen that the strength parameters of salt rock decrease with the increase in temperature cycles, and the rate of decrease gradually slows down. This is because as the number of temperature cycles increases, the damage caused by thermal stress to the internal structure of the specimen gradually accumulates, and microcracks initiate and gradually propagate inside the specimen, leading to a deterioration of the mechanical properties of the specimen. When the upper limit temperature remains constant and the number of temperature cycles increases, the number of microcracks generated inside the specimen gradually reaches the limit of this upper limit temperature, and the rate of decrease in compressive strength also slows down accordingly.

From Figure 7b it can be seen that the compressive strength and elastic modulus of the salt rock sample without periodic temperature action are 36.8 MPa and 1.7 GPa, respectively. When the number of cycles is 10 and the upper limit temperatures are 60, 70, and 80 °C, the compressive strength of the salt rock decreases by 6.9%, 11.9%, and 22.9%, and the elastic modulus decreases by 4.8%, 5.6%, and 6.5%, respectively. It can be seen that the strength parameters of salt rock decrease with the increase in the upper limit temperature, and the rate of decrease gradually accelerates. This is because different mineral particles have different coefficients of thermal expansion, which causes each mineral particle to be constrained by adjacent particles when deformed by periodic temperature effects, resulting in thermal stress [20]. Thermal stress often concentrates at the boundaries between mineral particles, causing stress concentration. When local stress exceeds the strength limit, microcracks are generated. As the upper limit temperature increases, the number and size of microcracks increase, eventually forming a widely distributed network of cracks, which gradually accelerates the deterioration of salt rock strength.

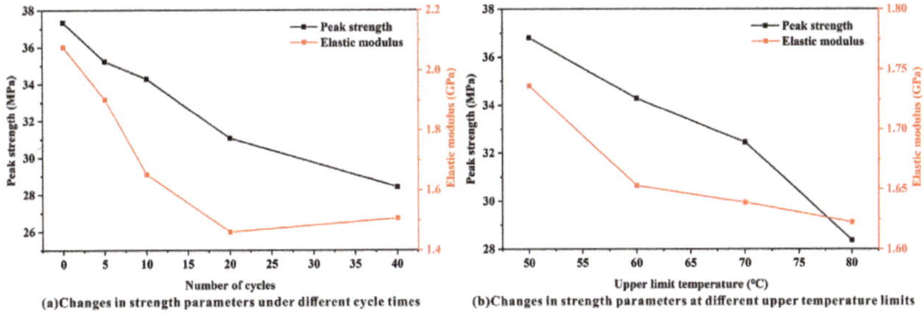

Figure 7. Strength parameters of salt rock under periodic temperature action. Note: ((**a**) is based on samples T1~T5, and (**b**) is based on samples T6~T9 in Table 1).

As shown in Figure 8, the peak strain energy (total strain energy U, elastic strain energy U_e, and dissipated strain energy U_d) of salt rock varies with the number of temperature cycles and the upper limit temperature. From the graph, it can be seen that as the number of cycles and upper limit temperature increase, the total strain energy, elastic strain energy, and dissipated strain energy at the peak point of the salt rock sample all decrease significantly. This indicates that salt rock absorbs more thermal energy under repeated thermal stress, leading to a large number of internal cracks, continuous accumulation of damage, and reduced structural integrity, making it more prone to failure under axial stress. Therefore, the energy required for the sample to reach peak strength is significantly reduced.

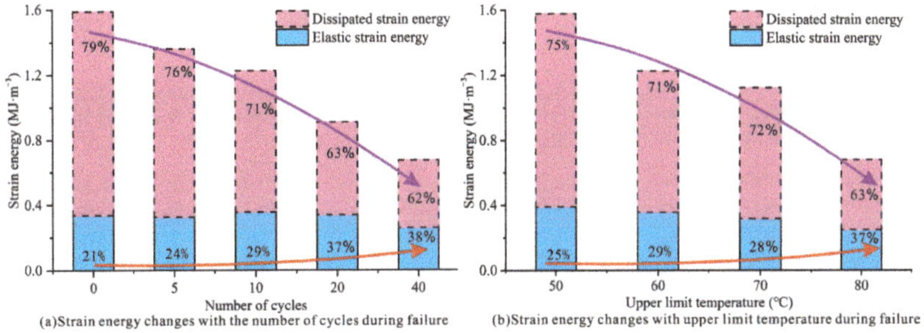

Figure 8. Characteristics of strain energy variation with cycle temperature during failure. Note: ((**a**) is based on samples T1~T5, and (**b**) is based on samples T6~T9 in Table 1).

However, it is worth noting that although all types of strain energy show a decreasing trend, the proportion of elastic strain energy in the total strain energy shows an increasing trend. This is because, unlike other rock masses, salt rock itself has a polycrystalline structure. After repeated temperature cycles, the internal structure of salt rock deteriorates significantly, and grain structure damage and microcrack accumulation inhibit the full development of plastic slip [15]. The strain process that should have entered the energy dissipation stage is "compressed", and its overall deformation tends to be more elastically dominated. This can also be confirmed by the strain values corresponding to the peak strength of the stress-strain curve in Figure 6. Therefore, the increase in the proportion of elastic strain energy essentially reflects the degradation trend of the plastic properties of salt rock under periodic temperature changes.

3.3. Acoustic Emission Characteristics

As shown in Figure 9, the evolution characteristics of ringing count (fracture activity) and total energy (damage scale) over time during the uniaxial compression test of salt rock under different temperature cycling times are presented [43]. From the figure, it can be seen that the process of sound emission ringing count and total energy change is divided into three stages:

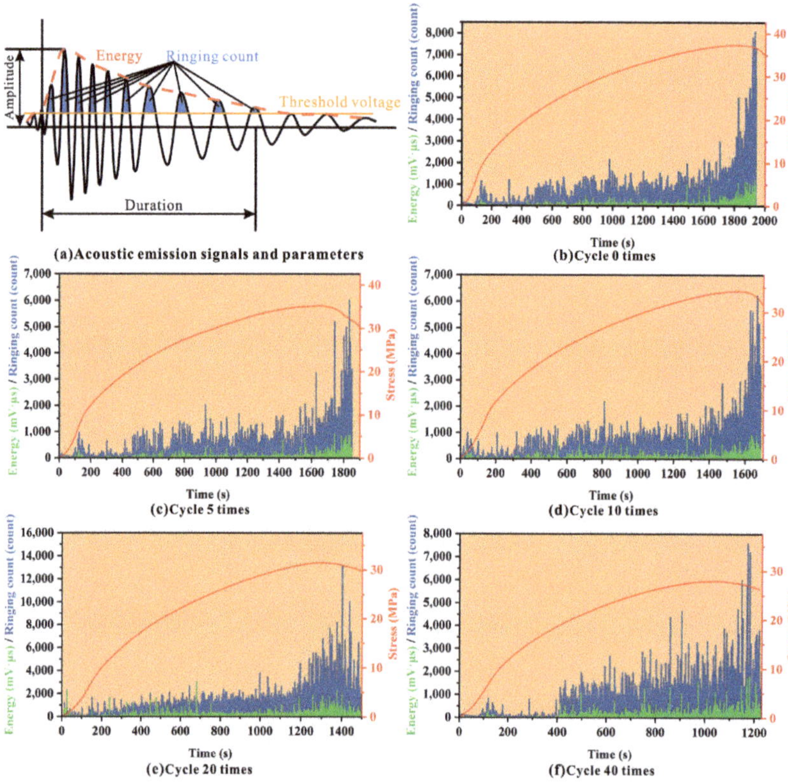

Figure 9. Acoustic emission counting, energy, and stress evolution under different temperature cycling cycles. Note: (The samples T1~T5 in Table 1).

① Active period: In the early stage of loading, the salt rock transitions from the compaction stage to the elastic stage. Due to the sudden loading of the salt rock (initial response triggered by fast loading), the initial cracks and pores inside the salt rock undergo instantaneous collision-type compression closure, resulting in impact failure. Therefore, during this stage, the sound emission signal is dense, and the ringing count and total energy continue to develop in a low-amplitude form.

② Calm period: After entering the elastic deformation stage, the acoustic emission ringing count and total energy begin to decrease, resulting in a sustained period of low signal, the "calm period". This is because after the similar impact effect ends, the salt rock is in the stage of elastic deformation, and the instantaneous accumulated elastic deformation will have a certain rebound recovery space due to the rapid generation of cracks and voids. During this stage, it mainly undergoes the release and recovery of elastic strain, and no new crack propagation has yet occurred. As the load continues to increase and enters the plastic deformation stage, microcracks begin to propagate, and acoustic emission signals reappear and remain at a relatively low stable level.

③ Destruction period: As the volume of salt rock expands and microcracks develop unstably, the acoustic emission ringing count and total energy rapidly increase when the peak strength of salt rock is about to be reached. When the peak strength is reached, the main cracks that cause salt rock failure have already formed. Subsequently, stress rapidly decreases and is insufficient to generate new cracks. At this point, the acoustic emission signal is mainly caused by the relative displacement and friction between the main cracks.

As shown in Figure 9, with the increase in the number of thermal cycles, both the ringing counts and total energy of salt rock during uniaxial compression failure exhibit a significant upward trend. After 20 thermal fatigue cycles, the ringing counts and total energy increased by 63.4% and 92.3%, respectively. This indicates that thermal fatigue enhances the activity of the fracturing process and enlarges the failure scale of salt rock.

Through extensive analysis of acoustic emission data, it was found that dividing the ringing count in the acoustic emission parameters by the duration is recorded as the average frequency AF, and dividing the rise time by the amplitude is defined as RA. The ratio of the two can be used to determine the type and proportion of cracks in the specimen after failure. Process the acoustic emission characteristic data of the salt rock sample and plot it as an AF-RA scatter density map, as shown in Figure 10. The slope of the boundary between tensile and shear cracks in the AF-RA scatter plot is AF_{max}/RA_{max} [43,44], where the scatter located on the left side of the boundary is defined as tensile cracks, and the scatter located on the lower right side of the boundary indicates the occurrence of shear cracks (Figure 10a).

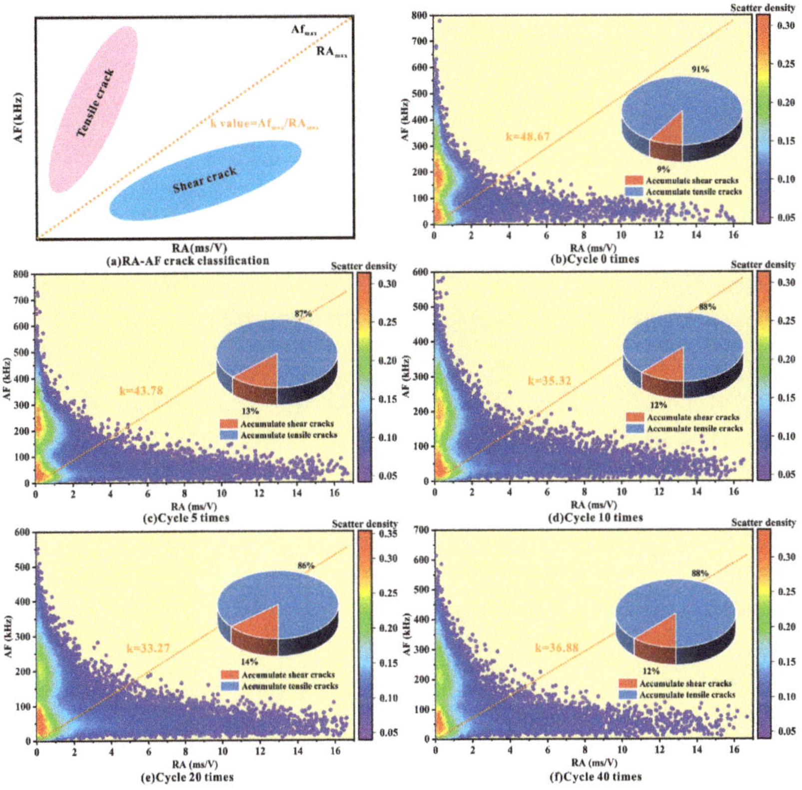

Figure 10. RA-AF probability density cloud map under different temperature cycle times. Note: (The samples T1~T5 in Table 1).

From Figure 10, it can be seen that the scatter points are distributed in a strip shape, closely following the RA axis and AF axis, and the distance from the scatter points to the density center is negatively correlated with the scatter density. From the density center to the outside, the scatter density decreases step by step. As the number of temperature cycles increases (Figure 10b–f), the position of the center of the scatter density changes significantly. In the absence of periodic temperature effects (Figure 10b), the number of data points for salt rock tensile failure is higher than that for shear failure [27], with tensile failure accounting for as high as 91%, indicating that the main form of sample failure is tensile failure. After 20 cycles of temperature action (Figure 10e), the proportion of salt rock tensile failure decreased to 86%, and the proportion of shear failure increased to 14%. Although tensile failure still dominated, the failure mode began to shift towards a combination of tensile and shear failure.

3.4. Compression Damage Characteristics Based on DIC

DIC stipulates that axial displacement is negative when it is vertically downward, and radial displacement is positive when it is horizontally rightward. As shown in Figure 11, the cloud map of the full field displacement of the salt rock surface after 40 temperature cycles is presented. For the convenience of analysis, three typical loading stages were selected as observation comparison points: elastic stage (①), plastic stage (②), and failure stage (③).

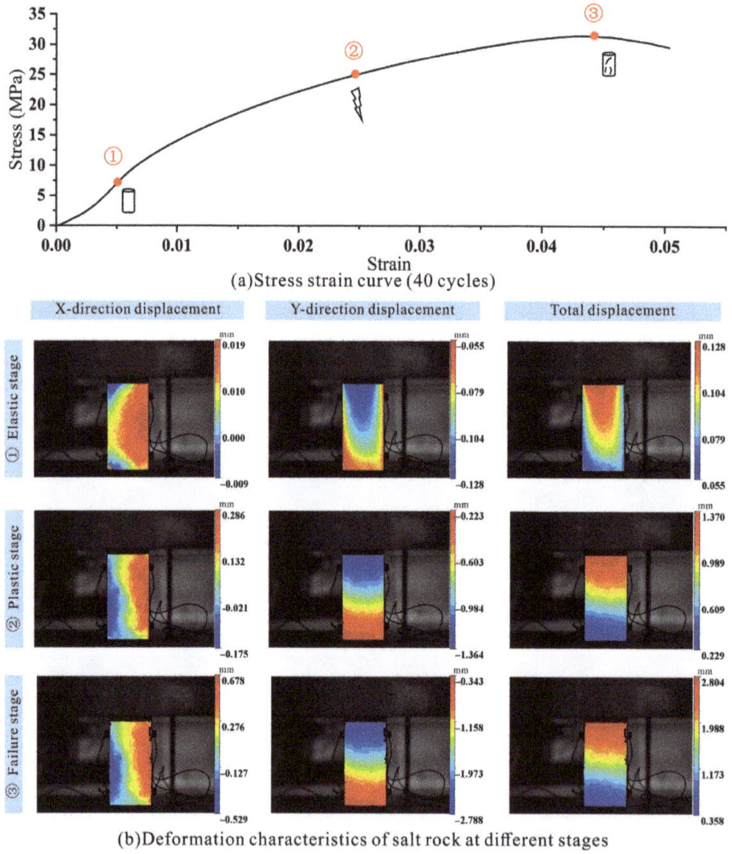

Figure 11. Full-field displacement cloud map of salt rock surface. Note: (The sample T5 in Table 1).

The left side of Figure 11 shows the radial displacement cloud map of the observed surface of the salt rock. In the elastic stage (①), the radial deformation of the salt rock is mainly concentrated in the lower half, and a significant local deformation state appears on the right side of the sample. As the load continues to be applied and enters the plastic stage (②), the radial displacement difference of the salt rock further intensifies, and the sample shows obvious expansion characteristics within a range of about 25–50 mm from the bottom. When entering the destruction stage (③), the displacement difference between the blue-green area on the left side and the red-yellow area on the right side of the salt rock significantly increases, forming a clear radial displacement mutation zone, exhibiting a strong Poisson effect. The deformation characteristics are consistent with the analysis results of AF-RA crack types and their proportions in Figure 10, indicating that the failure of salt rock is mainly caused by the tensile effect caused by the expansion on both sides.

The central image in Figure 11 shows the axial displacement cloud map of the observed surface of the salt rock. In the elastic stage (①), the displacement in the axial direction of the salt rock is negative, exhibiting a transfer-type compression state. After entering the plastic stage (②), the axial deformation of salt rock shows a clear gradient distribution characteristic, and the deformation of the upper part of the sample is significantly greater than that of the lower part, which is related to the loading end being located at the upper end of the sample. When entering the failure stage (③), the trend of axial displacement change remains unchanged, and the deformation of the upper part of the specimen continues to increase, without changing its evolution law. By comparing the numerical increments of axial and radial deformation of salt rock after entering the plastic stage, it was found that the final failure of the specimen was mainly caused by radial deformation.

On the right side of Figure 11 is the total displacement cloud map of the salt rock observation surface. In the elastic stage, the trend of total displacement variation is similar to axial deformation, indicating that the overall deformation of salt rock in this stage is mainly axial compression. After entering the plastic stage, although the total displacement and axial displacement cloud maps still show similar gradient distribution characteristics, there are significant differences between the two, mainly due to the gradual increase in radial deformation. According to the distribution characteristics of total displacement, the sample can be divided into three regions: A, B, and C. Region A is mainly red and has the highest total displacement. The B area is mainly yellow-green with a total displacement range of 0.6~0.9 mm, while the C area is mainly blue with a total displacement of about 0.2~0.6 mm. As the sample transitions from the plastic stage to the failure stage, the values in the red and yellow-green areas increase significantly, and a local concentration of total displacement appears in the lower left corner, showing a strain accumulation trend in the early stage of failure.

By comparing the stress-strain curves of salt rock under different temperature cycles (Figure 12a), it can be seen that with the increase in temperature cycles, the axial stress-strain curve shifts to the lower left, indicating that the axial deformation capacity of salt rock gradually weakens and the compressive strength also gradually decreases. To further analyze this trend, samples under the same strain conditions during the plastic stage were selected to compare the displacement cloud maps of the salt rock surface under different temperature cycling times.

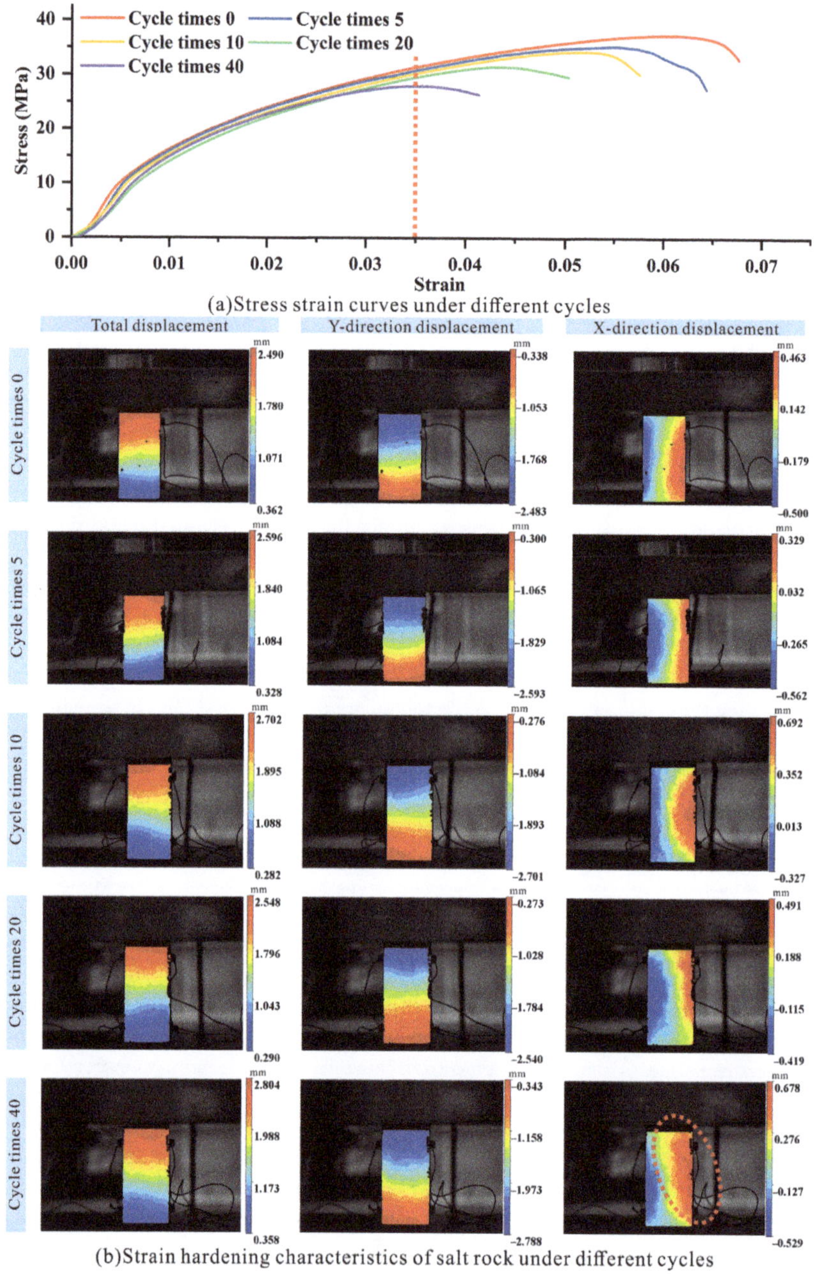

Figure 12. Displacement cloud map of salt rock plastic stage under cyclic temperature action. Note: (The samples T1~T5 in Table 1).

From Figure 12b, it can be seen that during the plastic stage under the same strain conditions, as the number of temperature cycles increases, the axial deformation of the salt rock is basically similar, but there is a clear displacement discontinuity zone in its radial deformation, resulting in deformation accumulation in local areas of the total displacement cloud map. This is because the temperature is constantly alternating states, and the salt rock will repeatedly expand and contract accordingly. Different areas inside the sample will produce uneven expansion rates, which will lead to the formation of local residual stresses. In addition, during the continuous temperature cycling process, thermal fatigue can promote the early development of cracks and damage within the polycrystalline structure of salt rock, and local damage limits the slip behavior of dislocations, resulting in a damage accumulation effect far greater than the plastic strengthening effect [15], ultimately leading to the specimen entering the failure stage earlier, manifested as a decrease in peak stress and peak strain. This can also be confirmed by the variation of peak strain energy (total strain energy U, elastic strain energy U_e, and dissipated strain energy U_d) of salt rock with the number of temperature cycles in Figure 8.

4. Discussion

When salt rock undergoes periodic temperature changes, thermal stress induces the continuous propagation of microcracks, leading to the accumulation of structural damage, increased radial deformation, and concentrated local displacement, causing it to enter the failure stage earlier, and the type of crack failure gradually shifts from tension-dominated to tension-shear composite. As a polycrystalline rock, salt rock exhibits heterogeneity and discontinuity between its crystal grains and typically demonstrates "elastic-plastic-creep" behavior under uniaxial compression. When the stress reaches a certain threshold, microcracks continuously initiate, expand, and penetrate, ultimately leading to a significant volume expansion phenomenon (Figures 13 and 14). To further reveal the influence of periodic temperature changes on the thermal expansion characteristics of salt rock during uniaxial compression, an analysis is conducted on the initial expansion stress and the volumetric expansion capacity changes of salt rock under different cycle times and upper temperature limits.

To obtain the radial deformation of the salt rock specimens, the virtual extensometer method was employed by leveraging the full-field displacement and strain measurement capabilities of DIC technology. This approach effectively overcomes the limitations of traditional strain gauges in measuring non-uniform deformation and allows for the calculation of volumetric strain in salt rock. In this study, three radial virtual extensometers were placed within the selected analysis area on the specimen surface at vertical positions of 25 mm, 50 mm, and 75 mm from bottom to top along the horizontal direction, as shown in Figure 13f. The strain curve in this article follows the regulations of rock mechanics tests; compressive strain is considered positive, and tensile strain is considered negative.

The radial strain of salt rock obtained based on DIC technology under periodic temperature changes is shown in Figures 13 and 14. It is evident from the figure that the radial strain measured by the virtual extensometers is greatest at the midsection of the specimen. This is primarily due to the end effects experienced by the upper and lower ends of the specimen during mechanical testing. Additionally, the mechanical testing system used in this study employs active loading at the top and passive loading at the bottom, resulting in radial deformation being concentrated mainly in the specimen's midsection, followed by the lower and upper ends. Therefore, to ensure the representativeness of the strain measurement, the radial strain at the specimen's midsection was selected for calculating volumetric strain.

The strain of the salt rock mass obtained based on DIC technology under periodic temperature changes is shown in Figures 13 and 14. From the figure, it can be seen that all samples exhibit significant volume expansion characteristics [20]. At the beginning of loading, the sample is in the compression and elastic deformation stage, and the internal pores gradually close. At this time, the radial strain is small, and the volumetric strain is mainly dominated by the axial strain. The sample volume is in the rapid shrinkage stage. After entering the initial stage of plastic deformation, as the radial strain gradually increases, the stress-strain curve enters a nonlinear stage, and the sample volume enters a short-term stable stage; that is, the volume is relatively stable. With the continuous development of plastic deformation, when the loading stress reaches the expansion starting stress, the stress-strain curve begins to reverse, and the volume of salt rock begins to expand, exhibiting expansion phenomenon. In the later stage of plasticity, the slope of the stress-strain curve decreases, and the specimen enters the accelerated expansion stage, eventually developing to failure.

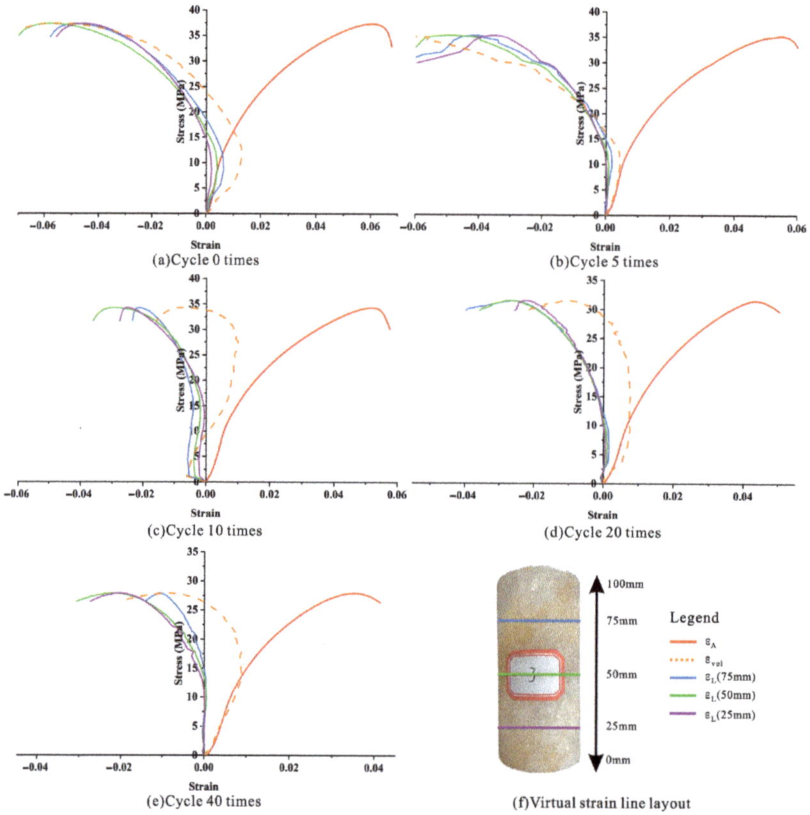

Figure 13. Comparison of strain curves of salt rock mass under different cycles. Note: (The samples T1~T5 in Table 1).

Under periodic temperature action, the initial stress of salt rock expansion and the change in volume expansion capacity at failure are shown in Figure 15. From the graph, it can be seen that as the number of temperature cycles and the upper limit temperature increase, the initial stress of salt rock expansion and the volume expansion capacity at failure both show a decreasing trend. The expansion boundary of salt rock under uniaxial compression is usually closely related to its elastic ultimate strength [20]. The thermal expansion and contraction effects caused by multiple temperature cycles form a complex thermal stress field inside the sample, which promotes the loosening and sliding of bonds between grains in the polycrystalline structure of salt rock, induces the continuous initiation and propagation of microcracks, and weakens the overall integrity of the structure, resulting in a significant decrease in the initial stress for expansion. In addition, under the influence of periodic temperature, the total strain energy of salt rock decreases throughout the compression process, and the sample enters the plastic stage earlier. As the intergranular slip intensifies and exceeds the mutual constraint range, energy is gradually released [20], ultimately exhibiting the thermal expansion characteristics of "premature initiation-weakened failure".

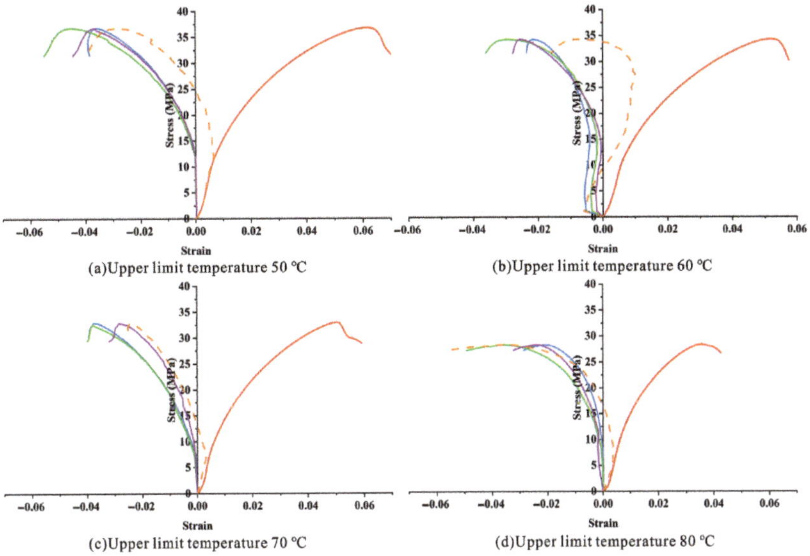

Figure 14. Comparison of strain curves of salt rock mass at different upper temperature limits. Note: (The samples T6~T9 in Table 1).

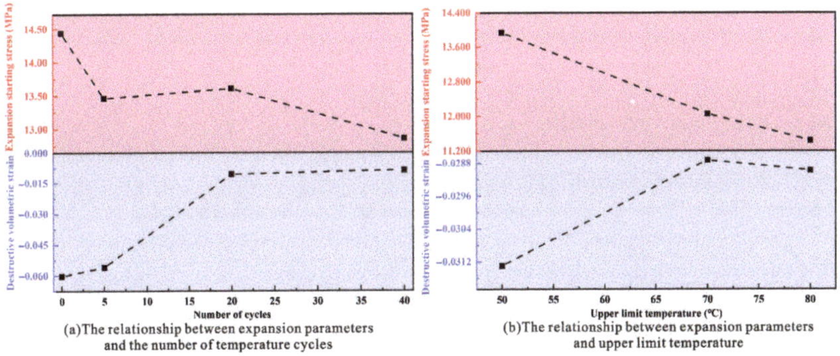

Figure 15. The relationship between the initial stress of expansion and the volume expansion capacity with periodic temperature changes. Note: ((**a**) is based on samples T1~T5, and (**b**) is based on samples T6~T9 in Table 1).

5. Conclusions

This article focuses on the study of the physical and mechanical properties of salt rock under periodic temperature changes and combines DIC and acoustic emission technology to qualitatively and quantitatively analyze the damage evolution process of salt rock. The influence of different upper limit temperatures and cycle times on the mechanical properties of salt rock is studied. The main conclusions are as follows:

(1) In the early stage of temperature cycling, the thermal expansion effect causes some cracks to close, and the wave velocity briefly increases. As the number of cycles increases, crack propagation gradually becomes dominant, and free water inside the salt rock evapo-

rates continuously. Impurity minerals such as gypsum undergo dehydration, resulting in an overall decrease in density, wave velocity, and hardness, and ultimately stabilizing. In terms of chromaticity, as the number of cycles and the upper limit temperature increase, the brightness of the salt rock gradually decreases, the color difference continues to increase, and eventually it tends to a new stable state.

(2) Salt rock exhibits typical "elastic-plastic-creep" characteristics under uniaxial compression. Periodic temperature changes induce microcrack propagation in salt rock, leading to accumulated structural damage, resulting in a 23.8% decrease in compressive strength and a 27.4% decrease in elastic modulus after 40 thermal cycles. As the number of cycles and upper limit temperature increase, the peak strain energy overall shows a decreasing trend, but the proportion of elastic strain energy increases from 21% to 38%. During the deformation process of salt rock, plastic slip is limited, showing an increasing trend of elastic-dominant characteristics.

(3) The acoustic emission response during uniaxial compression of salt rock is divided into an active period, a calm period, and a failure period, corresponding to the initial crack closure, elastic strain accumulation, and unstable propagation of microcracks, respectively. Under periodic temperature action, the crack types of salt rock change during compression. When not subjected to thermal cycling, tensile cracks accounted for 91%. After 20 temperature cycles, the proportion of tensile cracks decreased to 86%, and the proportion of shear cracks increased to 14%. The failure mode of salt rock gradually shifted from tension-dominated to tension-shear composite.

(4) The strain of salt rock gradually shifts from axial compression to radial expansion control during uniaxial compression, exhibiting a typical Poisson effect. As the number of cycles and upper limit temperature increase, the uneven expansion and thermal fatigue effects caused by thermal cycling accelerate the propagation of microcracks and suppress plastic slip, resulting in a decreasing trend in both the initial stress of expansion and the volume expansion capacity at failure, exhibiting the thermal expansion characteristics of "premature initiation-weakened failure". The sample structure fails prematurely, the bearing capacity decreases, and the peak stress and strain decay synchronously.

Author Contributions: Conceptualization, Y.G. and Y.Q.; methodology, Y.G. and Y.Q.; software, Y.G.; validation, Y.G.; formal analysis, Y.G.; investigation, Y.G., N.X., H.L., J.X., S.F. and L.C.; resources, Y.Q., N.X., H.L. and B.Z.; data curation, Y.G.; writing—original draft preparation, Y.G.; writing—review and editing, Y.G., Y.Q. and L.C.; visualization, Y.G. and L.C.; supervision, Y.Q., N.X., H.L., J.X., S.F. and B.Z.; project administration, Y.Q., N.X., J.X. and B.Z.; funding acquisition, N.X. All authors have read and agreed to the published version of the manuscript.

Funding: This research was funded by the National Key R&D Program of China, grant number 2023YFB4005500.

Institutional Review Board Statement: Not applicable.

Informed Consent Statement: Not applicable.

Data Availability Statement: Data are contained within the article. Data will be made available on request.

Acknowledgments: We sincerely thank the editors and all reviewers for their constructive and excellent reviews that helped improve the manuscript.

Conflicts of Interest: Authors Junhui Xu and Liuping Chen were employed by the company China National Salt Industry Group Co., Ltd. The remaining authors declare that the research was conducted in the absence of any commercial or financial relationships that could be construed as a potential conflict of interest.

References

1. Liang, W.; Yang, C.; Zhao, Y. Experimental investigation of mechanical properties of bedded salt rock. *Int. J. Rock Mech. Min. Sci.* **2007**, *44*, 400–411. [CrossRef]
2. Peng, H.; Fan, J.; Zhang, X.; Chen, J.; Li, Z.; Jiang, D.; Liu, C. Computed tomography analysis on cyclic fatigue and damage properties of rock salt under gas pressure. *Int. J. Fatigue* **2020**, *134*, 105523. [CrossRef]
3. Zeng, Z.; Ma, H.; Yang, C.; Zhao, K.; Wang, X.; Zheng, Z. Characterizing imbibition and void structure evolution in damaged rock salt under humidity cycling by low-field NMR. *Eng. Geol.* **2024**, *328*, 107371. [CrossRef]
4. Ozarslan, A. Large-scale hydrogen energy storage in salt caverns. *Int. J. Hydrogen Energy* **2012**, *37*, 14265–14277. [CrossRef]
5. Tarkowski, R. Underground hydrogen storage: Characteristics and prospects. *Renew. Sustain. Energy Rev.* **2019**, *105*, 86–94. [CrossRef]
6. Liu, W.; Zhang, Z.; Chen, J.; Jiang, D.; Wu, F.; Fan, J.; Li, Y. Feasibility evaluation of large-scale underground hydrogen storage in bedded salt rocks of China: A case study in Jiangsu province. *Energy* **2020**, *198*, 117348. [CrossRef]
7. Bérest, P. Heat transfer in salt caverns. *Int. J. Rock Mech. Min. Sci.* **2019**, *120*, 82–95. [CrossRef]
8. Yin, H.; Yang, C.; Ma, H.; Shi, X.; Zhang, N.; Ge, X.; Li, H.; Yue, H. Stability evaluation of underground gas storage salt caverns with micro-leakage interlayer in bedded rock salt of Jintan, China. *Acta Geotech.* **2020**, *15*, 549–563. [CrossRef]
9. Hunsche, U.; Albrecht, H. Results of true triaxial strength tests on rock salt. *Eng. Fract. Mech.* **1990**, *35*, 867–877. [CrossRef]
10. Senseny, P.; Hansen, F.; Russell, J.; Carter, N.; Handin, J. Mechanical behaviour of rock salt: Phenomenology and micromechanisms. *Int. J. Rock Mech. Min. Sci. Geomech. Abstr.* **1992**, *29*, 363–378. [CrossRef]
11. Liang, W.; Xu, S.; Zhao, Y. Experimental Study of Temperature Effects on Physical and Mechanical Characteristics of Salt Rock. *Rock Mech. Rock Eng.* **2006**, *39*, 469–482. [CrossRef]
12. Gao, R.; Wu, F.; Chen, J.; Zhu, C.; He, Q. Accurate characterization of triaxial deformation and strength properties of salt rock based on logarithmic strain. *J. Energy Storage* **2022**, *51*, 104484. [CrossRef]
13. Aldakheel, F. Micromorphic approach for gradient-extended thermo-elastic-plastic solids in the logarithmic strain space. *Contin. Mech. Thermodyn.* **2017**, *29*, 1207–1217. [CrossRef]
14. Urai, J.; Kukla, P.; Schléder, Z.; Spiers, C. *Dynamics of Complex Intracontinental Basins: The Central European Basin System*; Springer: Berlin/Heidelberg, Germany, 2008; pp. 277–290. [CrossRef]
15. Guo, W.; Li, J.; Wang, T.; He, T.; Xie, D.; Liao, Y.; Liu, C. Experimental Study on the Evolution Law of Permeability Characteristics of Salt Rocks Under Different Temperatures and Different Pore Pressures. *Rock Mech. Rock Eng.* **2025**, *58*, 1–23. [CrossRef]
16. Popp, T.; Kern, H.; Schulze, O. Evolution of dilatancy and permeability in rock salt during hydrostatic compaction and triaxial deformation. *J. Geophys. Res. Solid Earth* **2001**, *106*, 4061–4078. [CrossRef]
17. Miller, R. Engineering Classification and Index Properties for Intact Rock. Ph.D. Thesis, University of Illinois, Urbana, IL, USA, 1965; pp. 1–332.
18. Neff, P.; Eidel, B.; Martin, J. Geometry of Logarithmic Strain Measures in Solid Mechanics. *Arch. Ration. Mech. Anal.* **2016**, *222*, 507–572. [CrossRef]
19. Li, W.; Feng, K.; Ma, H.; Jiang, W.; Li, J.; Lu, Y. Microscopic mechanism analysis of temperature influence on rock salt and thermal damage evolution of surrounding wall in underground salt cavern. *Bull. Eng. Geol. Environ.* **2024**, *83*, 412. [CrossRef]
20. Zhai, S.; Wu, G.; Zhang, Y.; Wu, X. Study on the mechanical properties of high-temperature salt rock under uniaxial compression. *J. Rock Mech. Eng.* **2014**, *33*, 105–111.
21. Li, Z.; Ma, H.; Yao, Y. Research on the Basic Mechanical Properties of Salt Rock under High Temperature and High Pressure. *J. Undergr. Space Eng.* **2013**, *9*, 981–985.
22. Wang, Z.; Chen, F.; Dong, Z.; Li, H.; Shi, X.; Xu, Z.; Meng, X.; Yang, C. Study on the influence of temperature on the damage evolution of hot dry rock in the development of geothermal resources. *Geoenergy Sci. Eng.* **2024**, *241*, 213171. [CrossRef]
23. Sheinin, V.; Blokhin, D. Features of thermomechanical effects in rock salt samples under uniaxial compression. *J. Min. Sci.* **2012**, *48*, 39–45. [CrossRef]
24. Li, H.; Dong, Z.; Yang, Y.; Liu, B.; Chen, M.; Jing, W. Experimental study of damage development in salt rock under uniaxial stress using ultrasonic velocity and acoustic emissions. *Appl. Sci.* **2018**, *8*, 553. [CrossRef]
25. Dong, Z.; Li, Y.; Li, H.; Wang, Z.; Shi, X.; Chen, X.; Lu, Q. Experimental Study on the Influence of Temperature on Rock Salt Creep. *Rock Mech. Rock Eng.* **2023**, *56*, 3499–3518. [CrossRef]
26. Pan, X.; Chen, J.; Rui, Y.; Chen, Z.; Li, Z.; Du, J. Experimental study on short-term creep crack evolution and acoustic emission characteristics of salt rock after heat treatment at different temperatures. *Bull. Eng. Geol. Environ.* **2025**, *84*, 253. [CrossRef]
27. Zhang, S.; Liang, W.; Xiao, N.; Zhao, D.; Li, J.; Li, C. Fractional order viscoelastic plastic creep damage model of salt rock considering temperature. *J. Rock Mech. Eng.* **2022**, *41*, 3198–3209.
28. Zhang, S.; Xu, S.; Xiao, N.; Li, J.; Li, C. Creep damage model of deep salt rock under temperature stress coupling. *Coal J.* **2024**, *49*, 3425–3438.

29. Klafki, M.; Wagler, T.; Grosswig, S.; Kneer, A. Long-term down hole fibre optic temperature measurements and CFD-modeling for investigation of different gas operating modes. In Proceedings of the SMRI Fall Meeting, Chester, UK, 5–8 October 2003; pp. 179–189.
30. Mahmutoglu, Y. Mechanical Behaviour of Cyclically Heated Fine Grained Rock. *Rock Mech. Rock Eng.* **1998**, *31*, 169–179. [CrossRef]
31. Johnston, H.; Toksöz, N. Thermal cracking and amplitude dependent attenuation. *J. Geophys. Res. Solid Earth* **1980**, *85*, 937–942. [CrossRef]
32. Zhou, S.; Xia, C.; Hu, Y.; Zhang, P. Damage modeling of basaltic rock subjected to cyclic temperature and uniaxial stress. *Int. J. Rock Mech. Min. Sci.* **2015**, *77*, 77163. [CrossRef]
33. Inada, Y.; Kinoshita, N.; Ebisawa, A.; Gomi, S. Strength and deformation characteristics of rocks after undergoing thermal hysteresis of high and low temperatures. *Int. J. Rock Mech. Min. Sci.* **1997**, *34*, e1–e140. [CrossRef]
34. Becattini, V.; Motmans, T.; Zappone, A.; Madonna, C.; Haselbacher, A.; Steinfeld, A. Experimental investigation of the thermal and mechanical stability of rocks for high-temperature thermal-energy storage. *Appl. Energy* **2017**, *203*, 373–389. [CrossRef]
35. Jin, P.; Hu, Y.; Shao, J.; Zhao, G.; Zhu, X.; Li, C. Influence of different thermal cycling treatments on the physical, mechanical and transport properties of granite. *Geothermics* **2019**, *78*, 118–128. [CrossRef]
36. *ISO 22282:2012*; Geotechnical Investigation and Testing. International Organization for Standardization: Geneva, Switzerland, 2012.
37. Jiang, X.; Zhang, F.; Huang, B.; Titi, H.; Polaczyk, P.; Ma, Y.; Wang, Y.; Cheng, Z. Full-scale accelerated testing of geogrid-reinforced inverted pavements. *Geotext. Geomembr.* **2024**, *53*, 511–525. [CrossRef]
38. Liu, Z.; Liu, Y.; Wang, Z. Comparative Study of Temperature and Pressure Variation Patterns in Hydrogen and Natural Gas Storage in Salt Cavern. *Appl. Sci.* **2024**, *14*, 9005. [CrossRef]
39. Deng, F.; Jiang, F.; Wan, J.; Ji, W.; Li, J. Analysis of the deformation characteristics of surrounding rock due to interlayers during cyclic injection-production period of salt cavern hydrogen storage. *Fuel* **2025**, *385*, 134115. [CrossRef]
40. Labaune, P.; Rouabhi, A. Dilatancy and tensile criteria for salt cavern design in the context of cyclic loading for energy storage. *J. Nat. Gas Sci. Eng.* **2019**, *62*, 314–329. [CrossRef]
41. Liu, L.; Li, S.; Jiang, X.; Tao, F. A new two-sensor non-destructive testing method of grouted rock bolts. *Constr. Build. Mater.* **2022**, *317*, 125919. [CrossRef]
42. Jitka, K.; Radomír, K.; Martin, K.; Lenka, S.; Alena, V. New insight into the phase changes of gypsum. *Mater. Struct.* **2024**, *57*, 128. [CrossRef]
43. Li, W.; Zhu, C.; Yang, C.; Duan, K.; Hu, W. Experimental and DEM investigations of temperature effect on pure and interbedded rock salt. *J. Nat. Gas Sci. Eng.* **2018**, *56*, 29–41. [CrossRef]
44. Xie, H.; Xie, H.; Zhang, Z.; Yao, Q.; Cao, Z.; Gao, H.; Shan, C.; Yan, Z.; Yin, R. Fatigue fracture behaviors and damage evolution of coal samples treated with drying–wetting cycles investigated by acoustic emission and nuclear magnetic resonance. *Int. J. Rock Mech. Min. Sci.* **2025**, *185*, 105976. [CrossRef]

applied sciences

MDPI

Article

Theoretical Prediction and Safety Evaluation of Adjacent Pipeline Deformation Caused by Connecting Channel Excavation Reinforced with Freezing Method

Jun Zhang [1,2,*], Jiguo Liu [1,2], Shuoren Fu [3] and Zequn Hong [3]

[1] CCCC Second Highway Consultant Co., Ltd., Wuhan 430056, China; liujiguo@ccccltd.cn
[2] Research and Development Center on Tunnel and Underground Space Technology, CCCC, Wuhan 430056, China
[3] School of Mechanics and Civil Engineering, China University of Mining and Technology, Xuzhou 221116, China; ts22030036a31@cumt.edu.cn (S.F.); zqhong@cumt.edu.cn (Z.H.)
* Correspondence: zhangjun98@ccccltd.cn

Abstract: Underground excavation by freezing method can ensure the safety of the surrounding structures. The influence of excavation of a connecting channel between two tunnels by freezing method on adjacent pipelines is studied in this paper. Combined with field measurement, numerical simulation, and theoretical analysis, the stress and deformation law of the whole process of channel excavation by freezing method is studied. Based on Euler–Bernoulli beam theory prediction, the influence of temperature field and excavation parameters on the longitudinal deformation of pipeline is analyzed. The results show that the excavation rate significantly affects the pipeline settlement, and the settlement surges when the excavation rate exceeds 1.0 m/d. At the same time, the thick frozen soil wall formed by low freezing temperatures enhances the supporting ability and effectively reduces the formation disturbance and settlement. The study focuses on the influence of connecting channel excavation on the pipelines under uniform formation conditions, and puts forward the evaluation method of pipeline safety to provide a theoretical reference for engineering practice.

Keywords: freezing method; connecting channel; theoretical prediction; safety evaluation

1. Introduction

With the acceleration of urbanization, the development and utilization of underground space are becoming more and more extensive, and the construction of subway tunnels and their affiliated communication channels has become an important part of urban infrastructure [1,2]. However, in the construction process of tunnels and communication channels, especially when freezing technology is used for geotechnical reinforcement [3,4], its impact on the existing upper buildings has become a key problem to be solved urgently. Although freezing technology can effectively control the groundwater flow and enhance the formation stability, the temperature field change caused by it, the frost-heaving effect of soil, and the deformation in the subsequent thawing process may cause adverse effects on the upper buildings, such as foundation settlement and structural cracking [5,6]. Therefore, it is of great significance to deeply study the influence law of freezing of contact channels on the existing upper buildings to ensure the construction safety of underground engineering and protect the structural stability of existing buildings [7,8].

In the research field of tunnel–soil–building interaction, scholars at home and abroad have carried out a lot of work, especially on the influence of tunnel construction on underground pipelines, and have made remarkable progress [9,10]. These research results provide valuable references and enlightenment for this paper to study the effect of freezing of contact channels on the upper existing buildings [11,12].

In recent years, soil deformation caused by tunnel construction and its influence on underground structures has become a research focus. The free displacement field theory based on Gaussian distribution has been proposed, which lays a theoretical foundation for evaluating the influence of tunnel construction on underground pipelines [13,14]. On this basis, the continuous medium elastic solution of pipeline longitudinal deformation has been further derived, and the nonlinear change of soil stiffness has been considered, which makes the evaluation method more perfect [15,16]. Scholars have not only paid attention to the soil deformation caused by tunnel construction but also deeply studied the interaction mechanism between pipelines and soil. Considering the influence of soil stiffness weakening on pipeline bending stress, it has been found through model tests and numerical simulation that the longitudinal deformation of the pipeline can be more accurately predicted by considering the out-of-plane shear caused by pipe–soil interaction. In addition, there is more and more research on the discontinuous interface pipeline, and the calculation methods of pipeline deflection considering the interface effect have been proposed with respect to this, so that the simulation results are more close to the actual situation.

To simulate the deformation characteristics of the foundation more accurately, scholars have improved the foundation model. The traditional Winkler foundation model neglects the continuity of foundation deformation, while new models, such as the Pasternak foundation model, can better reflect the actual deformation of the foundation. Based on the Pasternak foundation model, the finite difference decomposition of discontinuous interface pipeline deflection has been constructed, which provides a new tool for evaluating the influence of freezing of the contact channel on superstructure. With the wide application of freezing technology in underground engineering, the changes of soil temperature field, seepage field, and stress field caused by freezing technology have been deeply studied by scholars at home and abroad. Remarkable achievements have been made in analytical solutions, the frost heave model, and experimental device development, which provide theoretical support for the application of freezing technology in the construction of subway connection passage. However, the research on the effect of freezing technology on the existing upper buildings is still insufficient, especially in the application of practical engineering validation, which needs to be strengthened. The improvement of the tunnel model itself has also been a research hotspot in recent years. The traditional tunnel model mainly simplifies the tunnel as an equal beam or Euler–Bernoulli infinite beam, but the urban shield tunnel is made of segments and bolts, and its physical and mechanical properties are weak. The Timoshenko beam model can better simulate the shear stiffness reduction effect of a tunnel, which provides a new idea for evaluating the influence of tunnel construction on underground structures. However, how to comprehensively consider the interaction between the tunnel, soil, and the superstructure in the concrete situation of freezing construction of the contact channel needs further research.

In summary, the study of tunnel–soil–building interaction has made some achievements, but there are still many deficiencies in the study of the influence of freezing of a contact passage on the existing upper structure. Through theoretical analysis, numerical simulation, and engineering measurement, this paper aims to deeply explore the changes in soil temperature field and stress field during the freezing process of a contact passage and its influence on the upper structure, to provide a scientific basis and technical support for the design and construction of underground engineering.

2. Project Overview

Dongge Road station–Binhu Road station is the 13th section of Nanning Rail Transit Line 3 from north to south, as shown in Figure 1. The depth of the tunnel is about 11.7~21.0 m. The starting point of this section is Dongge Road station, the end is Binhu Road station, the starting and ending mileage YDK14 + 778.298~YDK15 + 736.098, ZDK14 + 778.298~ZDK15 + 736.088, the short chain 4.485 m. The length of the left line is 953.306 m, and the length of the right line is 957.800 m. The interval tunnel is 6 m in diameter, and the line spacing is 4 m. The maximum slope of the line is 28.168%, and this section is a "V" shaped slope. The left line of this section adopts an earth pressure shield construction. The outer diameter D of the shield tunnel is φ6.0 m, and the thickness of the segment is 0.5 m. The construction (structure) in this section is mainly a water pipeline with a diameter of 3 m, and the concrete strength grade of the concrete pipeline is C40.

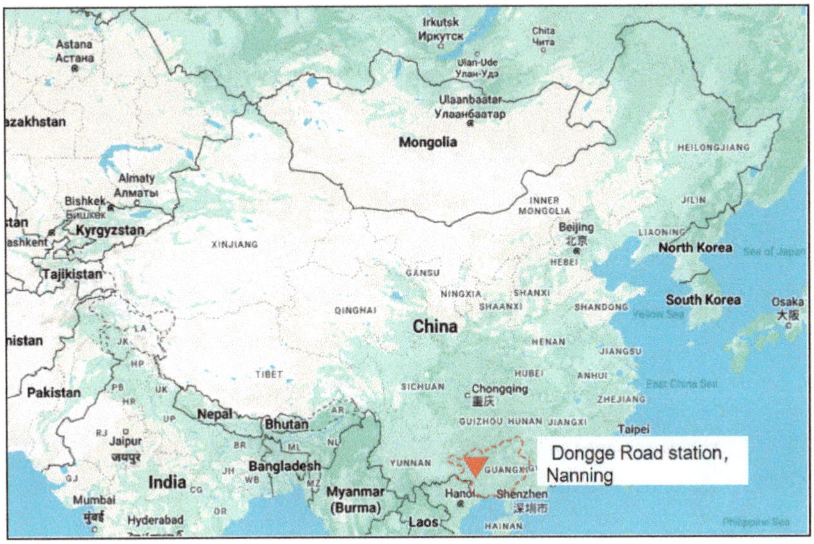

Figure 1. Location map of the shield tunnel in the Dongbin section.

The Dongbin section adopts the shield machine produced by Herrenkrenk, and the tunnel adopts the earth pressure shield tunneling. The tunnel in this section mainly passes through the boulder layer and soft plastic silty clay layer and partially passes through the silt layer and silty sand layer. The tunnel's buried depth is about 11.7 m.

3. The Establishment of a Numerical Model

3.1. Establishment of a Calculation Model of Freezing Body Temperature Field

To facilitate the calculation, the following assumptions are made for the finite element model:

- The soil is continuous and homogeneous during freezing;
- When the soil is frozen, the latent heat of phase change is continuously released at the freezing interface;
- Assume that the moisture in the soil is completely frozen when it is cold; that is, the content of unfrozen water is zero [17,18];
- The freezing temperature load on the freezing hole wall of the model changes with the change in temperature;
- The wall of the freezing tube is negligible relative to the whole model, so the material characteristics of the freezing tube itself are not considered.

To analyze the influence of horizontal freezing of the contact channel on the existing tunnel in the upper part, according to the general situation of the project and the symmetry of the contact channel, this paper adopts the FLAC3D finite difference software to establish a structural model, and then realize the three-dimensional temperature field calculation of the frozen body. The position relationship between the frozen tube and the existing tunnel is shown in Figure 2. The total size of the model is 44 m long × 15 m wide × 35 m high, as shown in Figure 3.

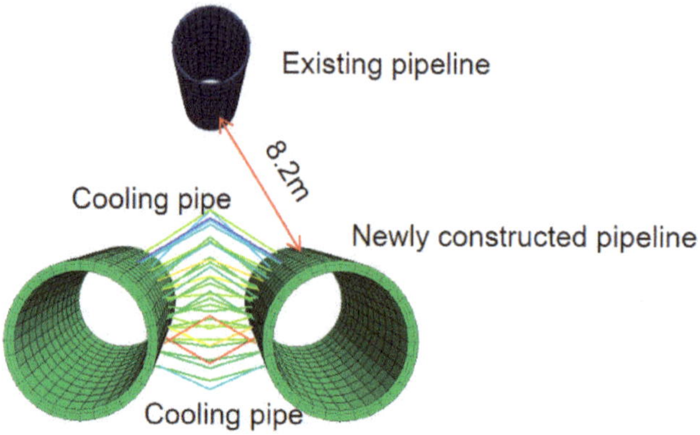

Figure 2. Position relationship between freezing tube and existing tunnel.

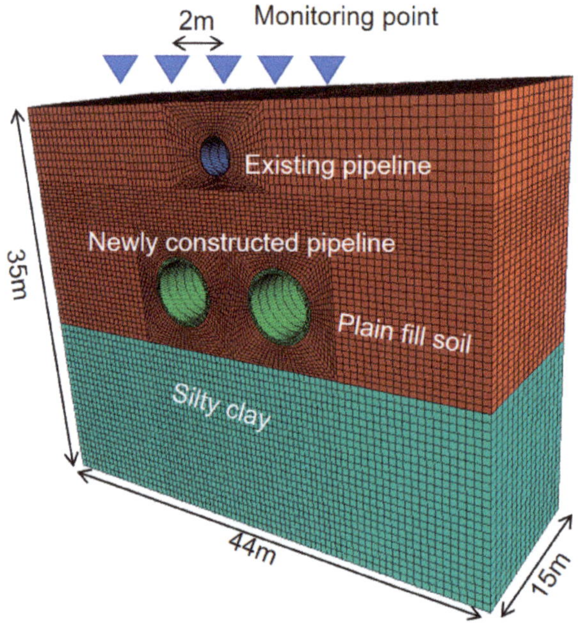

Figure 3. Numerical simulation grid model.

Application steps of FLAC3D in freezing method tunnel excavation calculation:

- According to the actual situation of the project, the three-dimensional geometric model, including tunnel, frozen soil wall, surrounding soil layer, and adjacent pipeline, is established in FLAC3D.
- Meshing ensures that the mesh density of the model is high enough at key locations, such as tunnel boundaries, the interface between the frozen wall, and the surrounding soil, to accurately capture stress concentration and deformation characteristics.
- The physical and mechanical parameters, such as elastic modulus, Poisson's ratio, density, cohesion and internal friction angle, are set according to laboratory tests or empirical formulas.
- Set the boundary conditions of the model, such as fixed boundaries, rolling boundaries, or free boundaries, to simulate the constraints in real situations.

- The freezing temperature field is set, the temperature gradient produced by the freezing tube in the soil is simulated, and the forming process of the frozen soil wall is calculated by a heat conduction equation.
- Monitoring the thickness, temperature distribution, and strength changes of the frozen soil wall, evaluating its supporting ability.
- According to the predetermined excavation sequence and excavation rate, the material of the tunnel is gradually removed to simulate the excavation process.
- In the process of excavation, the stress, strain, and displacement of soil around the tunnel are monitored in real time, especially the longitudinal deformation of adjacent pipelines.
- The simulation results are compared with the field-measured data to ensure the accuracy and reliability of the simulation results.
- The effects of different excavation rates, freezing temperatures, and other parameters on pipeline settlement are analyzed, and the conclusions mentioned in the abstract are verified.

When calculating the model numerically, the vertical tunnel axis direction in the water intake plane is the X axis, the direction along the tunnel axis is Y axis, and the vertical upward direction is the Z axis. In order to avoid the influence of a boundary effect on the numerical calculation results of the temperature field, the model range (X, Y, Z) was selected as (44, 15, 35 m), and considering the aspects of calculation accuracy and solution time, through the grid sensitivity analysis, we chose a set of grid configurations that can guarantee the calculation accuracy and control the calculation cost as the final simulation settings. The simulation results in this grid configuration are convergent and reliable enough to support our research conclusions.

The upper surface of the tunnel and frozen tube model is free, and the displacement of the lower surface and four side surfaces is completely constrained. The load is generated from top to bottom according to the gradient of gravity field considering only the action of gravity. The Mohr–Coulomb elastoplastic model was used as the soil element in the model. To make the calculation results more realistic, the model was subdivided into two layers, namely plain-filled soil and silty clay. The obtained rock and soil samples were brought back to the laboratory for various physical and mechanical tests, such as a density test, water content test, compression test, shear test, etc., in order to obtain the physical and mechanical parameters of rock and soil bodies, such as density, water content, compression modulus, internal friction angle, cohesion force, etc. The physical and mechanical parameters and thermophysical parameters of each soil layer are shown in Table 1. The elastic model was adopted for shield tunnel segment elements. The elastic modulus of the shield segment was set at 33.5 GPa, Poisson's ratio at 0.2, thermal conductivity at 0.5 W/(m °C), specific heat capacity at 1 kJ/(kg °C), and density at 2500 kg/m^3. Thermal conductivity of surface air and air in tunnel was 0.025 W/(m °C), and specific heat capacity was 1 kJ/(kg °C).

Table 1. Physical and mechanical parameters of rock and soil layer.

Rock and Soil Stratification	Natural Density (g/cm^3)	Cohesion (kPa)	Angle of Friction (°)	Void Rating	Permeability Coefficient (m/d)	Coefficient of Lateral Pressure	Modulus of Elasticity (MPa)	Poisson's Ratio	Thermal Conductive (W/m/°C)	Specific Heat Capacity (J/kg/°C)
Plain fill soil	1.96	15	10	0.752	3	0.35	4.5	0.35	/	/
Silty clay	1.98	38	13	0.710	0.006	0.33	6.5	0.33	1.85	1.46

In the actual freezing method construction process, the active freezing period of the interval contact channel is 40 d, and the numerical calculation time of the temperature field is also 40 d. Because the freezing pipe is arranged in an inclined radial shape, the freezing pipe layout on the upper and lower lines is not the same. It is a common and important step to arrange temperature measuring holes on site to monitor the temperature development of the freezing curtain during the construction of a freezing method of a subway contact passage. For the measurement of the temperature hole, there are usually repeated measurements to ensure the accuracy and reliability of the data. The instruments used to collect temperature data are usually high-precision temperature sensors or thermometers. These instruments can be installed in a temperature measuring hole and transmit temperature data wired or wirelessly to a data acquisition system or monitoring center. The temperature sensors in the temperature measurement holes must be installed and arranged according to the design requirements. At the same time, in order to ensure the accuracy of the measurement data, the front end of the temperature measurement tube should be welded and sealed, and no water should be permeated in the tube.

For the convenience of analysis, a typical cross-section in the direction of Y = −7.5 was selected, respectively, and a total of five temperature measuring holes were arranged, as shown in Figure 3. The data of the same point in the temperature measuring holes C1 and C3 in the tunnel were, respectively, taken as representatives, and the temperature change value was recorded according to the actual temperature measuring point in the numerical simulation. After 40 days of active freezing, the comparison between the numerical simulation of the temperature curves of the two temperature measuring holes and the field-measured temperature is shown in Figure 4.

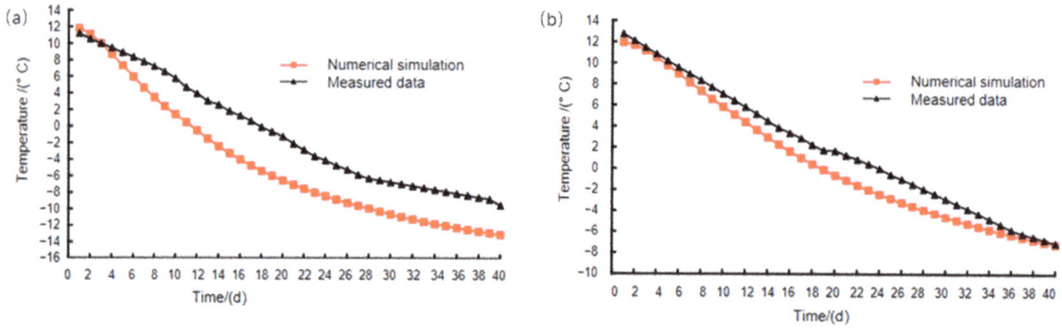

Figure 4. Comparison of temperature curves: (**a**) C1, (**b**) C3.

As shown in Figure 4, the average cooling velocity of the C1 temperature measuring hole is about 0.54 °C/d, and the numerical result is about 0.63 °C/d. After 40 days of active freezing, the measured temperature of the C1 temperature measuring hole is −9.50 °C, and the numerical simulation temperature is −13.10 °C. As shown in Figure 4, the average cooling speed of the C10 temperature measuring hole is about 0.48 °C/d, and the numerical result is about 0.48 °C/d. After active freezing for 40 days, the measured temperature of the C10 temperature measuring hole is −7.06 °C, and the numerical simulation temperature is −7.18 °C. It can be seen that the temperature cooling trends of the numerical simulation results at the same temperature measuring points are the same as that of the field-measured results. On the whole, the numerical simulation results are slightly lower than the measured temperature value, because the numerical calculation assumes that the groundwater is static. In the actual project, because the groundwater is flowing, the heat exchange of the frozen soil curtain is more complicated, and the flow of the groundwater will take away part of the cold amount, so the numerical simulation result is slightly lower than the measured temperature value. In summary, it can be shown that the transient freezing temperature field obtained by the numerical simulation of the finite difference method software FLAC 3D can reflect the actual situation of the project.

3.2. Pipeline Stress and Deformation Law during the Excavation of Contact Passage

After the excavation of the left and right sides of the tunnel, the change law of pipe stress and displacement is shown in Figures 5 and 6. It can be seen from Figures 5 and 6 that the breaking of the segment will lead to the redistribution of the stress field and displacement field in a certain surrounding area. Compared with the intact segment on the right, local stress concentration will occur after the segment is broken. As shown in Figure 5, the maximum stress value reaches 12.5 MPa, which is mainly concentrated in the bottom corner and the center area of the bottom edge of the segment opening. For shield tunnel segments using concrete with C50 strength grade, the safety factor is 4, the pipeline stress is only 0.78 MPa, and the safety factor is 50, which is within the safe range. At the same time, the breaking of the segment will cause the frozen soil curtain near the opening to shift into the tunnel, as shown in Figure 5. The displacement is very small, and the maximum displacement value is less than 10 mm. Whether in terms of stress or displacement, the breaking of the segment is within the safe range, and the construction can be carried out normally.

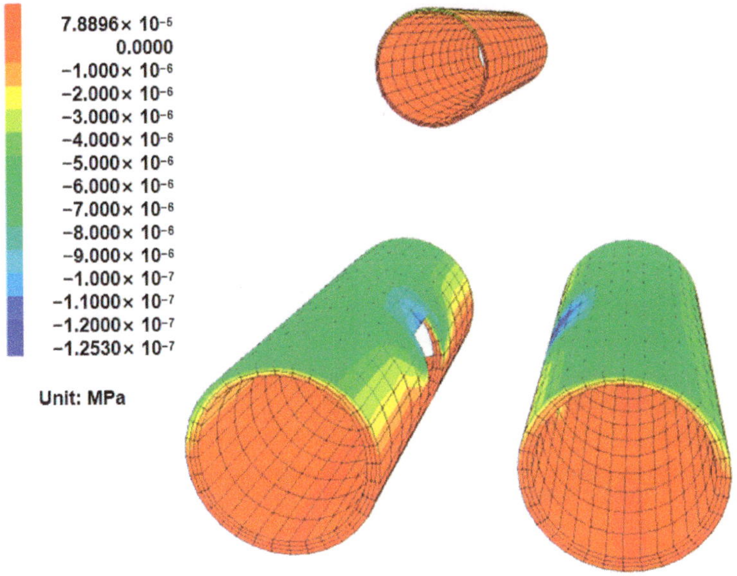

Figure 5. Stress program after fragment breaking.

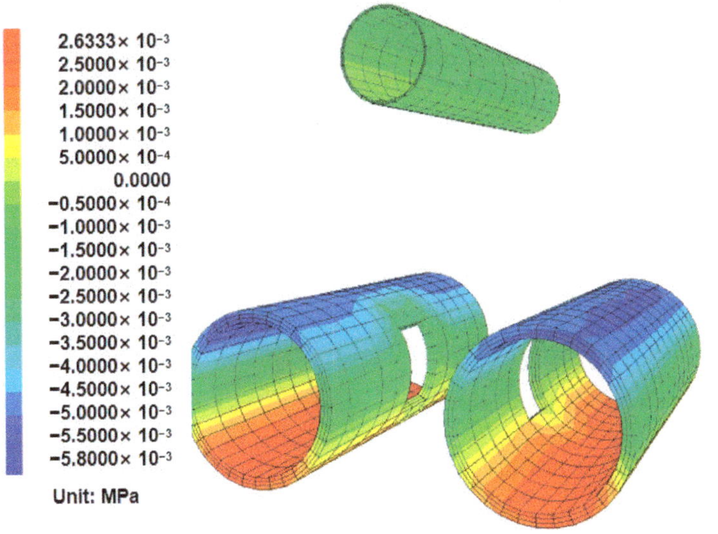

Figure 6. Displacement equivalent cloud map after segment break.

After the tunnel excavation, the deformation diagram of the tunnel and pipeline is shown in Figure 7. Without the initial support, the maximum subsidence value of the vault displacement is 11.8 mm, the maximum uplift value of the floor displacement is 13.2 mm, and the pipeline displacement is 10 mm, which is within the safe range.

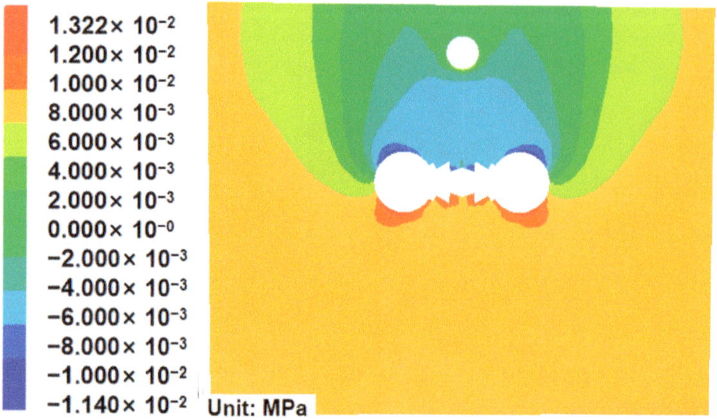

Figure 7. Z-displacement cloud image of the model after tunnel excavation.

4. Theoretical Derivation

4.1. Mechanical Model and Basic Assumptions

Figure 7 shows the engineering problem to be solved. The formation loss caused by the excavation of the circular shield tunnel causes the overlying strata to displace and then causes the underground pipeline to bend. As shown in Figure 8, the x coordinate is used to mark the position of each section along the pipeline. The theoretical derivation of this paper is based on the following basic assumptions and simplifications: tunnels and pipelines are buried horizontally in homogeneous isotropic strata, and their axes are perpendicular to each other. The foundation around the pipeline is simplified to the Pasternak foundation.

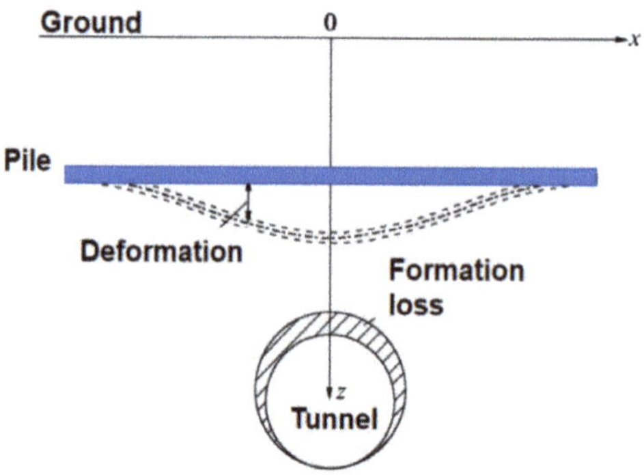

Figure 8. Schematic diagram of pipeline deflection caused by shield tunnel excavation.

4.2. Construction of Pipeline Deflection Differential Equation

As shown in Figure 9, the left and right ends of any joint position are intercepted with dx length for force analysis. It is assumed that they act on the line axis position and are evenly distributed along the y *direction.*

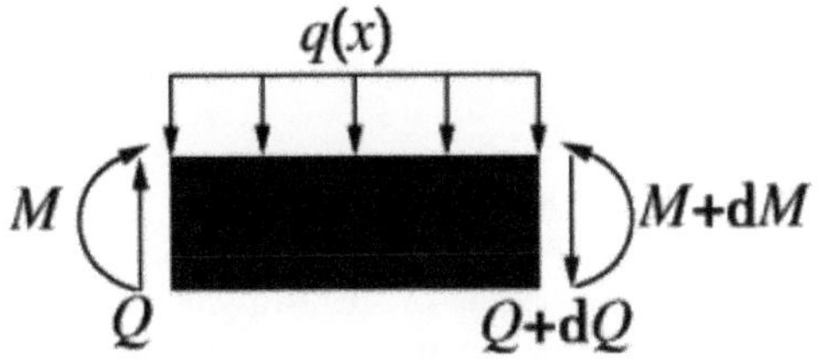

Figure 9. Element force analysis: micro force analysis of the left end of the joint.

Firstly, the vertical force balance of the microelement can be obtained using

$$(Q - \mathrm{d}Q) - q(x)b\mathrm{d}x - Q = 0 \tag{1}$$

where dx is the element width (m), q is the cross-section shear force (N), is the increment of cross-section shear force (N) along the length of dx of the pipeline, q(x) is the foundation reaction force *acting* on the pipeline (N/m^2), and b is the diameter of the pipeline (m).

Equation (1) can be obtained by simplification

$$\frac{\mathrm{d}Q}{\mathrm{d}x} = -q(x)b \tag{2}$$

Secondly, find the bending moment of the section of the left end of the element, which can be obtained by balancing the bending moment

$$(M - \mathrm{d}M) + Q\mathrm{d}x + q(x)b\frac{(\mathrm{d}x)^2}{2} - M = 0 \tag{3}$$

where *M* is the cross-section bending moment (N·m), and dM is the increment of the cross-section bending moment (N·m) along the dx length of the pipeline.

Equation (3) simplifies and omits the second-order trace, which can be obtained by

$$\frac{\mathrm{d}M}{\mathrm{d}x} = Q \tag{4}$$

Equation (2) is obtained in conjunction with (4):

$$q(x) = -\frac{1}{b}\frac{\mathrm{d}^2M}{\mathrm{d}x^2} \tag{5}$$

Suppose the force of the pipeline on the surrounding formation is *p*(*x*), and its interaction with *q*(*x*) is the interaction force, then

$$p(x) = -q(x) \tag{6}$$

Assuming that the bending stiffness of any cross-section of the pipeline along the direction is infinite, the pipeline will produce an overall translation along the *y* direction when it interacts with the soil. Combined with the assumption that the pipeline is always in contact with the surrounding strata, the vertical displacement u(*x*) of the soil on the contact surface of any cross-section of the pipeline will not change along the direction. Therefore, when the pipeline interacts with the soil, the relationship between the pipeline force *p*(*x*) and the ground displacement u(x) caused by the pipeline can be expressed as Equation (7).

$$p(x) = ku(x) - G_\mathrm{p}\frac{\mathrm{d}^2u(x)}{\mathrm{d}x^2} \tag{7}$$

where *k* is the foundation reaction coefficient (N/m^3) and *Gp* is the foundation shear stiffness (N/m).

Simultaneous Equation as (5)–(7) can be obtained

$$\frac{d^2M}{dx^2} = kbu(x) - G_pb\frac{d^2u(x)}{dx^2} \tag{8}$$

Similarly, for the right end of the joint and the standard pipe section, Equation (8) can be deduced.

The formation displacement around the pipeline originates from two aspects, one is the formation settlements (x) caused by the formation loss in tunnel excavation, and the other is the foundation displacement u(x) caused by the interaction between the pipeline and the soil. Assuming that the axial deflection along the pipeline is $w(x)$, the Equation (9) can be obtained based on the assumption that the pipeline is always in contact with the surrounding strata.

$$w(x) = s(x) + u(x) \tag{9}$$

In the Equation, $s(x)$ and $u(x)$ are taken from the position of the pipeline axis. By combining Equation as (8) and (9), we can obtain

$$-\frac{d^2M}{dx^2} + kbw(x) - G_pb\frac{d^2w(x)}{dx^2} = kbs(x) - G_pb\frac{d^2s(x)}{dx^2} \tag{10}$$

The deflection w(x) satisfies the Equation (11) of the extramural differential equation.

$$\frac{d^2w(x)}{dx^2} = -\frac{M}{EI} \tag{11}$$

where E is the elastic modulus (N/m^2) of the pipeline and I is the moment of inertia of the pipeline cross-section. By substituting Equation (11) into Equation (10), we can obtain

$$-\frac{d^2M}{dx^2} + kbw(x) + G_pb\frac{M}{EI} = kbs(x) - G_pb\frac{d^2s(x)}{dx^2} \tag{12}$$

The cross-sectional ground settlement s(x) perpendicular to the tunnel axis caused by ground loss in tunnel excavation conforms to the distribution pattern of the Gaussian curve, namely Equation (13).

$$s(x) = \frac{\pi R^2 V_1}{\sqrt{2\pi}K(z_0 - z_p)} \exp\left[-\frac{x^2}{2K^2(z_0 - z_p)^2}\right] \tag{13}$$

where R is the outer radius of the tunnel (m), Vl is the ground layer loss caused by tunnel excavation (dimension is 1), K is the width parameter of the settlement tank at the position of the pipeline axis (dimension is 1), and $z0$ and zp are divided into the axis burial depths of the tunnel and pipeline (m).

$$\frac{d^2M}{dx^2} = -EI\frac{d^4w(x)}{dx^4} \tag{14}$$

Substitute Equation (14) into Equation (10) to obtain

$$\begin{aligned} EI\frac{d^4w(x)}{dx^4} + kbw(x) - G_pb\frac{d^2w(x)}{dx^2} \\ = kbs(x) - G_pb\frac{d^2s(x)}{dx^2} \end{aligned} \tag{15}$$

The calculation Equations of foundation shear stiffness Gp are as follows: Equations (16) and (17).

$$G_p = \frac{E_sH_t}{6(1+v_s)}\psi_t \tag{16}$$

$$\psi_t = \frac{3}{2\gamma_p H_t} \frac{\sinh(\gamma_p H_t)\cosh(\gamma_p H_t) - \gamma_p H_t}{\sinh^2(\gamma_p H_t)} \tag{17}$$

where *Es* is the elastic modulus of the soil mass (N/m^2), vs. is the Poisson ratio of the soil mass, Ht is the thickness of the foundation shear layer (m), and Yp is the empirical *parameter* (M-1).

The algebraic equations applicable to each difference node are integrated into the following matrix form:

$$[K_p]\{w\} + [K_r]\{w\} - [K_{s1}]\{w\} = [K_r]\{s\} - [K_{s2}]\{s\} \tag{18}$$

where [*Kp*] is the stiffness matrix of the pipeline, [*Kr*] is the foundation reaction stiffness matrix, [*Ks1*] and [Ks2] are two different foundation shear stiffness matrices, {w} is the *pipeline* flexure vector, and {s} is the formation settlement vector generated by formation loss in tunnel excavation at each difference junction point.

Each matrix and vector can be obtained by Equation (18) and determined by the algebraic equations of the above-mentioned difference nodes.

4.3. Verification of the Theoretical Model

In order to verify the accuracy of the theoretical model, the calculation results of the theoretical model were compared with the Equation calculation results in Yao, Y. et al., (2023) [16], field monitoring data, and simulation data generated by the numerical model, as shown in Figure 10. The comparison results are shown in Figure 10. It can be seen from Figure 10 that the simulation results are highly consistent with the measured data, which proves the feasibility of the numerical simulation method. However, due to the limitation of simplified conditions, the influence of the Timoshenko beam theory on the longitudinal deformation of the pipeline is not fully reflected in the numerical simulation, resulting in a small width of the predicted settlement curve.

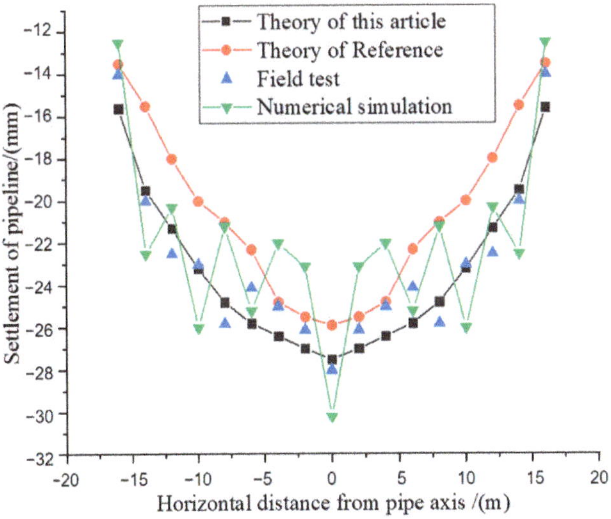

Figure 10. Comparison of a theoretical model with numerical simulation and measured data [16].

4.4. Influence of Tunnel Excavation Rate on Pipeline Settlement

Statistics on the influence of tunnel excavation rate on pipeline settlement are shown in Figure 11. The influence of daily tunnel excavation length (DEL) on pipeline settlement is analyzed. As can be seen from Figure 11, the faster the excavation rate of the tunnel, the greater the disturbance to the surrounding soil mass, which may lead to rapid changes in the formation stress and strain state, and then lead to more significant surface deformation. This deformation has an adverse effect on underground pipelines. When the excavation rate is less than 1.0 m per day, the settlement of the pipeline is less than 30 mm, while when the excavation rate is greater than 1.0 m per day, the settlement of the pipeline increases rapidly. Therefore, in the tunnel design and construction stage, it is necessary to comprehensively consider the geological conditions, excavation methods, and supporting measures, and reasonably control the excavation rate to reduce the impact on surface deformation to ensure construction safety and stability of the surrounding environment.

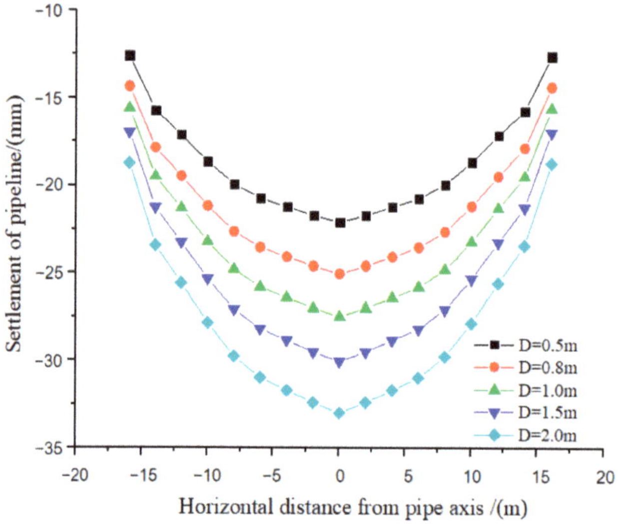

Figure 11. Influence of tunnel excavation rate on pipeline settlement.

4.5. *Influence of Tunnel Freezing Temperature on Pipeline Settlement*

The statistical relationship between tunnel freezing temperature and pipeline settlement shown in detail in Figure 12 deeply reveals the key role of temperature factor in the stability control of tunnel construction. Specifically, when a lower freezing temperature (such as −50 °C) is used, this extremely low-temperature environment significantly promotes the formation of frozen soil walls or frozen soil bodies, which not only increase in thickness, but also significantly improve in strength. This strengthening effect greatly enhances the support ability of the frozen soil structure to the surrounding soil, which is like building a solid barrier, effectively limiting the range and degree of formation disturbance during tunnel excavation, thereby reducing the formation settlement caused by construction activities.

On the other hand, the phenomenon of frost heave induced by low temperature is particularly significant in the soil above the tunnel, showing obvious uplift phenomenon. Although this uplift may seem unfavorable at the early stage of freezing, because it changes the original formation shape, it actually provides a "pre-compression" effect for the subsequent excavation process. However, with the gradual excavation of the tunnel, the soil uplifted by frost heave quickly loses its support and is transformed into a significant settlement. It is worth noting that this settlement often has a large magnitude and a certain lag; that is, the initial settlement may not be obvious, but with the passage of time and stress redistribution, the settlement will gradually increase.

Specific to the numerical value, comparing the pipe settlement under the two freezing conditions of −50 °C and −15 °C, we can clearly see the significant impact of temperature difference. At a low temperature of −50 °C, thanks to the stronger frozen soil support effect, the settlement of the pipeline is effectively controlled above −20 mm, showing a better construction control effect. In contrast, when the freezing temperature is increased to −15 °C, the strength and thickness of the frozen soil wall or frozen soil body are weakened, and the supporting effect on the surrounding soil is reduced, resulting in a significant increase in pipeline settlement to −33.2 mm, highlighting the important role of low-temperature freezing in controlling the settlement of tunnel construction.

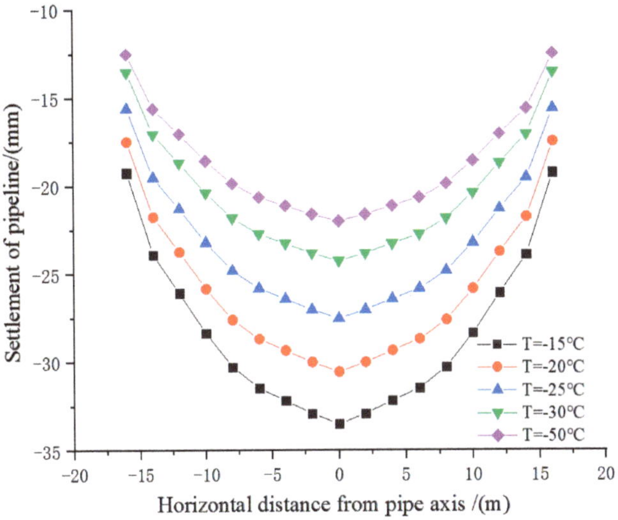

Figure 12. Influence of tunnel freezing temperature on pipeline settlement.

4.6. Influence of Thickness of the Frozen Curtain on Pipeline Settlement

To delve deeper into the complex relationship between the thickness of the frozen curtain and pipe settlement presented in Figure 13, we can explore the underlying physical mechanisms and engineering effects. With the increase in the thickness of the frozen curtain, its performance as a support structure is significantly enhanced, which is comprehensive in nature. Firstly, the thicker frozen curtain acts as a more solid barrier, and its enhanced support capacity is directly reflected in the effective resistance to the surrounding soil pressure and deformation. This enhanced resistance not only limits the scope and intensity of soil disturbance during tunnel excavation, but also significantly reduces the resulting soil settlement, providing important assurances for the safety and stability of tunnel construction.

During another critical stage of tunnel construction—the thawing process—the thicker frozen curtains also exhibit their unique advantages. With higher strength and stability, these thick frozen soil walls can release internal stress more slowly and contentedly in response to the challenge of melting, thus reducing the risk of sudden subsidence caused by rapid melting. Smaller melting volume and slower melting speed mean that the stability of the subsurface structure is better maintained, which is conducive to the smooth progress of construction and the improvement of final engineering quality.

It is particularly noteworthy that when the thickness of the frozen curtain is increased from 0.5 m to 1 m, the subsidence phenomenon above the tunnel shows a significant reduction. This transition indicates that the 1 m thick frozen curtain has been able to exert a strong support role during excavation, effectively alleviating the negative impact of ground disturbance. However, as the thickness of the frozen curtain is further increased to 2 m, although the subsidence volume continues to decrease, the rate of decrease is not as significant as before. This phenomenon can be explained from the perspective of mechanics and the saturation effect of materials: when the frozen curtain reaches a certain thickness, the growth of its bearing capacity will gradually become flat, i.e., reaching a "saturated" state. At this point, the marginal benefits of further increasing the thickness for subsidence control will gradually decrease.

In summary, the rational design of tunnel freezing curtain thickness is one of the key factors in controlling pipe settlement, ensuring construction safety and engineering quality. By scientific calculation and engineering practice, determining the optimal freezing curtain thickness can effectively reduce ground disturbance and settlement while optimizing construction costs and time, achieving a win–win situation of economic benefits and social benefits.

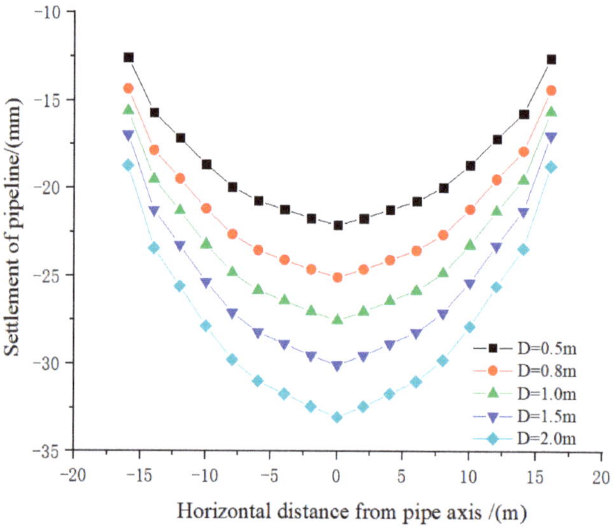

Figure 13. Influence of tunnel freezing temperature on pipeline settlement.

4.7. Safety Assessment of Buried HDPE Bellows Affected by Excavation

The pipe used in this paper is the HDPE corrugated pipe for water supply. According to the above results, the maximum circumferential allowable strain of the pipe can be determined according to article 4.4.6 in reference [19]; that is, the maximum vertical deformation of the polyethylene pipe under the combined action should be calculated as follows:

$$\omega_{d,max} \leq 0.05D_0 \quad (19)$$

where $\omega_{d,max}$ is the maximum vertical deformation of the polyethylene pipeline under the combined action, and D0 is the calculated diameter of the pipeline. The maximum allowable compressive and tensile strains are 5%.

According to the full text analysis, under different working conditions, the maximum settlement of the pipeline is less than 55 mm, while the allowable deformation of the pipeline is 150 mm. Therefore, the research in this paper can provide a basis for the excavation of refrigerated contact channels, and it can provide ideas for the safety criteria of pipelines.

5. Conclusions

- The breaking of the segment will lead to the redistribution of the stress field and displacement field in a certain surrounding area. Compared with the complete segment, the phenomenon of local stress concentration will occur after the segment is broken. Whether in terms of stress or displacement, the broken segment is within the safe range, and the construction can be carried out normally.
- The prediction theory of stratum displacement caused by tunnel excavation has an important impact on the estimation of pipeline longitudinal deformation. The tunnel

longitudinal deformation trend obtained based on the Euler–Bernoulli beam longitudinal prediction theory is consistent with the measured results.

- The faster the tunnel excavation speed, the greater the disturbance to the surrounding soil. When the excavation rate is less than 1.0 m/d, the pipeline settlement is less than 30 mm, and when the excavation rate is greater than 1.0 m/d, the pipeline settlement increases rapidly. The lower the freezing temperature, the thicker and stronger the frozen soil wall or frozen soil body, and the stronger the supporting effect on the surrounding soil. With the increase of freezing curtain thickness, the supporting ability of the freezing curtain is obviously enhanced. Thicker freezing curtains can resist the pressure and deformation of surrounding soil more effectively, thus reducing the formation disturbance and settlement caused by tunnel excavation.
- This paper mainly considers the influence of tunnel excavation on the existing pipeline when the formation is relatively uniform, and also provides readers with a way to judge the safety of the pipeline. If the stratification between new and old tunnels is obvious, the stratification effect should be further considered. In general, the analytical solution of pipe settlement caused by tunnel excavation presented in this paper can provide theoretical support for analyzing the influence of tunnel excavation on existing pipes.

Author Contributions: Conceptualization, J.Z., J.L., S.F. and Z.H.; methodology, S.F.; software, J.Z.; validation, J.Z.; writing—original draft preparation, S.F. and J.L.; writing—review and editing, J.L. and J.Z.; project administration, Z.H. and J.Z.; funding acquisition, J.Z. and J.L. All authors have read and agreed to the published version of the manuscript.

Funding: This research was supported by National Natural Science Foundation of China (NO. 52108386). The authors are deeply indebted to the financial support.

Data Availability Statement: Data is contained within the article.

Conflicts of Interest: Author J.Z. was employed by the company CCCC Second Highway Consultant Co., Ltd. The author declares no conflict of interest.

References

1. Li, Y.; Yao, A.; Gao, X.; Li, H.; Gong, Y. Rapid Prediction of Pile Friction Resistance Based on Soil Deformation Law Caused by a Super Large Diameter Tunnel: A Case Study. *KSCE J. Civ. Eng.* **2024**, *28*, 3033–3041. [CrossRef]
2. Zhang, J.; Pan, T.; Ma, K.; Xu, Q.; Kong, C. Modified stochastic medium prediction model for the deformation response of concealed underground stations under existing pipelines. *Sci. Rep.* **2023**, *13*, 9843. [CrossRef] [PubMed]
3. Li, Y.; Huang, C.; Lu, H.; Mou, C. Deformation Prediction Model of Existing Tunnel Structures with Equivalent Layered Method–Peck Coupled under Multiple Factors. *Mathematics* **2024**, *12*, 1479. [CrossRef]
4. Yamada, Y. A theoretical prediction method of inclined crack width in evaluation model for capacities of shear resistance mechanisms of RC beams. *J. Jpn. Soc. Civ. Eng. Ser. E2 (Mater. Concr. Struct.)* **2022**, *78*, 264–273. [CrossRef] [PubMed]
5. Guo, J. An Evaluation Model of the Stuck Risks Based on Remote Sensor Network and Fuzzy Logic. *IEEE Sens. J.* **2021**, *21*, 23658–23666. [CrossRef]
6. Wang, L.; He, C.; Cui, S.; Wang, F. Numerical Simulation of Surface Movement and Deformation Caused by Underground Mining with Complex Stratigraphic Boundary. *Adv. Civ. Eng.* **2021**, *2021*, 9967071. [CrossRef]
7. Huang, J.; Liu, J.; Guo, K.; Yang, S.; Lu, Y.; Wang, Y.; Wu, C. Evaluation Method of the Impact of Twin Shield Tunneling Construction on Elevated Bridges: Case Study. *Symmetry* **2024**, *16*, 1113. [CrossRef]
8. Liu, Q.; Shao, B.; Wang, B.; Zhang, W.; Ju, Z.; Ma, X.; Wang, P. Research on Prediction and Prevention of Nonuniform Deformation in Unsymmetrical Loading Tunnel. In Proceedings of the Engineering Geology for a Habitable Earth: IAEG XIV Congress 2023, Chengdu, China, 21–27 September 2023; Springer: Singapore, 2024. [CrossRef]
9. Li, F.; Tang, Q.; Yan, W.; Li, J.; Li, W.; Yang, T.; Lin, L.; Jiang, K.; Wang, K.; Zhang, W. Research on the failure analysis and prediction method of casing deformation under the synergistic effect of sand production and corrosion. *Eng. Fail. Anal.* **2023**, *154*, 107706. [CrossRef]
10. Ma, J.; Huang, K.; Zou, B.; Li, X.; Deng, Y. The influence of tunnel insulation measures on the temperature spatiotemporal variation of frozen soil during artificial ground freezing. *Cold Reg. Sci. Technol.* **2023**, *214*, 103942. [CrossRef]
11. Xu, X.; Wu, Z.; Weng, L.; Chu, Z.; Liu, Q.; Zhou, Y. Numerical investigation of geostress influence on the grouting reinforcement effectiveness of tunnel surrounding rock mass in fault fracture zones. *J. Rock Mech. Geotech. Eng.* **2024**, *16*, 81–101. [CrossRef]
12. Wu, X.; Gong, M.; Wu, G.W.S. Vibration reduction technology and the mechanisms of surrounding rock damage from blasting in neighborhood tunnels with small clearance. *Int. J. Min. Sci. Technol.* **2023**, *33*, 625–637. [CrossRef]

13. Qin, Z.; Liu, Y.; Liu, J.; Cao, B.; Zhang, F.; Li, W. Application of pretensioned grouting support in repairing surrounding rock damaged by dynamic load. *Arab. J. Geosci.* **2022**, *15*, 922. [CrossRef]
14. Zhang, S.; Lai, Y.; Zhang, X.; Pu, Y.; Yu, W. Study on the damage propagation of surrounding rock from a cold-region tunnel under freeze–thaw cycle condition. *Tunn. Undergr. Space Technol.* **2004**, *19*, 295–302. [CrossRef]
15. Zhang, X.; Lai, Y.; Yu, W.; Zhang, S. Non-linear analysis for the freezing–thawing situation of the rock surrounding the tunnel in cold regions under the conditions of different construction seasons, initial temperatures and insulations. *Tunn. Undergr. Space Technol.* **2002**, *17*, 315–325. [CrossRef]
16. Yao, Y.; Zhu, Y.; Shen, D.; Zhang, Z.; Wang, W. Fuzzy random evaluation of creep model of frozen soft soil in metro tunnel construction using artificial ground freezing technique. *Sci. Rep.* **2023**, *13*, 9468. [CrossRef] [PubMed]
17. Chen, J.H.; Zhang, J.; Chen, B.J.; Lu, G. The influence of the underpassing frozen connecting passage on the deformation of the existing tunnel. *Res. Cold Arid. Reg.* **2022**, *14*, 258–266. [CrossRef]
18. Chen, B. *Study on Mechanism of Freezing of Contact Channel on Existing Upper Tunnel*; Fujian College of Engineering: Fujian, China, 2021. (In Chinese)
19. *DBJ52/T039-2017*; Technical Regulations for Outdoor Buried Polyethylene (PE) Water Supply Pipeline Engineering. Guizhou Provincial Department of Housing and Urban-Rural Development: Guizhou, China, 2018.

applied sciences

MDPI

Article

Experimental Study on Damage Constitutive Model of Rock under Thermo-Confining Pressure Coupling

Yonglai Zheng [1], Zhengxie Zhang [1], Chenyu Hou [1,*], Tanbo Pan [2], Xin Lan [1] and Yifan Ji [1]

1 Department of Hydraulic Engineering, College of Civil Engineering, Tongji University, Shanghai 200092, China; 00016@tongji.edu.cn (Y.Z.); 2330610@tongji.edu.cn (Z.Z.); 2410024@tongji.edu.cn (X.L.); 1732622@tongji.edu.cn (Y.J.)

2 Department of Bridge Engineering, College of Civil Engineering and Architecture, East China Jiaotong University, Nanchang 330013, China; 1830149@tongji.edu.cn

* Correspondence: 2210368@tongji.edu.cn

Abstract: Underground engineering frequently encounters the challenges of high temperatures and high confining pressures. The combined effects of temperature and confining pressure can significantly alter the mechanical properties of rock. Evaluating the impact of these factors on rock is a crucial aspect of engineering. This study investigates the mechanical properties of rocks under various temperature and stress conditions, focusing on the deformation and failure characteristics of four typical rock types: granite, red sandstone, gray sandstone, and shale under temperature-confining pressure coupling. The results indicate that high temperatures cause internal structural damage and crack propagation in rocks, leading to a reduction in compressive strength and elastic modulus. Conversely, high confining pressures can inhibit crack propagation and enhance rock deformation capacity. Additionally, significant differences were observed in the mechanical responses of different rocks; red sandstone and shale predominantly exhibited shear failure under the coupled effects of temperature and confining pressure, whereas granite and gray sandstone exhibited bulging failure. Based on the experimental results, an elastic modulus fitting model considering the temperature-confining pressure coupling effect was proposed, and the parameters of the Drucker–Prager criterion were modified. A constitutive model was constructed to accurately reflect the stress–strain state of rocks under the coupled effects of temperature and confining pressure. The constitutive model results show good agreement with the experimental findings.

Keywords: confining pressure; damage constitutive model; elastic modulus; rock; temperature

1. Introduction

In geological environments, confining stress and temperature are two primary factors. Deep underground engineering often operates under high confining stress and elevated temperatures. The mechanisms of disaster formation in underground engineering are highly complex and differ significantly from the basic properties of rock under normal conditions. On the one hand, high temperatures typically cause damage to microstructures and expansion of fractures within the rock, which reduces the rock's stiffness, leading to a decrease in compressive strength and elastic modulus [1]. On the other hand, confining pressure alters the pore structure of the rock, induces new cracks, and significantly changes the properties of the rock.

Several scholars have studied the effects of temperature on the physical and mechanical properties of various rocks, summarizing and analyzing the changes in these properties with temperature variation [2–4]. Some researchers have used X-ray computed tomography (CT) technology to study the effects of temperature on the heterogeneity and anisotropy of granite. They observed that the porosity of granite increases with temperature after high-temperature treatment, and its heterogeneity and anisotropy also increase with temperature, with initial cracks below 200 °C having the most significant impact [5]. Moreover, as the temperature rises, the mineral particles in rock samples fracture and decompose, causing significant changes in the elastic modulus [1,6,7]. Under low-temperature conditions, the elastic modulus of rocks increases linearly with decreasing temperature, while Poisson's ratio remains unchanged [8]. Compared to high temperatures, high confining pressures can inhibit crack propagation and damage evolution, causing the failure characteristics of rocks to gradually transition from brittle to ductile failure [9,10]. Triaxial compression tests reveal that different confining pressure conditions significantly affect the mechanical and deformation characteristics of rocks [11]. Compressive strength and elastic modulus increase with confining pressure [12], primarily because increased confining pressure suppresses rock expansion, alters crack propagation paths [13], and thereby affects mechanical parameters, microcrack distribution, and energy evolution laws [14]. Based on these foundations, researchers have used analytical methods to calculate the equivalent elastic modulus of fractured rocks with random discrete fractures and regular fractures under confining pressure, providing a better evaluation of the deformation behavior of fractured rock masses [15]. Considering the combined effects of temperature and confining pressure, triaxial compression tests under different confining pressures and temperatures explore the impacts of these factors on the mechanical and deformation characteristics of rocks [16]. Increased confining pressure significantly enhances the compressive strength and elastic modulus of rocks [17], while the elastic modulus exhibits nonlinear changes with temperature [18]. The effects of temperature and confining pressure on rocks are coupled [19], causing significant changes in the microstructure of rocks [20]. Microcracks develop differently in various rock types [21,22], leading to significant differences in rock properties.

Based on the principles and results of field tests, the relationship between rock elastic modulus and other geotechnical parameters (such as joint distribution, continuity, deformation mode, and stiffness [23]) can be analyzed. By combining the strain energy theory of continuum mechanics and fracture mechanics [24], relevant parametric models can be established to describe changes in the elastic modulus [25], thereby determining the typical range of rock elastic deformation capacity [26]. Using continuum damage theory and introducing a Weibull distribution to describe the evolution of damage variables [27], the rock stress–strain curve can be divided into four stages: undamaged loading, stable damage propagation, intensified damage propagation, and stable damage propagation to saturation. Based on the analysis of the elastic modulus, a thermo-mechanical coupled damage constitutive model for rocks can be developed [28]. Researchers have also established constitutive models for different types of rocks based on this foundation [29,30], which hold significant implications for practical engineering and rock mechanics research.

The aforementioned research has made significant contributions to the analysis of the mechanical properties of rocks under the coupled effects of temperature and confining pressure. However, there are still some limitations. The current studies involve a relatively limited range of rock types, and research on rocks under the coupled effects of temperature and confining pressure often employs high-temperature pretreatment followed by conventional triaxial tests in the laboratory. This approach makes it difficult to accurately reflect the real-time stress–strain response of rocks under high-temperature and high-pressure conditions. Moreover, the failure structure models of rocks under the coupled effects of temperature and confining pressure remain underexplored, and the selected probabilistic microelement strength models and strength criteria are relatively simplistic. In addition, in the existing structural models, the elastic modulus is often fitted using polynomial functions, which hinders a detailed description of the elastic modulus development trend for specific rock masses. Furthermore, the comprehensive effects of temperature and confining pressure on the elastic modulus have not been fully considered.

Therefore, this study employs a combined approach of laboratory experiments and theoretical calculations to investigate the room temperature uniaxial compression tests and conventional triaxial compression tests of four typical rocks under the coupled effects of temperature and confining pressure: granite, red sandstone, gray sandstone, and shale. The research aims to explore the deformation evolution and failure characteristics of these four types of rocks under the combined influence of temperature and confining pressure. A fitting model for the elastic modulus of rocks under these coupled conditions is proposed, and the parameters of the Drucker–Prager criterion are revised to reflect the effects of temperature and confining pressure. Additionally, a constitutive model is constructed to accurately capture the stress–strain state of rocks under these combined effects, providing insights into rock deformation characteristics that align with practical engineering conditions.

2. Experimental Program

2.1. Specimen Preparation

The primary focus of this study is on four typical types of rocks: granite, red sandstone, gray sandstone, and shale. Before conducting triaxial compression experiments, uniaxial compression tests were first performed on these four rock types to obtain their peak strengths and average uniaxial compressive strengths.

For the uniaxial compression experiments, the surface of the rock samples and the instrument joints were cleaned. The rock samples were then encased in FEP heat shrink tubing. Due to its high toughness, temperature resistance, and immunity to chemicals such as oil, this material can effectively prevent hydraulic silicone oil from infiltrating the rock samples during the experiment when heated with a heat gun, as shown in Figure 1a. After placing the rock samples, circumferential and axial strain gauges were sequentially installed on the samples, and loading was applied, as illustrated in Figure 1b.

(a) (b)

Figure 1. Installation of conventional uniaxial test rock samples and sensors: (**a**) rock sample encased in FEP heat shrink tubing; (**b**) installation of circumferential and axial strain gauges on the rock sample.

The compressive strength of rock exhibits dispersion and follows a normal distribution. Therefore, based on the relevant theories of material mechanics, the standard compressive strength values for the four types of rocks were calculated. In order to ensure that the rock samples are within the elastic phase during the loading process under different working conditions, 70% of the standard compressive strength value was chosen as the stress limit for this experiment. The results of the uniaxial compression tests and the selected stress limits for the four types of rocks are shown in Table 1.

Table 1. Uniaxial Compression Test Results.

Rock Specimen	Peak Strength/MPa	Average Uniaxial Compressive Strength/MPa	Standard Value of Compression Strength/MPa	Upper Limit of Stress/MPa
G-JZ-1	119.96			
G-JZ-2	133.72	133.73	115.49	80.84
G-JZ-3	147.50			
RS-JZ-1	25.40			
RS-JZ-2	21.98	25.04	21.36	14.95
RS-JZ-3	27.75			
GS-JZ-1	148.26			
GS-JZ-2	151.24	158.17	138.47	96.93
GS-JZ-3	175.03			
SH-JZ-1	60.72			
SH-JZ-2	60.74	60.29	59.28	41.49
SH-JZ-3	59.42			

2.2. Testing System

The experiment used the GCTS high-temperature and high-pressure true triaxial testing servo apparatus at Tongji University, as depicted in Figure 2. This apparatus comprises an operational control system and a triaxial load application module, equipped with sensors for displacement, load, and temperature. These sensors accurately capture data on axial stress and strain, radial strain, confining pressure, and temperature. The primary parameters are as follows: a maximum confining pressure of 60 MPa, a maximum axial pressure of 500 MPa, a stress accuracy of 0.01 MPa, and a deformation accuracy of 0.001 mm. The apparatus can record various mechanical parameters of rock under temperature-stress coupling in real-time and simultaneously plot stress–strain curves.

Figure 2. GCTS high-temperature and high-pressure rock triaxial apparatus.

2.3. Test Procedure

This experiment involves conventional triaxial tests under various temperature and confining pressure conditions. Based on the studied stratum distribution, the test temperatures are set at five levels: 20 °C, 40 °C, 60 °C, 80 °C, and 120 °C, and the confining pressures are set at six levels: 0 MPa, 5 MPa, 10 MPa, 15 MPa, 20 MPa, and 25 MPa. Relevant studies indicate that thermal cracking in granite within the range of 100 °C to 500 °C does not cause structural damage [31], and the effects of temperature and cyclic loading can reinforce rock samples under certain conditions [32]. This reinforcement phenomenon is present in most rock types [31,33]. Some studies suggest that mild temperatures [31] or cyclic loading [33,34] have a hardening effect on rocks. Triaxial loading and unloading tests on some rocks reveal minimal strain differences during loading and unloading, with a maximum difference of only 6.96% [35]. Moreover, the peak stress ($\sigma_f{}'$) obtained from the triaxial cyclic loading and unloading tests is similar to the peak stress (σ_f) obtained from the triaxial compression tests. Before reaching the peak stress, the number of loading and unloading cycles at high-stress levels is relatively low. As a result, cyclic loading before peak stress has minimal impact on the peak stress of the rock [36]. Based on these research results, the following basic steps are used in this experiment:

(1) The standard granite sample was securely enclosed in a specialized fluoro rubber sleeve to prevent hydraulic oil infiltration, ensuring test accuracy. After clearing debris from the steel block connections, the rock core was placed flat with end blocks at both ends. The axial strain gauge was carefully adjusted to its proper position, and the circumferential strain gauge was installed at the center of the specimen. A slight touch confirmed proper connection by displaying minimal displacement on the computer. Initial values were then set on the computer.

(2) Using the GCTS high-temperature and high-pressure rock triaxial apparatus, the rock samples were subjected to an initial confining pressure state ($\sigma_2 = \sigma_3$ of 0 MPa) and maintained consistently. In this experiment, the loading was controlled by a fixed axial displacement rate. The loading rate was set to 0.03 mm/min in the computer software control system, and the stress limit was set to 70% of the standard uniaxial compressive strength of the rock samples at room temperature [37]. When the axial stress reached the preset stress limit, the system automatically stopped loading. Subsequently, the confining pressure and stress were unloaded to 0 MPa. During the loading process, the system continuously collects and records the stress and strain values of the rock sample in real-time. Peak strength, residual strength, axial strain, and circumferential strain are directly obtained from the recorded data. The real-time stress–strain curve is monitored to ensure that the test data meet the requirements of this study.

(3) Repeat step (2), applying initial confining pressures of $\sigma_2 = \sigma_3$ at 5 MPa, 10 MPa, 15 MPa, 20 MPa, and 25 MPa. Complete the triaxial compression tests at 20 °C under different confining pressure conditions.

(4) After unloading the confining pressure and stress to 0 MPa, the hydraulic oil in the pressure chamber was heated to the preset temperature of 40 °C using the temperature control system of the GCTS high-temperature, high-pressure rock triaxial apparatus. The temperature was maintained for a period of time to ensure uniform heating of the specimen. Once the temperature stabilized, steps (2) and (3) were repeated to conduct the triaxial compression test at 40 °C under different confining pressure conditions.

(5) Repeat step (4), heating the samples to the predetermined temperatures of 60 °C, 80 °C, and 120 °C. When the experiment reaches the condition of 25 MPa confining pressure and 120 °C temperature, the loading continues until the sample fails. Complete the triaxial compression tests at 60 °C, 80 °C, and 120 °C under different confining pressure conditions.

(6) After the sample failure, turn off the loading system and temperature control system of the triaxial apparatus. Once the temperature of the hydraulic oil in the pressure chamber drops to room temperature, unload the confining pressure, drain the hydraulic oil, and remove the samples. Observe and record the failure characteristics of the samples and organize and analyze the experimental data.

3. Results and Discussion

3.1. Failure Modes

The crack characteristics of the four types of rocks, after conventional triaxial compression failure and uniaxial compression failure under temperature-stress coupling, are shown in Figure 3. It can be observed that the failure modes of red sandstone and shale under uniaxial stress and temperature-stress coupling are primarily characterized by shear failure. Under uniaxial stress, a macroscopic through-crack typically appears on the specimen surface, accompanied by several small micro-cracks. However, under the coupling of temperature and stress, the number of micro-cracks on the specimen surface decreases, with the failure characteristics focusing on a single main crack. Additionally, the angle of the crack relative to the principal stress direction increases. This phenomenon is primarily due to the generation of thermal stress, where temperature changes within the rock induce thermal stress. The action of thermal stress can make the internal stress field of the rock more uniform, reducing stress concentration and, consequently, the number of micro-cracks. Simultaneously, the complex stress state of thermal stress and confining pressure alters the propagation direction of the main crack, increasing its angle relative to the principal stress direction. These findings indicate that the coupling effect of temperature and confining pressure significantly influences the failure modes and characteristics of red sandstone and shale.

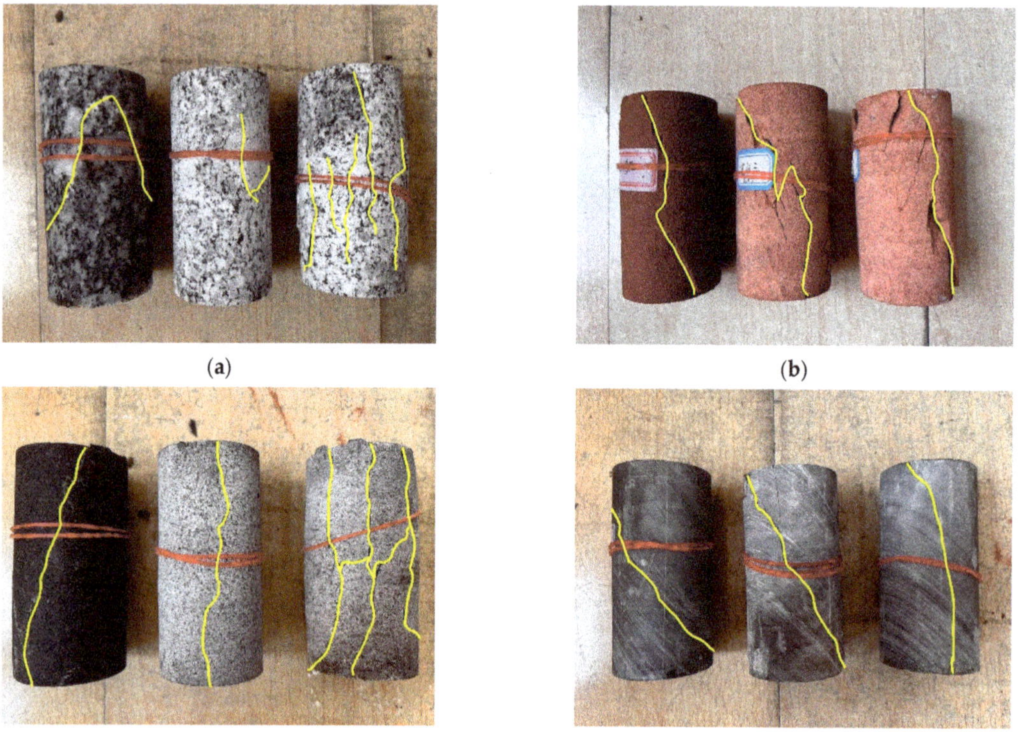

(**a**) (**b**) (**c**) (**d**)

Figure 3. Failure patterns of various rock types under triaxial and uniaxial compression: (**a**) failure patterns of granite under triaxial and uniaxial compression; (**b**) failure patterns of red sandstone under triaxial and uniaxial compression; (**c**) failure patterns of gray sandstone under triaxial and uniaxial compression; (**d**) failure patterns of shale under triaxial and uniaxial compression.

Additionally, the failure modes of granite and gray sandstone exhibit bulging failure, with most cracks parallel to the principal stress direction under uniaxial stress. However, under the coupling of temperature and stress, the failure mode transitions to shear failure, and the shear plane of the specimen displays a "Y"-shaped shear surface. The primary reason for this is that uniaxial compression creates stress concentration zones within the rock perpendicular to the direction of the compressive stress. This stress concentration easily leads to bulging failure, characterized by expansion and cracking along the perpendicular direction. Conversely, under the combined effects of temperature and confining pressure, the internal stress field of the rock becomes more complex. Confining pressure inhibits bulging failure, while the changes in the rock at high temperatures make shear failure the primary failure mode. In this scenario, the rock is likelier to experience sliding and failure along weak planes under shear stress.

When subjected to temperature-stress coupling, the fracture surface of the specimen during failure is relatively smooth and flat. In contrast, the fracture surface under uniaxial stress appears rough and uneven. Additionally, around the fracture surface under uniaxial conditions, multiple secondary cracks parallel to the main fracture surface and a few spalled fragments can be observed.

3.2. *Analysis of Mechanical Properties of Rocks*

3.2.1. Analysis of Stress–Strain Curves for Different Rocks

Figures 4 and 5 illustrate the stress–strain curves of granite under different temperatures and confining pressures. From Figure 4a,f, it can be observed that the yield stage of granite is not distinct, the crack development stage is short, and the stress immediately drops after reaching the peak, resulting in failure accompanied by a noticeable cracking sound, indicative of brittle failure. When granite samples are loaded to the stress limit, they remain in the elastic stage. Under the same confining pressure conditions, the slope of the curve increases with rising temperature, while the separation between the curves at different temperatures decreases as the confining pressure increases. This is primarily due to the reduction of pores between mineral particles within the microstructure of the rock sample as the confining pressure increases, and the confining effect induced by the increase in confining pressure surpasses the thermal expansion effect caused by the rise in temperature.

As illustrated in Figure 5, when the temperature remains constant, the stress–strain curves progressively converge with the increase in confining pressure. This indicates that granite is more sensitive to changes in confining pressure under lower confining conditions. However, once a certain confining pressure threshold is reached, the slope of the curve stabilizes, indicating that the deformation behavior of the rock tends to become consistent. Comparing Figure 5a,e, it is evident that, at a confining pressure of 25 MPa and a temperature of 120 °C, the compressive strength and peak strain of granite are both greater than those under normal conditions. It indicates that the coupled effects of temperature and stress can enhance the density of the mineral substructure within the rock sample, thereby increasing its load-bearing capacity and improving its ductility.

Figure 6 shows the stress–strain curves of red sandstone under different temperature and confining pressure conditions. From Figure 6a,b, it can be observed that the curves of red sandstone mainly exhibit four stages: the pore crack compaction stage, the linear elastic deformation stage, the microcrack development stage, and the post-failure stage. The nonlinear characteristics of the crack closure and compaction stage in red sandstone are not obvious, and the curve quickly enters the elastic deformation stage, showing linear variation characteristics. Comparing Figure 6c,d, it is found that the slope of the curve in the elastic stage gradually increases. The confining pressure restricts the lateral deformation of the rock sample and the expansion of internal cracks, enhancing the closure degree of internal cracks and thus increasing the load-bearing capacity of the sample. Moreover, the compressive strength and peak strain of red sandstone decrease under high-temperature and high-pressure conditions. Higher temperatures cause a crack expansion in red sandstone, reducing its density. The influence of plastic deformation at elevated temperatures outweighs the confining pressure's restrictive effect, thereby decreasing the sample's load-bearing capacity. Consequently, temperature becomes the primary factor determining the compressive strength of red sandstone.

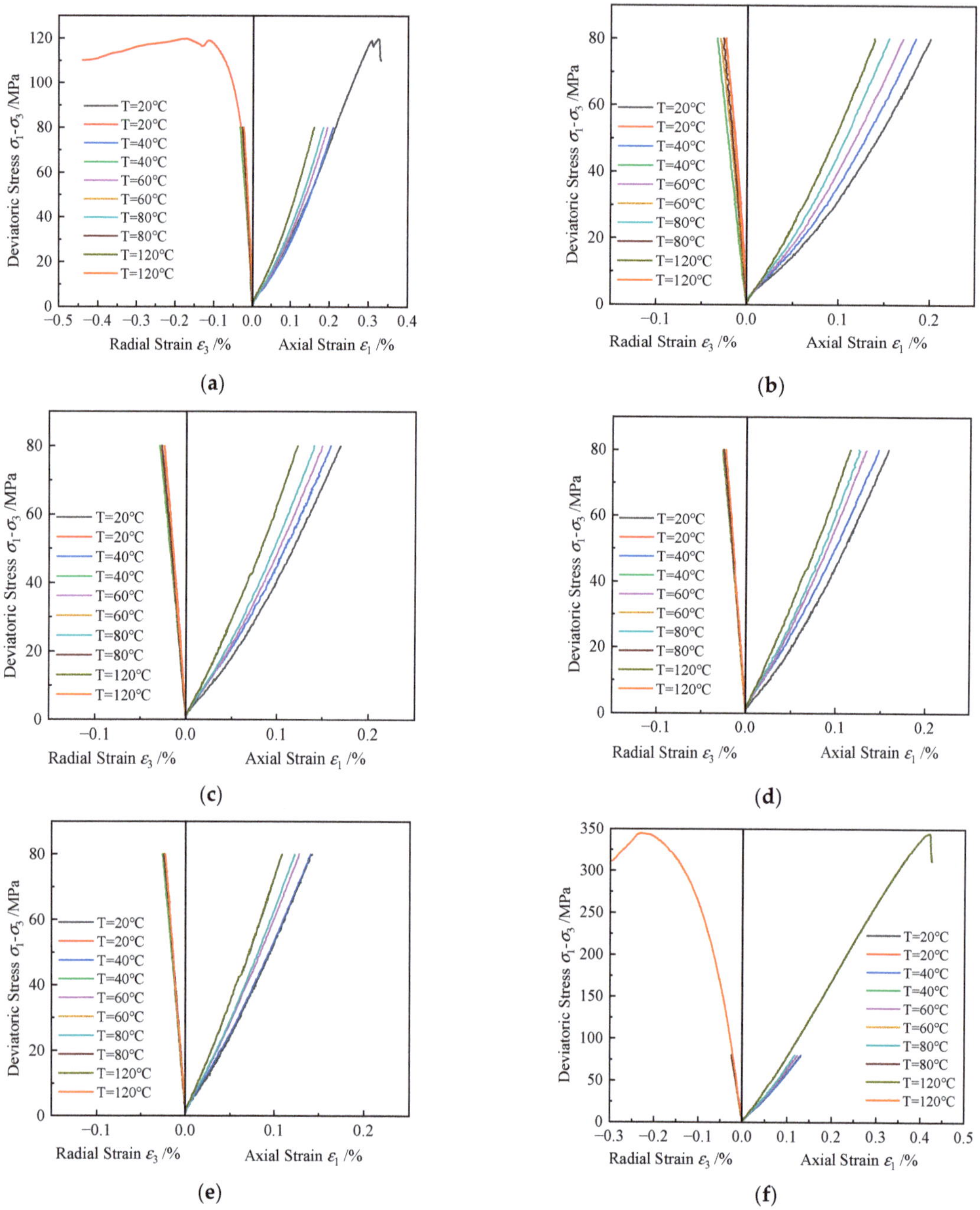

Figure 4. Stress–strain curves of granite under different temperature conditions: (**a**) granite σ_3 = 0 MPa; (**b**) granite σ_3 = 5 MPa; (**c**) granite σ_3 = 10 MPa; (**d**) granite σ_3 = 15 MPa; (**e**) granite σ_3 = 20 MPa; (**f**) granite σ_3 = 25 MPa.

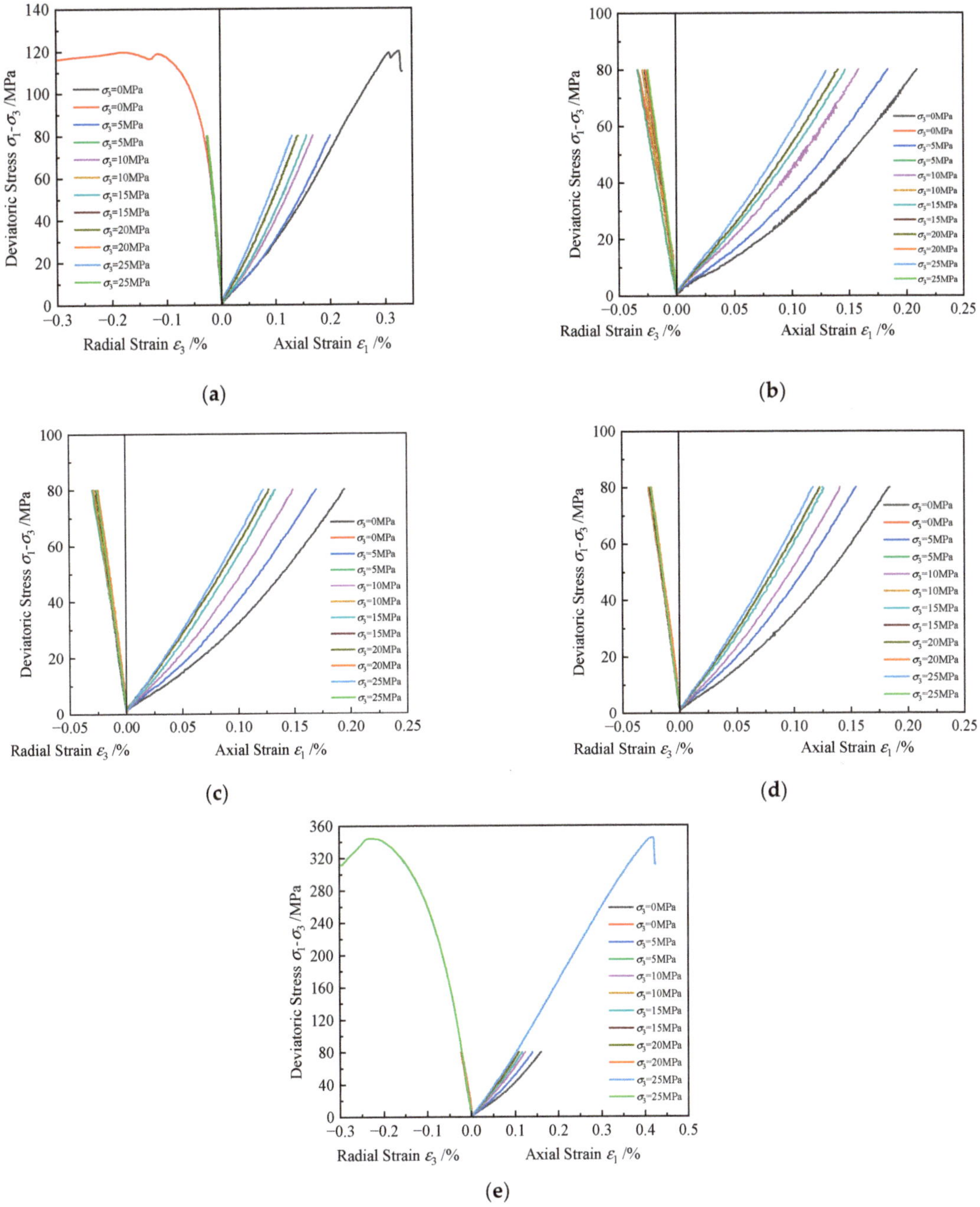

Figure 5. Stress–strain curves of granite under different confining pressures: (**a**) granite T = 20 °C; (**b**) granite T = 40 °C; (**c**) granite T = 60 °C; (**d**) granite T = 80 °C; (**e**) granite T = 120 °C.

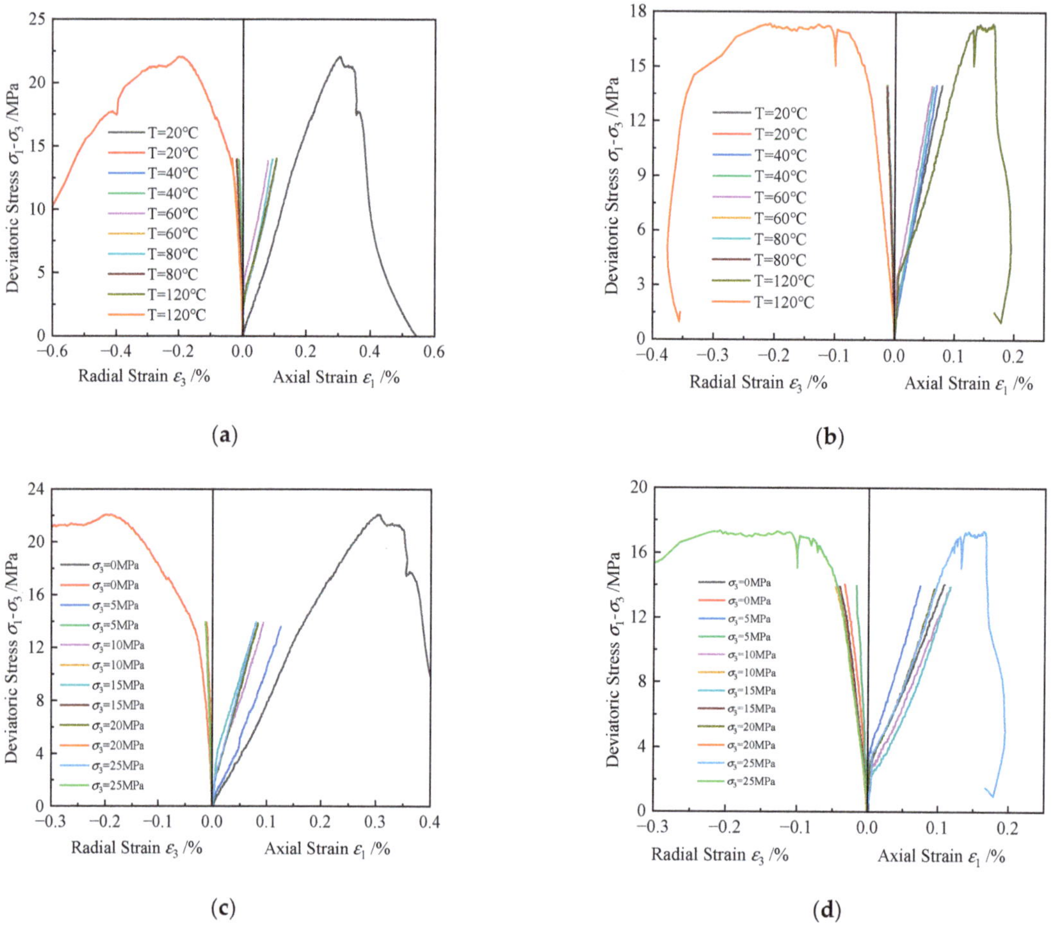

Figure 6. Stress–strain curves of red sandstone under different temperature and confining pressure conditions: (**a**) red sandstone σ_3 = 0 MPa; (**b**) red sandstone σ_3 = 25 MPa; (**c**) red sandstone T = 20 °C; (**d**) red sandstone T = 120 °C.

Figure 7 illustrates the stress–strain curves of gray sandstone under different temperature and confining pressure conditions. It can be observed that the curves of gray sandstone similarly exhibit the four stages of pore crack compaction, linear elastic deformation, microcrack development, and post-failure. After reaching a certain strain, the curve begins to show nonlinear growth, indicating plastic deformation and the formation of a distinct deformation region, characteristic of ductile failure. Additionally, the compressive strength and peak strain of gray sandstone significantly increases under high-temperature and high-pressure conditions, indicating that the coupled effects of temperature and stress enhance the mineral density within the gray sandstone sample, thereby improving its load-bearing capacity and ductility.

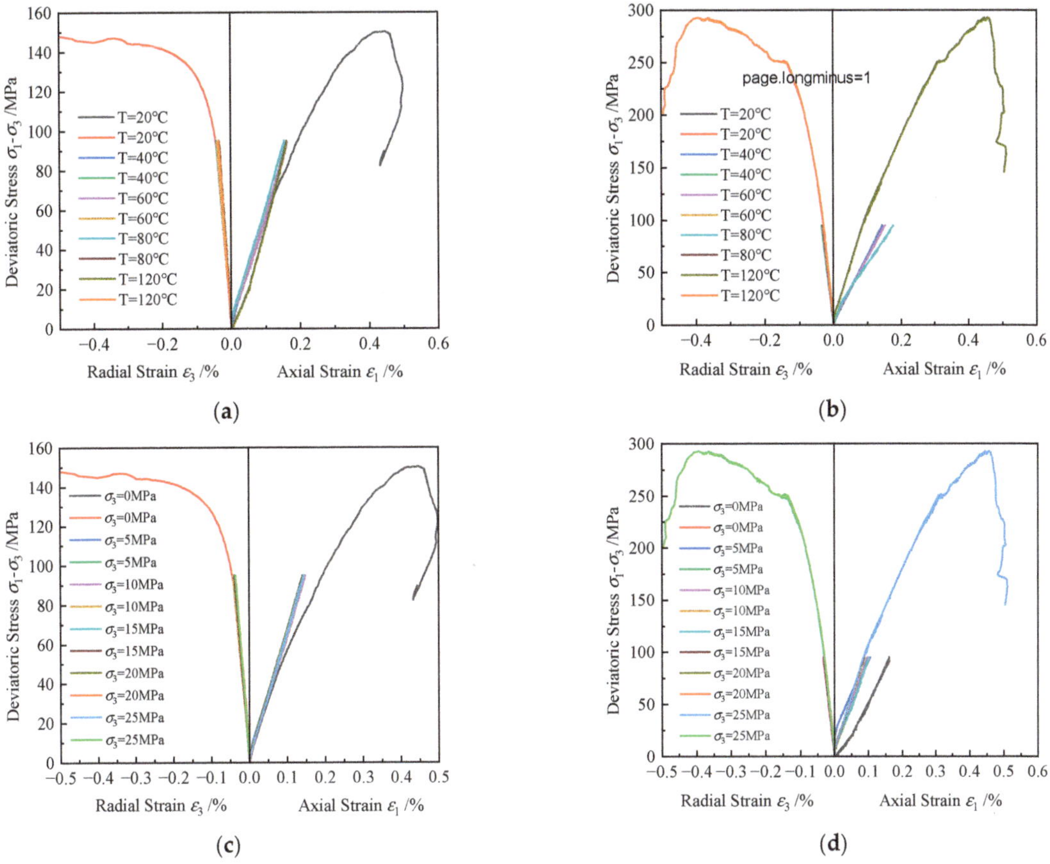

Figure 7. Stress–strain curves of gray sandstone under different temperature and confining pressure conditions: (**a**) gray sandstone σ_3 = 0 MPa; (**b**) gray sandstone σ_3 = 25 MPa; (**c**) gray sandstone T = 20 °C; (**d**) gray sandstone T = 120 °C.

The stress–strain curves of gray sandstone under different temperature and confining pressure conditions are shown in Figure 8. It can be observed that the curve of shale is similar to that of red sandstone and calcareous sandstone, also exhibiting four stages. As the temperature increases, the elastic modulus of the shale increases, the lateral deformation decreases, and the resistance to elastic deformation and the load-bearing capacity are improved. The compressive strength and peak strain of shale also improves under high-temperature and high-pressure conditions.

Therefore, granite demonstrates pronounced brittle failure characteristics under high temperature and pressure, with the effect of confining pressure on its deformation behavior being greater than that of temperature changes. Compared to granite, red sandstone, gray sandstone, and shale exhibit a steeper slope during the elastic phase. The elastic modulus and stiffness of these rocks increase significantly with rising temperature and confining pressure. The nonlinear increase in the curve of gray sandstone is more pronounced, demonstrating characteristics of ductile failure, which is distinctly different from the brittle failure observed in granite.

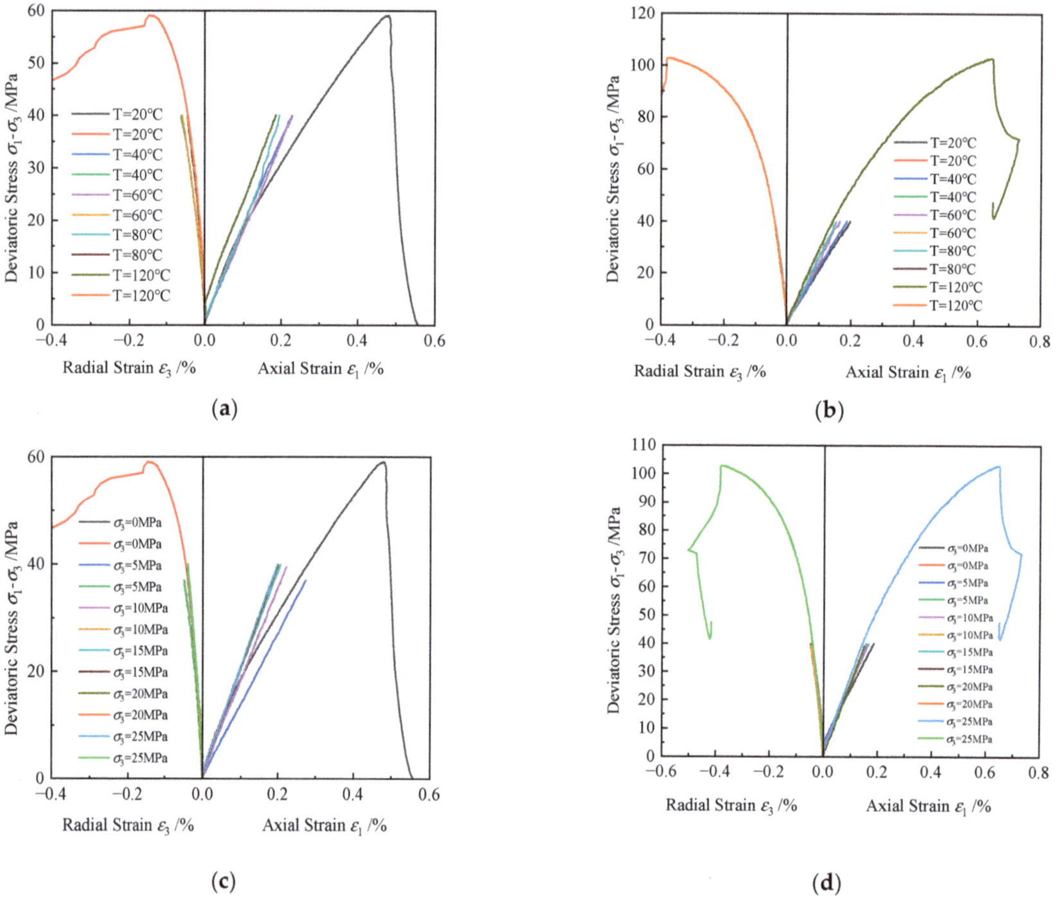

Figure 8. Stress–strain curves of shale under different temperature and confining pressure conditions: (**a**) shale σ_3 = 0 MPa; (**b**) shale σ_3 = 25 MPa; (**c**) shale T = 20 °C; (**d**) shale T = 120 °C.

3.2.2. Analysis of Elastic Modulus for Different Rocks

Granite exhibits distinct stages of pore crack compaction and elastic deformation during loading. The transition from a concave upward shape to an approximately linear shape in the stress–strain curve serves as the boundary, allowing a separate discussion of the deformation characteristics of these two stages. The ratio of stress to strain at this boundary point is used as the deformation modulus E of the pore crack compaction stage, determined by Equation (1). The tangent elastic modulus E_t of the elastic deformation stage is represented by the tangent slope of the linear segment of the stress–strain curve, determined by Equation (2). The determination of E and E_t is shown in Figure 9.

$$E = \frac{\sigma_{01}}{\varepsilon_{01}} \tag{1}$$

$$E_t = \frac{\sigma_{02} - \sigma_{01}}{\varepsilon_{02} - \varepsilon_{01}} \tag{2}$$

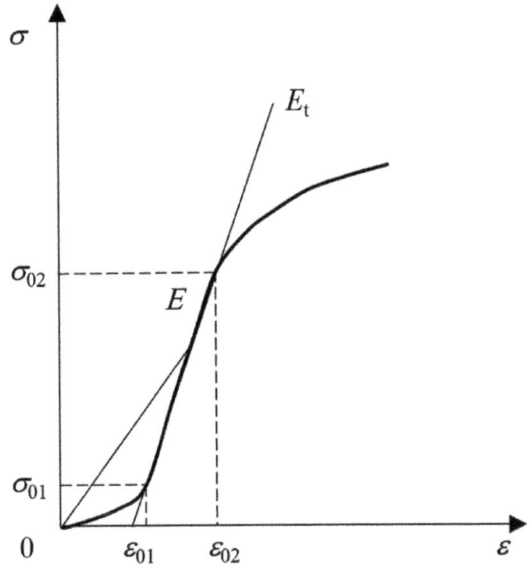

Figure 9. Schematic diagram for determining the E and the E_t.

Figure 10a indicates that, as the temperature increases, the deformation modulus E increases. In the temperature range of 40 °C to 80 °C, the slopes of the curves remain relatively low, particularly under lower stress conditions (e.g., 0 MPa and 5 MPa). This suggests that, at lower temperatures, the impact of temperature on the elastic modulus is minimal, and the stiffness of the rock does not increase significantly. Conversely, in the temperature range of 100 °C to 120 °C, the slopes of the curves increase markedly, especially under higher stress conditions (e.g., 25 MPa). During this phase, the elastic modulus exhibits a rapid upward trend as temperature rises. This behavior is primarily attributed to the enhanced interactions between mineral particles, the repair or rearrangement of crystal lattice defects, and the resulting increase in rock density and hardness under high-temperature and high-pressure conditions, leading to a significant increase in the elastic modulus. At confining pressures of 0 MPa, 5 MPa, 10 MPa, 15 MPa, 20 MPa, and 25 MPa, E increases by 38.17%, 59.31%, 42.46%, 41.53%, 33.73%, and 34.76%, respectively, and the increment in rock deformation modulus initially rises and then levels off. At lower confining pressures, the increase in pressure compresses and closes microcracks and pores, enhancing the overall stiffness and load-bearing capacity of the rock sample. However, as confining pressure further increases, the effect of microcrack closure saturates, and the increase in deformation modulus stabilizes. Figure 10b illustrates that the tangent elastic modulus (E_t) increases as the temperature rises. At confining pressures of 0 MPa, 5 MPa, 10 MPa, 15 MPa, 20 MPa, and 25 MPa, E_t increases by 39.09%, 40.17%, 28.82%, 21.17%, 26.92%, and 25.06%, respectively. The significant rise in E_t when the confining pressure increases from 0 MPa to 5 MPa is attributed to the compaction of microcracks and pores during the crack closure phase. As the confining pressure reaches 15 MPa, the rate of increase in E_t declines due to the formation of new microcracks or the expansion of existing ones. However, when the confining pressure increases to 25 MPa, the rate of increase in E_t stabilizes, indicating that new cracks are rapidly compressed under high confining pressure, with the rock potentially nearing its density limit.

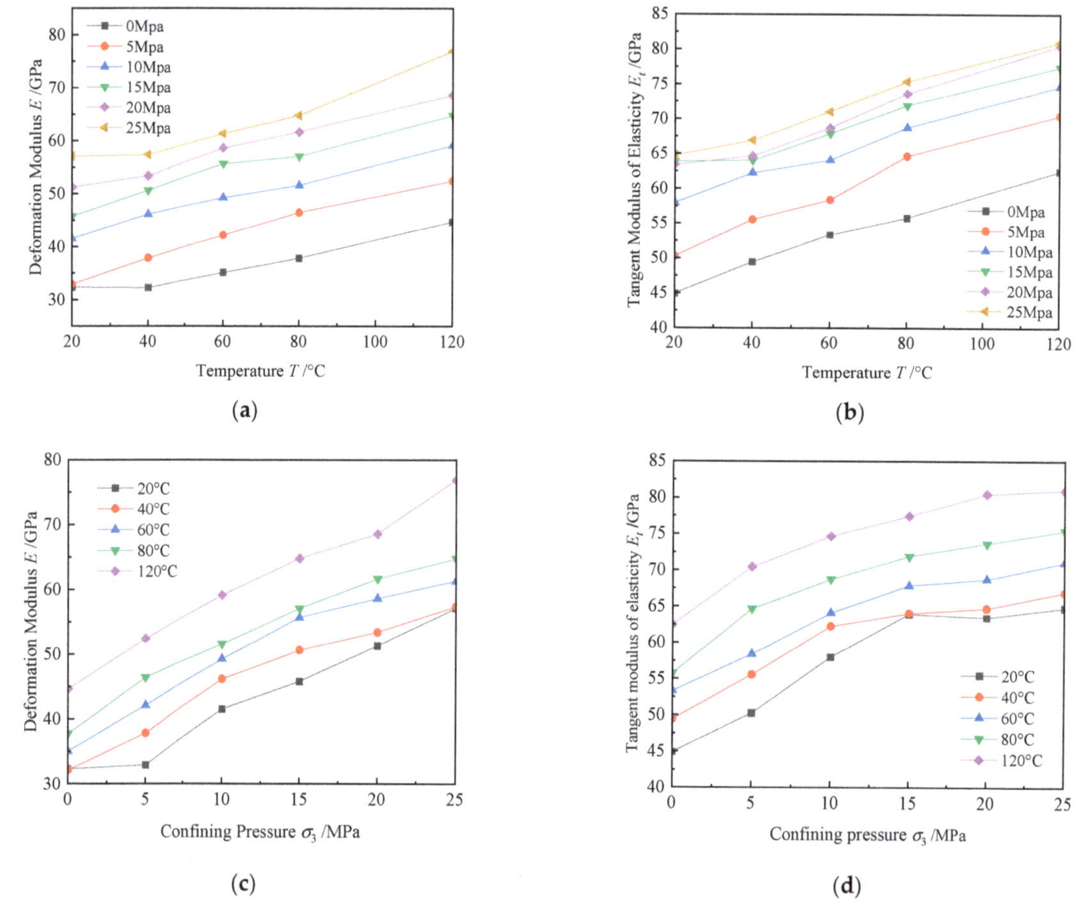

Figure 10. Relationship diagrams of granite deformation modulus, tangent elastic modulus with temperature, and confining pressure: (**a**) relationship between granite deformation modulus and temperature; (**b**) relationship between granite tangent elastic modulus and temperature; (**c**) relationship between granite deformation modulus and confining pressures; (**d**) relationship between granite tangent elastic modulus and confining pressures.

Figure 10c shows a gradual increase in the deformation modulus *E* as the confining pressure rises from 0 MPa to 25 MPa. The confining pressure causes internal pores and microcracks to gradually close, enhancing the load-bearing capacity of the rock sample and leading to an increase in deformation modulus. At temperatures of 20 °C, 40 °C, 60 °C, 80 °C, and 120 °C, *E* increases by 76.51%, 78.03%, 74.74%, 71.47%, and 72.15%, respectively. When the temperature increases from 20 °C to 40 °C, *E* slightly increases due to the thermal expansion effect, where the crack closure effect outweighs the expansion effect. As the temperature rises to 80 °C, the rate of increase in *E* slightly declines due to the predominant crack expansion effect caused by continuous heating. When the temperature reaches 120 °C, the increase in *E* stabilizes as the thermal expansion and crack closure effects balance out. Figure 10d demonstrates a gradual rise in E_t as the confining pressure increases from 0 MPa to 25 MPa. At temperatures of 20 °C, 40 °C, 60 °C, 80 °C, and 120 °C, the increases of E_t are 44.18%, 35.33%, 33.15%, 35.18%, and 29.64%, respectively. As the temperature rises from 20 °C to 40 °C, the increase in E_t slightly declines, possibly due to most cracks and pores being compressed and closed during the crack closure phase while, in the elastic deformation phase, new microcracks induced by temperature rise affect structural integrity and reduce the rate of increase in elastic modulus. When the temperature rises to 80 °C, the increase in E_t stabilizes due to the balanced effect of thermal expansion on cracks. At 120 °C, high temperature causes the expansion of cracks and pores, influencing the trend of E_t changes and potentially altering mineral structures, leading to a decrease in the rate of increase in E_t. Obviously, confining pressure has a more significant impact on the deformation modulus during the pore crack compaction phase and on the tangent elastic modulus during the elastic phase of granite. The increase in deformation modulus with rising confining pressure is greater than that caused by temperature rise, indicating that confining pressure plays a major role in the changes in deformation modulus and elastic modulus.

The stress–strain curves of red sandstone, gray sandstone, and shale exhibit a single slope, indicating that the elastic modulus remains constant. The nonlinear characteristics of the crack closure compaction stage are not evident and, after loading, the stress–strain curves quickly transition into the elastic deformation stage, showing linear variation characteristics. Therefore, the tangent elastic modulus E_t is used to represent the elastic modulus for the entire deformation stage. This value is defined as the slope of the linear segment in the stress-axial strain relationship.

Figure 11 illustrates the relationship between tangent elastic modulus and temperature, confining pressure for different rocks. From Figure 11a,b, it can be observed that the tangent elastic modulus E_t of red sandstone exhibits significant trends with changes in temperature and confining pressure. As the temperature rises, E_t first increases and then decreases. Between 20 °C and 40 °C, the increase in temperature causes thermal expansion of minerals, which closes cracks and increases the sample's density, leading to an increase in E_t. In the range of 60 °C to 80 °C, at low confining pressures (0–15 MPa), the increase in temperature still causes E_t to increase, while at high confining pressures (15–25 MPa), E_t remains relatively unchanged and tends to stabilize. When the temperature reaches 120 °C, E_t significantly decreases, primarily due to high temperatures causing mineral particle expansion, generating thermal stress, weakening internal bonds, and intensifying crack propagation, which reduces the elastic modulus.

Figure 11c,d indicate that the trend of E_t of gray sandstone differs from that of red sandstone. As temperature increases, E_t initially decreases and then increases. At room temperature (20 °C), when the confining pressure increases, E_t increases from 62.39 GPa to 65.18 GPa, an increase of 4.47%; at 40 °C, the increase is 7.74%; at 60 °C, the increase is 3.29%. The E_t of gray sandstone increases with confining pressure, but the rate of increase gradually decreases. This indicates that the effect of confining pressure on the closure of internal cracks in gray sandstone gradually reaches saturation. At 80 °C, E_t first decreases and then increases, reflecting the significant thermal expansion effect at high temperatures, leading to pore and crack expansion, but at high confining pressures, the closure effect dominates. At 120 °C, E_t increases with confining pressure, indicating that thermal expansion causes mineral particles to contact more closely.

Figure 11e,f illustrate that the tangent elastic modulus E_t of shale is significantly affected by temperature and confining pressure. In the range of 20 °C to 60 °C, at low confining pressures (0–5 MPa), E_t first rises and then falls, while at high confining pressures (10–25 MPa), E_t gradually rises. High temperatures cause the closure effect of internal pores and cracks in shale to exceed the expansion effect, leading to an increase in E_t.

The differing trends in the development of elastic modulus for the four rock types under varying temperature and confining pressure conditions primarily stem from differences in their internal structures and mineral compositions. Granite and shale generally exhibit an increase in elastic modulus under high temperature and pressure, reflecting the thermal stability of their structures and the effect of confining pressure on fracture closure. In contrast, red sandstone and gray sandstone show a more complex response, particularly under high-temperature conditions. Their porosity and diverse mineral compositions can lead to a decrease in elastic modulus under certain conditions.

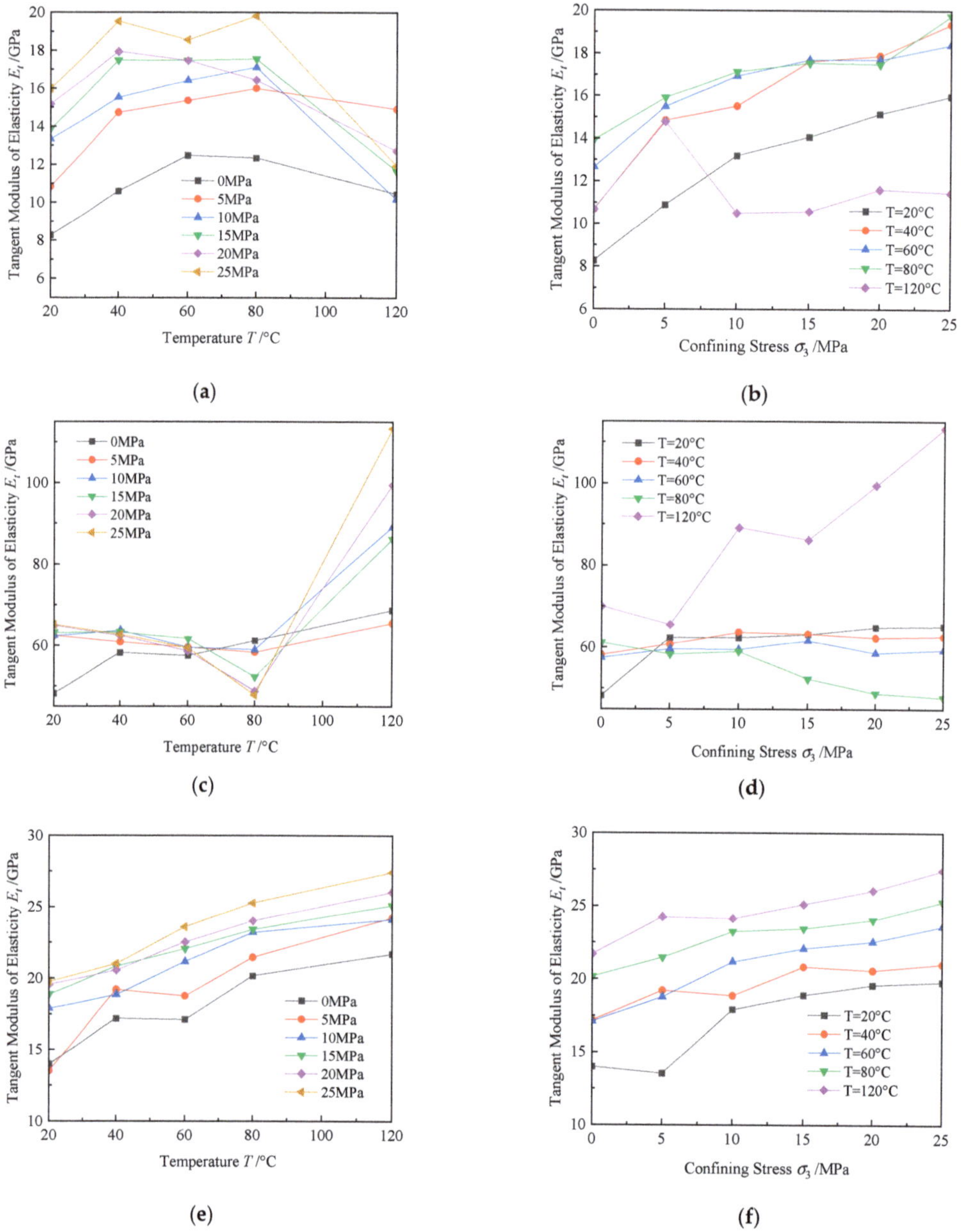

Figure 11. Relationship between tangent elastic modulus and temperature, confining pressure for different rocks: (**a**) relationship between red sandstone tangent elastic modulus and temperature; (**b**) relationship between red sandstone tangent elastic modulus and confining pressures; (**c**) relationship between gray sandstone tangent elastic modulus and temperature; (**d**) relationship between gray sandstone tangent elastic modulus and confining pressures; (**e**) relationship between shale tangent elastic mod-ulus and temperature; (**f**) relationship between shale tangent elastic mod-ulus and confining pressures.

3.3. Model Analysis of Elastic Modulus of Rock

3.3.1. Modeling the Elastic Modulus of Rock under the Influence of Temperature

Clearly, at low temperatures microcracks within the rock tend to close, while at high temperatures these microcracks may expand or new cracks may form, leading to a nonlinear variation in the rock's elastic modulus. Additionally, the differing thermal expansion coefficients of various minerals cause extra internal stress and strain within the rock, thereby affecting its mechanical properties. Under the influence of temperature, the tangent elastic modulus of some rock masses may also exhibit a decreasing trend. Considering these factors, the relationship between the rock's elastic modulus and temperature can be well-fitted using a polynomial function:

$$E(T) = AT^3 + BT^2 + CT + E_0 \tag{3}$$

where E_0 represents the elastic modulus of the rock under uniaxial compression at room temperature (with room temperature taken as T = 20 °C), and A, B, and C are the fitting parameters to be determined.

Figure 12 shows the fitting results for different types of rocks. By integrating the data from the four rock types, the overall goodness of fit (R^2) for the polynomial model is above 0.90, indicating a good fit. It is observed that the gray sandstone has a lower goodness of fit for the temperature model at a confining pressure of 5 MPa. This is likely due to the higher porosity of gray sandstone causing the internal pores to close during the initial stages of confining pressure changes, which leads to abrupt changes in the tangent elastic modulus. This behavior is challenging for the model to accurately describe. However, in other developmental stages, the model effectively fits the trend of the tangent elastic modulus with temperature changes for all four rock types. This demonstrates that the polynomial model can accurately describe the nonlinear changes in the tangent elastic modulus of rocks under varying temperatures.

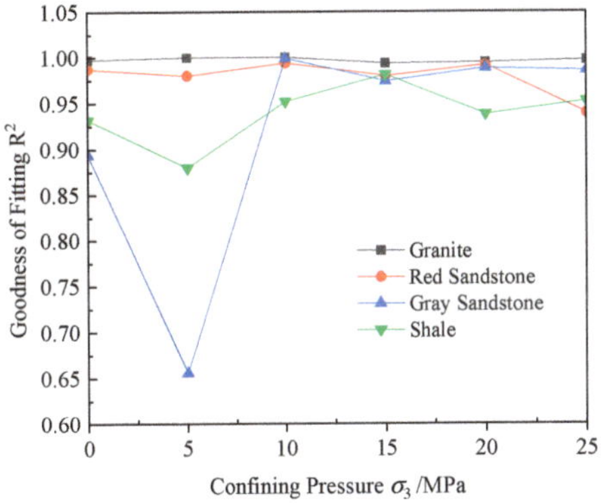

Figure 12. Comparison of model fitting goodness for four rock types under temperature influence.

3.3.2. Modeling the Elastic Modulus of Rock under the Influence of Confining Pressure

Through the mechanical properties of the rock samples, it can be observed that the tangent elastic modulus of the four types of rocks studied in this paper exhibits nonlinear growth with increasing confining pressure. Researchers have proposed many regression formulas or empirical formulas based on experimental results and most models use quadratic curves for fitting [6], which can efficiently fit the data within a limited range. However, these models fail to accurately describe the part where the growth rate of the elastic modulus decreases with increasing confining pressure. The experiments indicate that the rate of increase in the tangent elastic modulus (E_t) gradually slows down as the confining pressure increases, suggesting the existence of a limiting value for E_t. Based on the micromechanical damage theory, we use a composite exponential formula to represent the relationship between E_t and σ_3. The formula is expressed as follows:

$$E(\sigma_3) = E_0[1 - (\frac{A}{E_0} - 1)(1 - e^{(-\frac{\sigma_3}{B})})] \tag{4}$$

where E_0 represents the elastic modulus of the rock under uniaxial compression; A and B are the parameters to be determined. The fitting results under different confining pressures are shown in Figure 13.

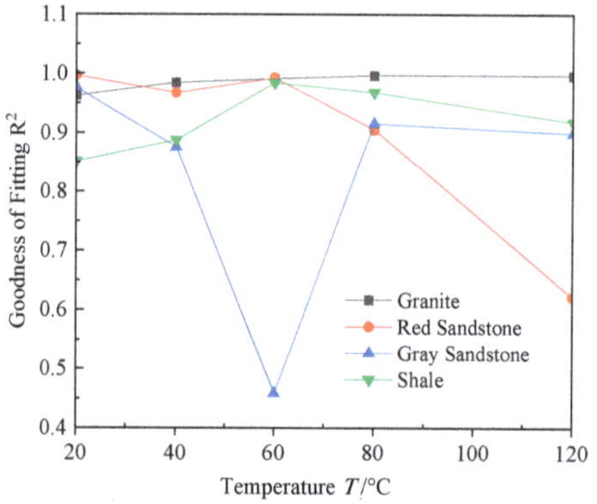

Figure 13. Comparison of model fitting goodness for four rock types under confining pressure influence.

By integrating the data from the four rock types, the overall R^2 value exceeded 0.90, indicating a good fit. This suggests that the model effectively describes the variation of the tangent elastic modulus of rocks under the influence of confining pressure. In particular, it accurately captures the phenomenon where the rate of change in the tangent elastic modulus decreases as the confining pressure reaches a certain value. This behavior is primarily attributed to the closure of internal cracks within the rock mass, leading to an increase in elastic modulus, which ultimately approaches its limit. The composite exponential model successfully reflects this trend. The fitting results for calcareous sandstone at 60 °C were less satisfactory, likely due to the higher porosity of calcareous sandstone compared to the other rock types, as discussed in Section 3.2.2. This higher porosity causes abrupt changes in the tangent elastic modulus within the 10 MPa to 20 MPa range, thereby reducing the effectiveness of the model fitting.

3.3.3. Modeling the Elastic Modulus of Rock under the Influence of Temperature and Confining Pressure

The thermo-mechanical coupling effect is not a simple additive relationship between temperature and confining pressure but rather a complex interaction. Therefore, drawing on the fitting models from Sections 3.3.1 and 3.3.2, both temperature and confining pressure factors are jointly considered. By referencing the development trend of the elastic modulus of rocks under thermo-mechanical coupling, the following formula is used for nonlinear fitting:

$$E(T, \sigma_3) = AT^3 + BT^2 + CT + E_0[1 - (\frac{D}{E_0} - 1)(1 - e^{(-\frac{\sigma_3}{E})})] \tag{5}$$

where E_0 represents the elastic modulus of the rock under uniaxial compression at room temperature (with room temperature taken as T = 20 °C), and A, B, C, D, E are the fitting parameters to be determined.

Table 2 demonstrates the results of fitting the tangent elastic modulus under temperature-confining pressure coupling. Figure 14 demonstrates the fitting results of the tangent elastic modulus in relation to temperature and confining pressure for the four types of rocks. The results indicate that the nonlinear fitting model effectively describes the development trend of the tangent elastic modulus under thermo-mechanical coupling for all four rock types. The fitting results for granite and shale are particularly accurate, whereas the fitting for red sandstone and gray sandstone is slightly less precise. This discrepancy is mainly due to the higher degree of fracture development and porosity in red sandstone and gray sandstone, which can cause abrupt changes in the tangent elastic modulus under high temperature and pressure conditions. Consequently, the model finds it challenging to accurately fit these abrupt trends. Future research could focus on investigating the specific characteristics of these rock types further.

Table 2. Fitting results of tangent elastic modulus under thermo-mechanical coupling.

	A	B	C	D	E	R^2
G	-1.5519×10^{-5}	0.00319	−0.01056	23.53821	11.65526	0.98819
RS	3.6669×10^{-6}	−0.00245	0.26988	2.36925	8.99786	0.81258
GS	1.7180×10^{-4}	−0.02891	1.35263	35.58206	15.28693	0.71489
SH	-1.5150×10^{-6}	-6.6366×10^{-5}	0.11000	6.38303	20.27230	0.95229

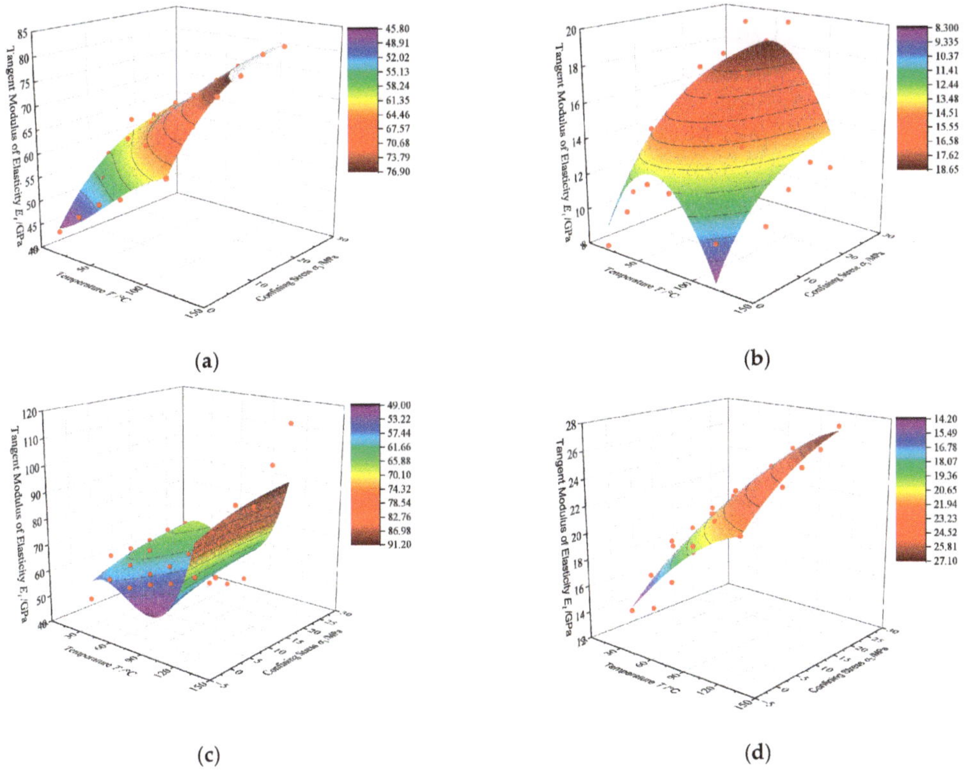

Figure 14. Fitting results of tangent elastic modulus in relation to temperature and confining pressure for rocks (The red dots in the graph indicate actual test data): (**a**) granite; (**b**) red sandstone; (**c**) gray sandstone; (**d**) shale.

4. Damage Constitutive Model of Rock

4.1. Rock Damage Theory Based on the Drucker–Prager Criterion

Based on the Drucker–Prager strength criterion, the microelement strength of the rock F^* is defined as:

$$F^* = f(\sigma^*) = \alpha I_1 + \sqrt{J_2} \tag{6}$$

where α represents the strength parameter of the rock microelement; I_1 denotes the first invariant of stress; J_2 signifies the second invariant of deviatoric stress.

Expressing the above I_1 and J_2 in terms of stresses gives:

$$I_1 = \frac{E_{\varepsilon_1}(\sigma_1 + 2\sigma_3)}{\sigma_1 - 2v\sigma_3} \tag{7}$$

$$\sqrt{J_2} = \frac{E\varepsilon_1(\sigma_1 + 2\sigma_3)}{\sqrt{3}(\sigma_1 - 2v\sigma_3)} \tag{8}$$

Substituting Equations (7) and (8) into Equation (6), the microelement strength of the rock under confining pressure can be obtained as:

$$F^* = \frac{E\varepsilon_1\left[\sin\varphi(\sigma_1 + 2\sigma_3) + \sqrt{3 + \sin^2\varphi}(\sigma_1 - \sigma_3)\right]}{\sqrt{9 + 3\sin^2\varphi}(\sigma_1 - 2v\sigma_3)} \tag{9}$$

Substituting Equation (9) into Equation (14), the rock damage evolution equation under the coupled effect of temperature and confining pressure is obtained as follows:

$$D_s = 1 - \frac{E_T}{E_0}\exp\left\{-\left\{\frac{E_T\varepsilon_1\left[\sin\varphi(\sigma_1 + 2\sigma_3) + \sqrt{3 + \sin^2\varphi}(\sigma_1 - \sigma_3)\right]}{F_0\sqrt{9 + 3\sin^2\varphi}(\sigma_1 - 2v\sigma_3)}\right\}^m\right\} \tag{10}$$

4.1.1. Rock Damage under Confining Pressure

The damage variable can be defined in various ways. Given that material damage results in alterations to the microstructure and certain macroscopic physical properties, the failure of rock materials can be viewed as a process of damage accumulation. Consequently, the benchmark for measuring damage can be selected from both micro and macro perspectives. The damage to microelements within the rock under load is generally random. The damage variable D can be defined as the ratio of the number of damaged units n at a certain stress level q to the total number of units N in the initial state, as follows:

$$D = \frac{n}{N} \tag{11}$$

According to the Krajcinovic model [38,39], the damage variable is the probability p of microelement failure. If the probability density of microelement failure is $\phi(x)$, then p is the cumulative distribution function F, expressed as:

$$D = p = \int_0^F \phi(x)dx \tag{12}$$

Based on the assumption that the strength of rock microelements follows a Weibull distribution, its probability density function can be expressed as:

$$\phi(F) = \frac{m}{F_0}\left(\frac{F}{F_0}\right)^{m-1}\exp\left[-\left(\frac{F}{F_0}\right)^m\right] \tag{13}$$

where F represents the microelement strength; the parameter m represents the shape parameter of the Weibull distribution, indicating the uniformity of the internal strength of the rock material. The parameter F_0 is the scale parameter of the Weibull distribution, reflecting the average strength level of the rock material.

Substituting Equation (13) into Equation (12), the damage variable of rock under external load is:

$$D = \int_0^F \phi(x)dx = 1 - \exp[-(\frac{F}{F_0})^m] \tag{14}$$

4.1.2. Rock Damage under Temperature

Based on the theory of macroscopic damage mechanics, the thermal damage of rock can be characterized using macroscopic mechanical parameters. Therefore, the elastic modulus can be used to define the thermal damage variable, which can be expressed as:

$$D_T = 1 - \frac{E_T}{E_0} \tag{15}$$

where E_T is the elastic modulus of the rock under the temperature T; E_0 is the elastic modulus of the rock at room temperature (20 °C).

4.1.3. Total Damage of Rock under Thermo-Confining Pressure Coupling

During triaxial compression tests under high temperature and high confining pressure, the primary causes of rock damage are the destructive effects of temperature and confining pressure on the internal structure of the rock. High temperatures cause thermal expansion of internal materials and cracks within the rock, usually leading to internal damage. Confining pressure compresses and closes the microcracks and pores within the rock, thereby enhancing its overall stiffness and density. Due to the uncertainty in the extent of rock damage caused by the coupling of temperature and stress, certain mathematical methods are needed to quantitatively analyze the total damage.

Assuming that the effective volumes of undamaged rock, thermally damaged rock, and rock damaged are V_0, V_1, and V_2 respectively, then we have:

$$D_T = 1 - \frac{V_1}{V_0} \tag{16}$$

$$D_m = 1 - \frac{V_2}{V_1} \tag{17}$$

By combining the above assumptions, we can establish the following relationship:

$$D_s = D_T + D - D_T D \tag{18}$$

By substituting Equations (14) and (15) into the overall equation, the total damage variable for rock under the coupled effects of temperature and confining pressure can be obtained as:

$$D_s = 1 - \frac{E_T}{E_0} \exp[-(\frac{F}{F_0})^m] \tag{19}$$

According to Lemaitre's strain equivalence hypothesis, the constitutive relationship of the rock damage can be expressed as follows:

$$\sigma_i = \sigma_i'(1 - D_s); \quad i = 1,2,3 \tag{20}$$

According to the generalized Hooke's law, we have:

$$\varepsilon_i' = \frac{\sigma_i' - v(\sigma_j' + \sigma_k')}{E_T} \tag{21}$$

where $\varepsilon_i{}'$ is the effective strain corresponding to the effective stress $\sigma_i{}'$.

According to the deformation compatibility conditions, we have:

$$\varepsilon_i' = \varepsilon_i \tag{22}$$

From Equations (20)–(22), the damage constitutive relation of rock can be derived as follows:

$$\sigma_1 = E_T\varepsilon_1(1 - D_s) + v(\sigma_2 + \sigma_3) \tag{23}$$

The initial axial strain generated under the initial confining pressure is:

$$\sigma_s = \sigma_1 - \sigma_3 \tag{24}$$

$$\varepsilon_0 = \frac{1 - 2v}{E_T}\sigma_3 \tag{25}$$

$$\varepsilon_1 = \varepsilon + \varepsilon_0 \tag{26}$$

By substituting Equations (10) and (24)–(26) into Equation (23), the damage constitutive model of rock under thermo-confining pressure coupling can be obtained as follows:

$$\begin{aligned} \sigma_1 &= [E_T\varepsilon_1 + (1 - 2v)\sigma_3]\tfrac{E_T}{E_0} \times \\ &\exp[-\left\{\tfrac{[E_T\varepsilon_1 + (1-2v)\sigma_3][(\sigma_1 + 3\sigma_3)\sin\varphi + \sigma_1\sqrt{3+\sin^2\varphi}]}{F_0\sqrt{9+3\sin^2\varphi}[\sigma_1 + (1-2v)\sigma_3]}\right\}^m] + (2v - 1)\sigma_3 \end{aligned} \tag{27}$$

In summary, the parameters that need to be determined are E_T, m and F_0. The elastic modulus under the influence of temperature, E_T, can be determined using the fitting model discussed in Section 3.3.1. The parameters m and F_0 can be determined from the peak points (ε_p,σ_p) of the rock stress–strain curves.

The peak stress σ_p and peak strain ε_p satisfy the following two geometric conditions:

$$\varepsilon = \varepsilon_p,\ \ \sigma_1 = \sigma_p \tag{28}$$

$$\varepsilon = \varepsilon_p,\ \ \frac{d\sigma_1}{d\varepsilon_1} = 0 \tag{29}$$

By substituting Equations (28) and (29) into Equation (27), two relationships regarding m and F_0 can be obtained, as shown below:

$$m = \frac{1}{\ln\frac{E_T}{E_0}[\frac{E_T\varepsilon_p + (1-2v)\sigma_3}{\sigma_p + (1-2v)\sigma_3}]} \tag{30}$$

$$F_0 = \frac{[E_T\varepsilon_p + (1 - 2v)\sigma_3][(\sigma_1 + 3\sigma_3)\sin\varphi + \sigma_p\sqrt{3 + \sin^2\varphi}]}{\sqrt{9 + 3\sin^2\varphi}[\sigma_1 + (1 - 2v)\sigma_3]}m^{\frac{1}{m}} \tag{31}$$

By adjusting the parameters m and F_0 based on experimental data and substituting them into Equation (27), a damage constitutive model can be constructed that accurately reflects the stress–strain state of rock under thermo-mechanical coupling.

4.2. Experimental Validation of the Damage Constitutive Model of Rock under Thermo-Confining Pressure Coupling

To validate the rationality and effectiveness of the method proposed in this paper, we evaluated the statistical damage Drucker–Prager constitutive model for granite, red sandstone, gray sandstone, and shale under conditions of 20 °C and 0 MPa, as well as 120 °C and 25 MPa. Based on the experimental results, we extracted mechanical parameters such as peak stress, peak strain, confining pressure, elastic modulus, and Poisson's ratio. These parameters were substituted into Equations (30) and (31) to calculate the Weibull distribution parameters m and F_0 for the four rock types under the conditions of 20 °C, 0 MPa, and 120 °C, 25 MPa. The specific results arc shown in Table 3.

Table 3. Mechanical parameters of rocks under different thermo-mechanical coupling conditions.

Rock Type	T/°C	σ_3/MPa	σ_p/MPa	ε_p/%	φ/°	E_T/GPa	v	m	F_0
G	20	0	119.02	0.31	45	44.93	0.15	6.56	44.42
	120	25	344.60	0.42	50	81.02	0.21	35.59	93.30
RS	20	0	22.06	0.29	45	8.27	0.23	14.24	6.92
	120	25	17.04	0.13	53	11.93	0.22	1.00	32.90
GS	20	0	150.22	0.45	45	48.16	0.07	2.80	68.49
	120	25	289.53	0.43	45	113.34	0.35	0.94	124.62
SH	20	0	59.09	0.48	45	13.99	0.30	8.07	20.98
	120	25	102.63	0.63	45	27.43	0.28	1.24	80.31

Where σ_p is the peak stress; ε_p is the peak strain; φ is the internal friction angle; v is the poisson ratio.

By substituting the calculated Weibull distribution parameters m and F_0 along with the values of the elastic modulus, internal friction angle, and other parameters into Equation (27), the statistical damage Drucker–Prager constitutive model for various types of rocks under the conditions of 20 °C, 0 MPa and 120 °C, 25 MPa can be obtained. The general form of the Drucker–Prager constitutive model under these conditions can be expressed in Table 4.

Table 4. Constitutive models for various rock types.

Rock Type	T/°C	σ_3/MPa	Constitutive Model
G	20	0	$\sigma_1 = (44.93\varepsilon_1)\exp[-\left\{\frac{(44.93\varepsilon_1)(2.58\sigma_1)}{143.94\sigma_1}\right\}^{6.56}]$
	120	25	$\sigma_1 = (105.04\varepsilon_1 + 18.80)\exp[-\left\{\frac{(81.02\varepsilon_1+14.5)(2.66\sigma_1+57.45)}{306.05\sigma_1+4437.78}\right\}^{35.59}] + 14.5$
RS	20	0	$\sigma_1 = (8.27\varepsilon_1)\exp[-\left\{\frac{(8.27\varepsilon_1)(2.58\sigma_1)}{22.42\sigma_1}\right\}^{14.24}]$
	120	25	$\sigma_1 = (11.93\varepsilon_1 + 10.45)\exp[-\left\{\frac{(15.98\varepsilon_1+14)(2.71\sigma_1+59.90)}{108.69\sigma_1+1521.62}\right\}^{1}] + 14$
GS	20	0	$\sigma_1 = (48.16\varepsilon_1)\exp[-\left\{\frac{(48.16\varepsilon_1)(2.58\sigma_1)}{221.93\sigma_1}\right\}^{2.8}]$
	120	25	$\sigma_1 = (114.26\varepsilon_1 + 13.04)\exp[-\left\{\frac{(65.17\varepsilon_1+7.5)(2.58\sigma_1+53.03)}{403.82\sigma_1+3028.61}\right\}^{0.94}] + 7.5$
SH	20	0	$\sigma_1 = (13.99\varepsilon_1)\exp[-\left\{\frac{(13.99\varepsilon_1)(2.58\sigma_1)}{67.98\sigma_1}\right\}^{8.07}]$
	120	25	$\sigma_1 = (38.04\varepsilon_1 + 15.25)\exp[-\left\{\frac{(27.43\varepsilon_1+11)(2.58\sigma_1+53.03)}{260.23\sigma_1+2862.58}\right\}^{1.24}] + 11$

Based on the damage constitutive models in Table 4, the theoretical stress–strain curves for granite, red sandstone, gray sandstone, and shale under conditions of 20 °C, 0 MPa and 120 °C, 25 MPa are compared with the experimental stress–strain curves obtained from tests.

Figure 15 compares the theoretical and experimental stress–strain curves, where it can be observed that the theoretical results obtained from the established constitutive model coincide well with the experimental results, indicating that the Drucker–Prager constitutive model for the four types of rocks under conditions of 20 °C, 0 MPa, and 120 °C, 25 MPa is reliable. By comparing the tangent elastic modulus $E_{t'}$ derived from the theoretical stress–strain curves with the tangent elastic modulus E_t obtained from the experimental stress–strain curves, the accuracy and applicability of the proposed models are further validated, as shown in Table 5.

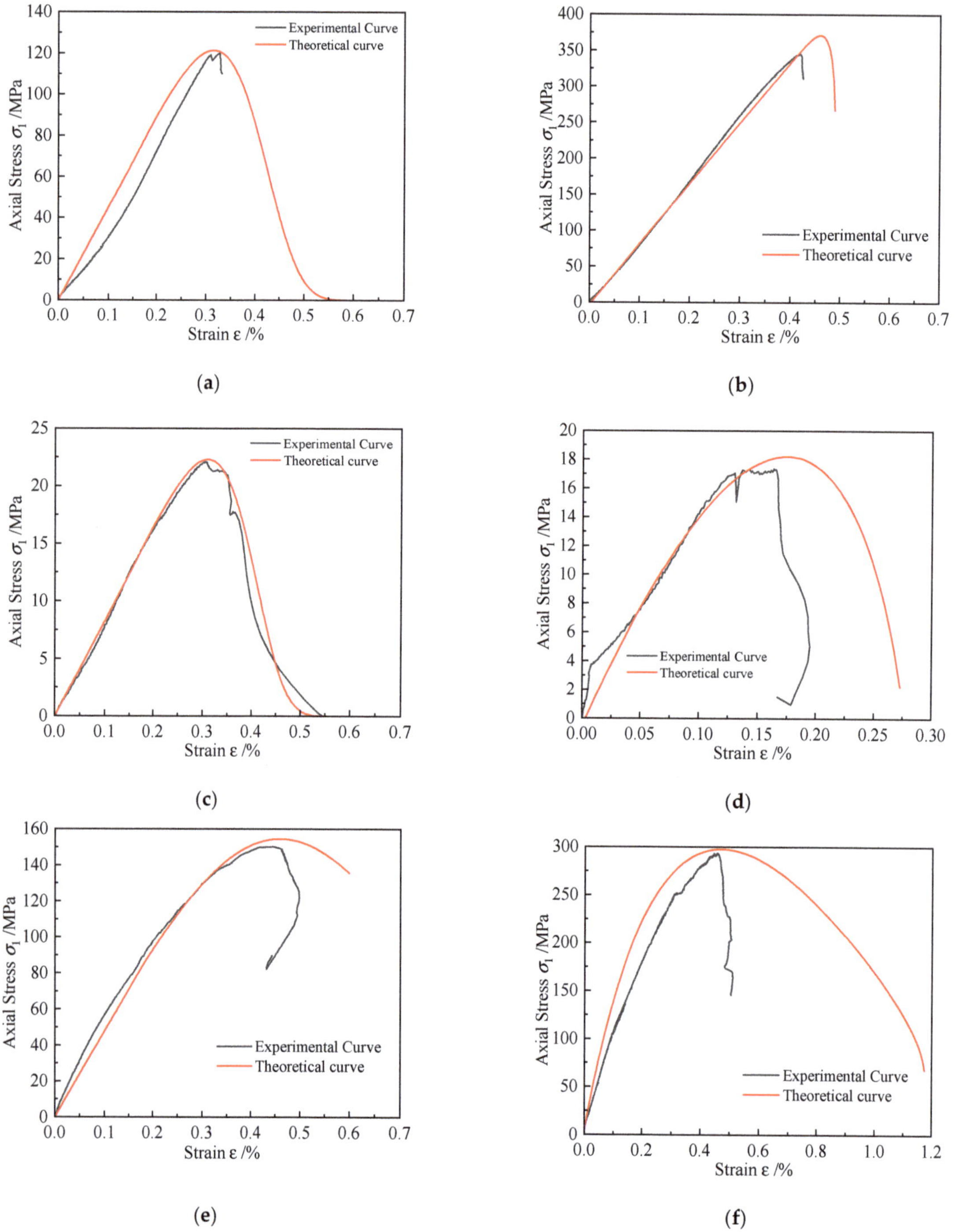

Figure 15. *Cont.*

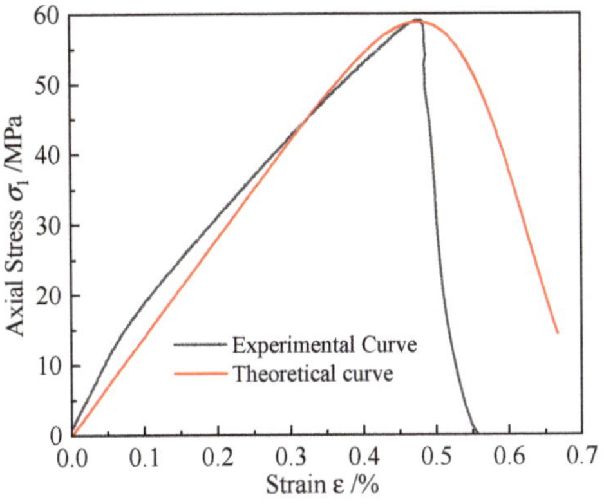

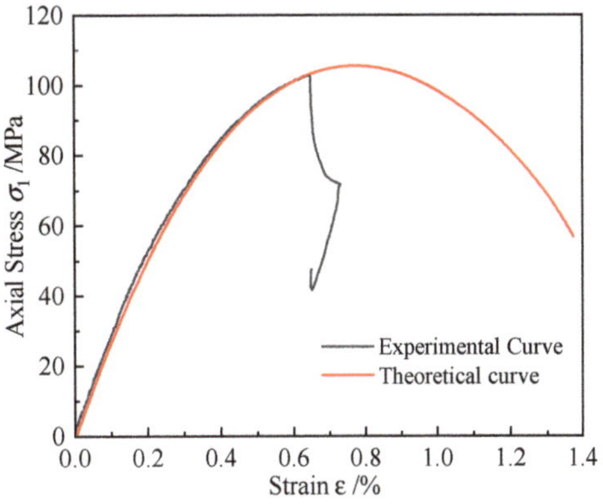

Figure 15. Comparison of theoretical and experimental stress–strain curves: (**a**) granite 20 °C, 0 MPa; (**b**) granite 120 °C, 25 MPa; (**c**) red sandstone 20 °C, 0 MPa; (**d**) red sandstone 120 °C, 25 MPa; (**e**) gray sandstone 20 °C, 0 MPa; (**f**) gray sandstone 120 °C, 25 MPa; (**g**) shale 20 °C, 0 MPa; (**h**) shale 120 °C, 25 MPa.

Table 5. Theoretical and experimental tangent elastic modulus for different rock types.

Rock Type	T/°C	σ_3/MPa	E_t/GPa	$E_{t'}$/GPa	Specific Value
G	20 120	0 25	44.93 81.02	43.92 80.68	0.98 1.00
RS	20 120	0 25	8.27 20.81	7.08 20.71	0.86 0.99
GS	20 120	0 25	48.16 65.71	47.58 63.29	0.99 0.96
SH	20 120	0 25	13.99 27.43	12.25 26.71	0.88 0.98

Figure 16 illustrates the comparison between $E_{t'}$ and E_t. The data points primarily cluster around the 45° line, indicating a high degree of correlation between the theoretical and actual values. This suggests that the theoretical model accurately predicts the experimental outcomes. Data analysis reveals that the average ratio of the theoretical to experimental tangent elastic modulus is 0.95, with a standard deviation of 0.003. This indicates that the constitutive model exhibits high precision and reliability, effectively predicting the actual values of the tangent elastic modulus.

The theoretical peak stress σ_p and peak strain σ'^{p} obtained from the stress–strain curves of different rock types are compared with the experimental peak stress ε_p and peak strain $\varepsilon_{p'}$ as shown in Table 6. Figure 17 shows the theoretical calculated values of peak stress and peak strain compared to the test values. Data analysis indicates that the average ratio of the theoretical peak strain to the experimental peak strain is 1.088, with a standard deviation of 0.0134. Similarly, the average ratio of the theoretical peak stress to the experimental peak stress is 1.03, with a standard deviation of 0.00083. These results demonstrate that the constitutive model exhibits high accuracy and reliability, effectively predicting the actual values of peak stress and peak strain.

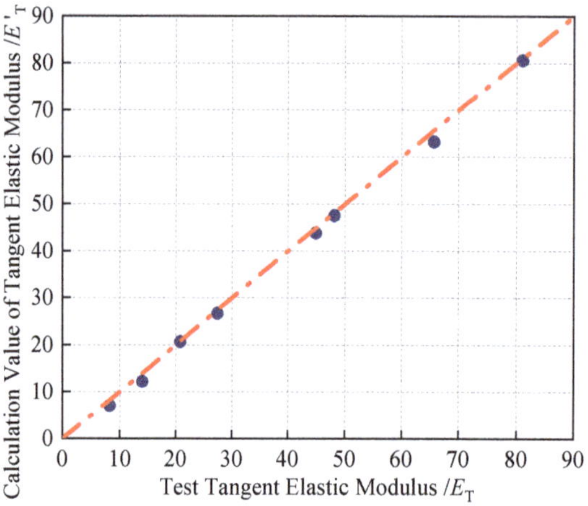

Figure 16. Comparison of theoretical and experimental tangent elastic modulus.

Table 6. Theoretical peak strain values for different rock types.

Rock Type	T/°C	σ_3/MPa	σ_p/MPa	σ'^{p}/MPa	Specific Value	ε_p	$\varepsilon_{p'}$	Specific Value
G	20	0	119.73	121.31	1.01	0.32	0.32	1.00
	120	25	344.60	371.13	1.08	0.42	0.46	1.10
RS	20	0	22.07	22.27	1.01	0.30	0.31	1.03
	120	25	17.04	18.20	1.09	0.13	0.17	1.31
GS	20	0	150.22	154.48	1.03	0.45	0.46	1.02
	120	25	293.08	297.46	1.01	0.46	0.47	1.02
SH	20	0	59.09	58.82	0.99	0.48	0.48	1.00
	120	25	102.43	105.59	1.03	0.63	0.77	1.00

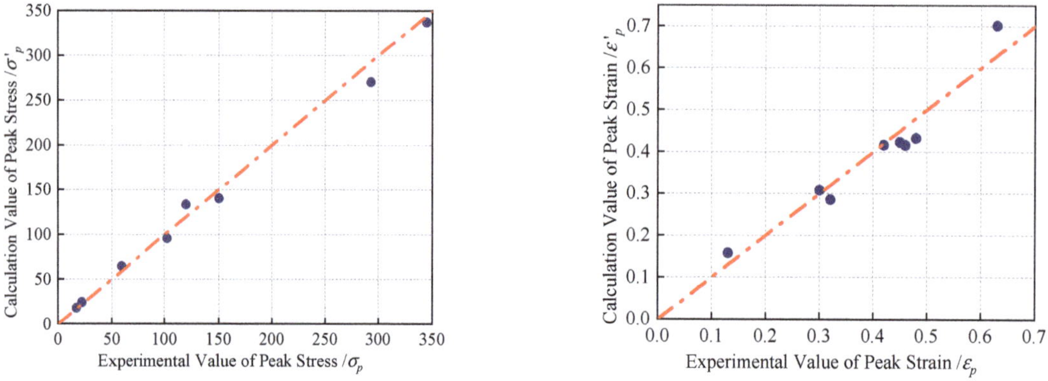

(**a**) (**b**)

Figure 17. Comparison of theoretical and experimental peak stress and peak strain: (**a**) comparison of theoretical and experimental peak stress values; (**b**) comparison of theoretical and experimental peak strain values.

In conclusion, the constitutive model for the temperature-confining pressure coupled damage evolution equation presented in this paper successfully represents the stress–strain behavior of rocks. The mechanical parameters derived show strong consistency. A comparison of the theoretical peak points predicted by the damage constitutive model with the experimental peak points demonstrates a high level of correlation. This confirms that the model accurately captures the stress–strain peak characteristics of rocks during triaxial compression, highlighting its important engineering significance.

5. Conclusions

This study utilizes triaxial compression tests to investigate the effects of temperature and confining pressure on the deformation and mechanical properties of four typical rock types: granite, red sandstone, gray sandstone, and shale. The primary conclusions are as follows:

(1) Temperature influences the elastic modulus of rocks differently. Granite and shale show an increase in elastic modulus with temperature. Red sandstone's elastic modulus increases initially, then decreases after a threshold, while gray sandstone exhibits the opposite trend. These differences are attributed to variations in internal structure, mineral composition, and the behavior of microcracks. Confining pressure significantly impacts elastic modulus and strain, with more pronounced changes under low confining pressure.

(2) The combined effects of temperature and confining pressure enhance the peak strength and strain of granite, gray sandstone, and shale at 120 °C and 25 MPa, compared to 20 °C and 0 MPa, indicating increased density, load-bearing capacity, and ductility. Conversely, red sandstone shows reduced strength at higher temperatures, likely due to thermal expansion and internal crack propagation. The coupling also affects rock failure characteristics, with smoother fracture surfaces under triaxial conditions compared to uniaxial conditions.

(3) A fitting model for the tangent elastic modulus (E_t) under temperature and confining pressure was developed. A polynomial model captures temperature-induced changes, particularly the decline in E_t, while a composite exponential model fits the confining pressure effects. The combined model accurately predicts E_t under dual influences, with a high goodness-of-fit (R^2 close to 1), offering a reliable reference for evaluating rock deformation.

(4) A constitutive model accounting for temperature and confining pressure coupling was formulated based on the Drucker–Prager criterion and Weibull statistical damage theory. Empirical formulas were derived for different rocks under varying conditions, and the model was validated against experimental results, showing strong consistency.

Author Contributions: Conceptualization, Z.Z.; methodology, Z.Z., Y.Z. and C.H.; software, C.H.; validation, T.P., Y.J. and X.L.; formal analysis, Z.Z.; investigation, Z.Z.; resources, Y.Z. and X.L.; data curation, Z.Z.; writing—original draft preparation, Z.Z.; writing—review and editing, Z.Z., Y.Z., C.H. and T.P.; visualization, Z.Z.; supervision, T.P. and Y.J.; project administration, Y.Z. and C.H.; funding acquisition, Y.Z. All authors have read and agreed to the published version of the manuscript.

Funding: This research received no external funding.

Institutional Review Board Statement: Not applicable.

Informed Consent Statement: Not applicable.

Data Availability Statement: The data presented in this study are available upon request from the corresponding author. The data are not publicly available due to privacy.

Acknowledgments: The authors would like to thank Cui Kun for his great help in this article, and his support has made it possible to present the article in a better way. The authors sincerely thank all the anonymous reviewers for their comments that improved the quality of the paper.

Conflicts of Interest: The authors declare no conflicts of interest.

References

1. Luo, N.; Liang, H.L.; Shen, T.; Yang, W.H. Temperature Dependence of Young's Modulus of Red Sandstone. *Therm. Sci.* **2019**, *23*, 1599–1606. [CrossRef]
2. Tian, H.; Kempka, T.; Xu, N.X.; Ziegler, M. Physical Properties of Sandstones After High Temperature Treatment. *Rock Mech. Rock Eng.* **2012**, *45*, 1113–1117. [CrossRef]
3. Zhang, W.Q.; Sun, Q.; Hao, S.Q.; Geng, J.S.; Lv, C. Experimental study on the variation of physical and mechanical properties of rock after high temperature treatment. *Appl. Therm. Eng.* **2016**, *98*, 1297–1304. [CrossRef]
4. Zhang, L.Y.; Zhu, H.H. Three-dimensional Hoek-Brown strength criterion for rocks. *J. Geotech. Geoenviron.* **2007**, *133*, 1128–1135. [CrossRef]
5. Fan, L.F.; Gao, J.W.; Wu, Z.J.; Yang, S.Q.; Ma, G.W. An investigation of thermal effects on micro-properties of granite by X-ray CT technique. *Appl. Therm. Eng.* **2018**, *140*, 505–519. [CrossRef]
6. Yin, T.B.; Chen, Y.J.; Li, X.B.; Li, Q. Effect of high temperature and strain rate on the elastic modulus of rocks: A review. *Int. J. Earth Sci.* **2021**, *110*, 2639–2660. [CrossRef]
7. Liu, W.; Zhang, L.; Luo, N. Elastic modulus evolution of rocks under heating-cooling cycles. *Sci. Rep.* **2020**, *10*, 13835. [CrossRef]
8. Suknev, S.V. Determination of elastic properties of rocks under varying temperature. *J. Min. Sci.* **2016**, *52*, 378–387. [CrossRef]
9. Yang, S.Q.; Jiang, Y.Z.; Xu, W.Y.; Chen, X.Q. Experimental investigation on strength and failure behavior of pre-cracked marble under conventional triaxial compression. *Int. J. Solids Struct.* **2008**, *45*, 4796–4819. [CrossRef]
10. Liu, J.-T.; Wang, J.-B.; Ge, H.-K.; Zhou, W.; Chen, B.-B.; Li, X.-D.; Xue, X.-J.; Luo, S.-L. Effect of gravel on rock failure in glutenite reservoirs under different confining pressures. *Pet. Sci.* **2023**, *20*, 3022–3036. [CrossRef]
11. Arzúa, J.; Alejano, L.R.; Walton, G. Strength and dilation of jointed granite specimens in servo-controlled triaxial tests. *Int. J. Rock Mech. Min.* **2014**, *69*, 93–104. [CrossRef]
12. Chen, M.; Zang, C.-W.; Ding, Z.-W.; Zhou, G.-L.; Jiang, B.-Y.; Zhang, G.-C.; Zhang, C.-P. Effects of confining pressure on deformation failure behavior of jointed rock. *J. Cent. South Univ.* **2022**, *29*, 1305–1319. [CrossRef]
13. Yang, H.; Dahlhaus, N.; Renner, J. On the Effect of Lateral Confinement on Rock-Cutting Tool Interactions. *Rock Mech. Rock Eng.* **2024**, *57*, 6063–6083. [CrossRef]
14. Meng, Y.; Jing, H.; Liu, X.; Yin, Q. Effects of initial unloading level on the mechanical, micro failure and energy evolution characteristics of stratified rock mass under triaxial unloading confining pressure. *Theor. Appl. Fract. Mech.* **2023**, *128*, 104161. [CrossRef]
15. Chen, D.; Chen, H.; Zhang, W.; Lou, J.; Shan, B. An analytical solution of equivalent elastic modulus considering confining stress and its variables sensitivity analysis for fractured rock masses. *J. Rock Mech. Geotech. Eng.* **2022**, *14*, 825–836. [CrossRef]
16. Wang, H.W.; Wang, Y.Y.; Fu, X. Experimental Study on Coupling Influence of Temperature and Confining Pressure to Deformation and Strength Characteristics of Rock-like Material with Pre-Existing Crack. *Materials* **2021**, *14*, 7572. [CrossRef]
17. Li, M.; Wu, Z.; Weng, L.; Ji, J.; Liu, Q. Numerical Investigation of Coupled Effects of Temperature and Confining Pressure on Rock Mechanical Properties in Fractured Rock Mass Using Thermal-Stress-Aperture Coupled Model. *Int. J. Geomech.* **2021**, *21*, 04021195. [CrossRef]
18. Yang, J.; Fu, L.-Y.; Fu, B.-Y.; Wang, Z.; Hou, W. High-temperature effect on the material constants and elastic moduli for solid rocks. *J. Geophys. Eng.* **2021**, *18*, 583–593. [CrossRef]
19. Demirci, A.; Görgülü, K.; Durutürk, Y.S. Thermal conductivity of rocks and its variation with uniaxial and triaxial stress. *Int. J. Rock Mech. Min.* **2004**, *41*, 1133–1138. [CrossRef]
20. Kumari, W.G.P.; Ranjith, P.G.; Perera, M.S.A.; Shao, S.; Chen, B.K.; Lashin, A.; Al Arifi, N.; Rathnaweera, T.D. Mechanical behaviour of Australian Strathbogie granite under in-situ stress and temperature conditions: An application to geothermal energy extraction. *Geothermics* **2017**, *65*, 44–59. [CrossRef]
21. Deng, S.; Zhang, Y.; Yan, X.; Yang, S.; Li, C.; Peng, H. Experimental study on P-wave response characteristics of hot dry rock under high temperature and high confining pressure. *Geomech. Energy Environ.* **2023**, *36*, 100514. [CrossRef]
22. Zhou, Q.; Zhu, Z.; Liu, W.; Lu, H.; Fan, Z.; Nie, X.; Li, C.; Wang, J.; Ren, L. Hydraulic fracturing behaviors of shale under coupled stress and temperature conditions simulating different burial depths. *Int. J. Min. Sci. Technol.* **2024**, *34*, 783–797. [CrossRef]
23. Yang, J.P.; Chen, W.Z.; Yang, D.S.; Tian, H.M. Estimation of Elastic Moduli of Non-persistent Fractured Rock Masses. *Rock Mech. Rock Eng.* **2016**, *49*, 1977–1983. [CrossRef]
24. Wu, F.Q.; Deng, Y.; Wu, J.; Li, B.; Sha, P.; Guan, S.G.; Zhang, K.; He, K.Q.; Liu, H.D.; Qiu, S.H. Stress-strain relationship in elastic stage of fractured rock mass. *Eng. Geol.* **2020**, *268*, 105498. [CrossRef]
25. Hua, D.J.; Jiang, Q.H.; Liu, R.Y.; Gao, Y.C.; Yu, M. Rock mass deformation modulus estimation models based on in situ tests. *Rock Mech. Rock Eng.* **2021**, *54*, 5683–5702. [CrossRef]
26. Malkowski, P.; Ostrowski, L. The Methodology for the Young Modulus Derivation for Rocks and Its Value. *Procedia Eng.* **2017**, *191*, 134–141. [CrossRef]
27. Zhao, H.; Shi, C.; Zhao, M.; Li, X. Statistical Damage Constitutive Model for Rocks Considering Residual Strength. *Int. J. Geomech.* **2017**, *17*, 04016033. [CrossRef]
28. Xu, X.L.; Gao, F.; Zhang, Z.Z. Thermo-mechanical coupling damage constitutive model of rock based on the Hoek-Brown strength criterion. *Int. J. Damage Mech.* **2018**, *27*, 1213–1230. [CrossRef]

29. Chen, Y.; Xiao, P.; Du, X.; Wang, S.; Wang, Z.; Azzam, R. Study on Damage Statistical Constitutive Model of Triaxial Compression of Acid-Etched Rock under Coupling Effect of Temperature and Confining Pressure. *Materials* **2021**, *14*, 7414. [CrossRef]
30. Li, X.; Yin, Z.; Falsone, G. A Time-Dependent Creep Constitutive Model of Deep Surrounding Rock under Temperature-Stress Coupling. *Math. Probl. Eng.* **2021**, *2021*, 9938869. [CrossRef]
31. Wong, L.N.Y.; Zhang, Y.; Wu, Z. Rock strengthening or weakening upon heating in the mild temperature range? *Eng. Geol.* **2020**, *272*, 105619. [CrossRef]
32. Zhao, G.; Guo, Y.; Chang, X.; Jin, P.; Hu, Y. Effects of temperature and increasing amplitude cyclic loading on the mechanical properties and energy characteristics of granite. *Bull. Eng. Geol. Environ.* **2022**, *81*, 155. [CrossRef]
33. Rao, G.M.N.; Murthy, C.R.L. Dual role of microcracks: Toughening and degradation. *Can. Geotech. J.* **2001**, *38*, 427–440. [CrossRef]
34. Taheri, A.; Royle, A.; Yang, Z.; Zhao, Y. Study on variations of peak strength of a sandstone during cyclic loading. *Geomech. Geophys. Geo-Energy Geo-Resour.* **2015**, *2*, 1–10. [CrossRef]
35. Lin, H.; Liu, J.; Yang, J.; Ran, L.; Ding, G.; Wu, Z.; Lyu, C.; Bian, Y. Analysis of damage characteristics and energy evolution of salt rock under triaxial cyclic loading and unloading. *J. Energy Storage* **2022**, *56*, 106145. [CrossRef]
36. Miao, S.; Liu, Z.; Zhao, X.; Ma, L.; Zheng, Y.; Xia, D. Plastic and damage energy dissipation characteristics and damage evolution of Beishan granite under triaxial cyclic loading. *Int. J. Rock Mech. Min.* **2024**, *174*, 105644. [CrossRef]
37. Tiang, Y.; Chen, H.; Xiong, L.; Xu, Z. Conventional triaxial loading and unloading test and PFC numerical simulation of rock with single fracture. *Arch. Civil. Eng.* **2024**, *70*, 233–254. [CrossRef]
38. Krajcinovic, D.; Fonseka, G.U. The Continuous Damage Theory of Brittle Materials, Part 1: General Theory. *J. Appl. Mech.* **1981**, *48*, 809–815. [CrossRef]
39. Fonseka, G.U.; Krajcinovic, D. The Continuous Damage Theory of Brittle Materials, Part 2: Uniaxial and Plane Response Modes. *J. Appl. Mech.* **1981**, *48*, 816–824. [CrossRef]

applied sciences

MDPI

Article

Adaptive Kriging-Based Heat Production Performance Optimization for a Two-Horizontal-Well Geothermal System

Haisheng Liu [1,2,3,4], Wan Sun [2], Jun Zheng [1,*] and Bin Dou [1]

[1] Engineering Faculty, China University of Geosciences, Wuhan 430074, China; liuhaish2024@126.com (H.L.); doubin@cug.edu.cn (B.D.)
[2] Shanghai Shallow Geothermal Energy Engineering Technology Research Center, Shanghai 200040, China; sunwan820224@126.com
[3] Qinghai Provincial Second Geological Exploration Institute, Xining 810003, China
[4] Qinghai Province Core Drilling Engineering Technology Research Center, Xining 810003, China
* Correspondence: junzheng@cug.edu.cn

Abstract: Optimizing heat generation capacity is crucial for geothermal system design and evaluation. Computer simulation is a valuable approach for determining the influence of various parameter combinations on a geothermal system's ability to produce heat. However, computer simulation evaluations are often computationally demanding since all potential parameter combinations must be examined, posing significant hurdles for heat generation performance evaluation and optimization. This research proposes an adaptive Kriging-based heat generation performance optimization method. Firstly, a two-horizontal-well geothermal system with rectangular multi-parallel fractures is constructed. The heat production performance optimization problem is then established, and the temperature and enthalpy of the outlet water are calculated using computer simulation and Kriging. A parameterized lower confidence bounding sampling scheme (PLCB) is developed to adaptively update Kriging in order to strike a compromise between optimization accuracy and computation burden. The outcomes of the optimization are compared to those of the Kriging-based optimization approach and other common infill options to demonstrate the efficiency of the proposed method. The outlet temperature curve obtained with PLCB-AKO-1 rose for a longer time and the heat generation power curve reached a stable output without a downward trend. According to the Friedman and Wilcoxon signed ranks tests, the PLCB-1-AKO technique is statistically superior to alternative strategies.

Keywords: heat production performance; Kriging; hot dry rock; geothermal system; numerical simulation

1. Introduction

Geothermal energy is one of the most economically sustainable, low-carbon, and renewable energy alternatives to traditional fossil fuels owing to its stability, concentrated quality, and purit [1,2]. Geothermal reservoirs, especially hot dry rock (HDR), are of particular interest to researchers due to their great potential for power generation [3]. Developing geothermal energy from HDR is made possible by reforming the reservoir to create a permeable fracture system and then pumping the working fluid into the fracture to extract heat from high-temperature HDR [4].

Due to the large capital expenditure required to develop an effective system, the geothermal system should be meticulously planned in advance to maximize its profitability. Figure 1 depicts a basic configuration of a geothermal facility. When designing a geothermal system, optimal heat extraction performance and low development costs are desired. As a result, the prediction and optimization of heat production performance is a key technology that affects the development cost and operational life of a specific geothermal system [5]. Conducting computer simulations to evaluate various parameter combinations to improve performance is a common practice [6]. The effects of fluid flow rate, temperature, lateral well spacing fracture number, fracture breadth, and production time on a horizontally fractured geothermal reservoir have been studied in the literature [7].

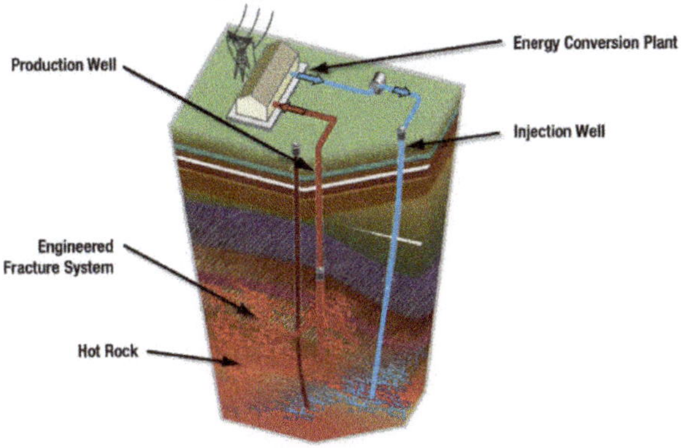

Figure 1. Type geothermal plant layout of HDR.

The previous studies help us to comprehend the importance of choosing the appropriate parameters. The heat production performance of a certain geothermal system is significantly influenced by the thermal reservoir characteristics, fracture conditions, and injection parameters of the circulation fluid. Particularly, the circulation fluid flow and, consequently, the effectiveness of thermal extraction are significantly influenced by the characteristics of the thermal reservoir, such as reservoir volume, permeability, and fracture characteristics, such as length, opening, roughness, and tortuosity [8]. In other words, a variety of factors impact heat production performance, and the underlying effect mechanism is complex, making its prediction extremely difficult [9]. Furthermore, the impact of various factors on geothermal performance has not been well examined. One solution to this problem is to investigate all potential parameter combinations and then select the optimal combination. However, because all potential combinations must be analyzed, it is frequently computationally demanding [10].

Using approximations instead of computationally expensive experiments/simulations is an effective solution to this computationally intensive problem [11]. By revealing the connections between the variables and objective functions, approximations can provide insights into the optimization problem [12]. Due to their ability to handle black-box and computationally demanding functions, approximation models could be used in practically all sorts of optimization problems [13]. When establishing an approximation model, samples are anticipated to be scattered over the whole design space in order to extract as much information as possible about the real experiments or simulations to be substituted. Additionally, the sampling technique has a significant impact on the approximation accuracy [14]. In general, a larger sample size allows approximations to explore the design/optimization area more thoroughly [15]. Large sample sizes of experiments/simulations, however, are impractical to implement, especially when computational costs are an overburden [16]. To enhance approximation approaches even further, the adaptive sampling methodology was created [17]. Researchers have lately switched from standard (or static) to adaptive techniques, known as dynamic or adaptive approximations, for approximation-based analysis and optimization [18].

Multiple fracturing technology and horizontal wells have significantly advanced during the last few years, improving gas and oil output [19]. The employment of multiple fracturing technology and horizontal wells have also attracted wide attention recently. Yu et al. developed a 3D thermal-hydrological-mechanical model with contacting cracks and assessed the efficacy of its heat extraction [20]. To examine the impact of numerous hydraulic fractures on geothermal energy extraction performance, geothermal systems with multiple fractured horizontal wells were analyzed [21]. Research shows that horizontal wells are preferable to vertical wells for connecting injection and production wells [19]. Additionally, a geothermal system with a horizontal well and several fractures might guarantee a constant in-situ stress and temperature field in the fractures at the same depth, preventing major flow short circuit between the fractures [22]. In addition to boosting heat extraction, it may also significantly reduce produced seismicity compared to a single fracture because of the decreased individual energy release from each fracture [6].

The adaptive Kriging approximation and numerical simulations are integrated in this study to optimize heat production performance for a geothermal system with two horizontal wells and rectangular multi-parallel fractures. In the beginning, a numerical model for a two-horizontal-well geothermal system with rectangular multi-parallel fractures is developed. The heat production performance optimization issue is then constructed using the heat generating power calculation. A substantial processing burden and a "black box" problem come from the computer simulation investigation. In order to offer production-specific enthalpy and outlet temperature responses, Kriging approximation is employed and adaptively updated to take the role of the computer simulation. We use a parameterized lower confidence bounding scheme (PLCB), a sequential sampling technique that was recommended in our earlier work, to update the Kriging approximation with dynamically identified sample sites in order to strike a compromise between optimization accuracy and efficiency. As a result, the approximation update is led by both the optimization accuracy and the approximation's expected uncertainty. The optimization outcome is compared to that of the Kriging-based optimization approach. Furthermore, the effectiveness of the PLCB system is proved through comparisons with other commonly used infill schemes.

2. Method

2.1. Heat Production Performance Optimization Problem Formulation

As a high temperature geothermal resource, HDR is primarily used for power production. The capacity of heat generation power must be addressed when developing a geothermal system. As a result, heat generating capacity is taken as an assessment indicator to evaluate heat production performance. The following equation is widely used to compute heat generating power according to the First Law of Thermodynamics:

$$W_e = 0.45q(h_{pro} - h_{inj})(1 - T_o/T_{pro}), q = \rho v A \tag{1}$$

where W_e stands for the heat generation power, MW; q represents the circulation mass flow rate, kg/s; h_{pro} and h_{inj} stand for the enthalpy of the producing water and injection specific enthalpy, J/kg, respectively; T_0 is the injection temperature, while T_{pro} is the outlet temperature; and 0.45 is the thermal energy conversion coefficient. The mass flow rate q could be calculated according to $q = \rho v A$, where ρ stands for the density of the circulating working fluid, kg/m^3; v represents the injection velocity of circulating working fluid, m/s; while A denotes the injection pipe's cross sectional area, m^2.

Therefore, the calculation formula of total power generation Q within a certain time is as follows:

$$Q = \int_0^t 0.45q(h_{pro} - h_{inj})(1 - T_o/T_{pro})dt \tag{2}$$

The total power generation Q is employed to reflect the heat production performance, which would be optimized to obtain the best performance. The production specific enthalpy h_{pro} and outlet temperature T_{pro} are obtained through the computer simulations described in Section 2.3, which is a black-box issue. A Kriging model is employed to replace the production specific enthalpy h_{pro} and outlet temperature T_{pro} in the optimization problem to benefit the whole optimization procedure with easy implementation and much less computational burden. In this work, the fracture space W, mass flow rate of injected circulating working fluid q, and well space L are taken as the design parameters to gain the maximum heat generation power. The well spacing L can be decomposed into L_1 and L_2 in the vertical and horizontal directions.

Therefore, the optimization problem could be formulated as follows:

$$\begin{array}{ll} max & Q \\ s.t. & 6 \le q \le 12, \mathrm{kg/s}, \ 40 \le W \le 120, \mathrm{m} \\ & 600 \le L \le 1000, \mathrm{m}, \ T_{pro} \ge 393.15 \ \mathrm{K} \end{array} \quad (3)$$

The optimization objective function is to maximize the total power generation Q. The fracture space W, mass flow rate of injected circulating working fluid q, and well space L are the variables, which are employed in the numerical simulation model to obtain the production specific enthalpy. There is a cut-off outlet temperature constraint (presented as $T_{pro} \ge 393.15$ K) considering the system lifetime; that is, an outlet temperature less than 393.15 K is considered to have no development and utilization value.

2.2. Adaptive Kriging-Based Heat Production Performance Optimization

To solve the total power generation maximization problem proposed in the preceding section, an adaptive Kriging-based heat production performance optimization is proposed in this section.

The design parameters for the heat production performance optimization problem are specifically fracture space, injection flow rate, and well space. The performance of heat production is taken into account while calculating the heat producing power. Figure 2 illustrates how the suggested adaptive Kriging-based heat generation performance optimization is carried out. Computer simulations based on the numerical model for a geothermal system with rectangular multi-parallel fractures conducted in Section 2.3 are employed to obtain the production specific enthalpy h_{pro} and outlet temperature T_{pro} at the sample points. The heat generation power in Equation (2) is then calculated based on the simulated response of production specific enthalpy h_{pro} and outlet temperature T_{pro}. A relationship is established between the heat generation performance and the input parameters (fracture space, injection flow, and well spacing) through the use of Kriging modeling. Additionally, a sequential sampling technique is used to update the Kriging model. The PLCB sampling method proposed in our previous work [15], which aims to adaptively balance the exploration/exploitation procedure, is employed to locate new samples to adaptively update the Kriging approximation.

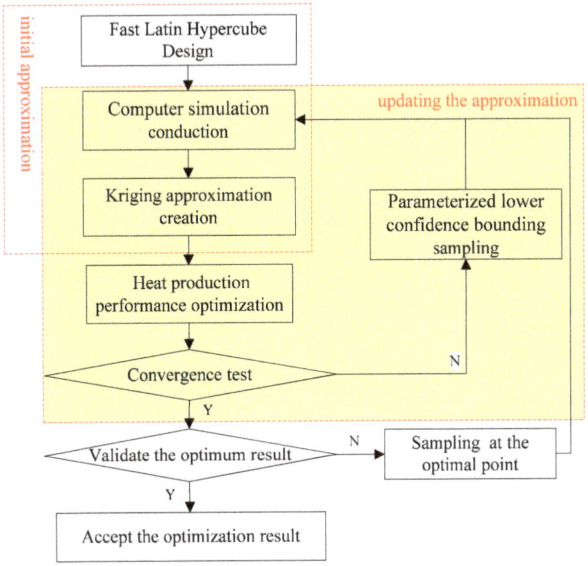

Figure 2. Procedure of the adaptive Kriging-based heat production performance optimization.

2.3. Numerical Simulation for a Two-Horizontal-Well Geothermal System

The numerical simulation technique for a two-horizontal-well geothermal system with rectangular multi-parallel fractures is detailed in this section. In this research, the computational fluid dynamics software, ANSYS Fluent 16, is used as the simulation solution. It is widely used for simulating and analyzing fluid flow and heat transfer through porous media, multiphase flow, and phase transition in various engineering applications [23]. The convergence condition of the numerical simulation in this work is that the residual value of energy is set to 10^{-6} and the other variables' residual values are set to 10^{-3}.

2.3.1. Numerical Model Description

The fractured zone and the impermeable rock mass zone are represented by a numerical model with rectangular multi-parallel cracks. Figure 3 depicts the conceptual model. As Figure 3b,c demonstrate, L_1, L_2, W and D represent the fracture length, the fracture height, the fracture aperture, and the fracture space, respectively. Taking the boundary impact into account, there is a reserve distance of 25 m between the inlet and outflow and the edge. Because the well diameter is just 0.3 m, which is significantly less than the dimension of the rock mass, the flow heat transfer within the well is considered insignificant. Therefore, fracture interfaces with the injection well are considered as inlets, while interfaces with the producing well are treated as the outlets. Thus, the model is periodic along the z axis. Figure 3c demonstrates one period, in which Z = 0 is the fracture's central axis.

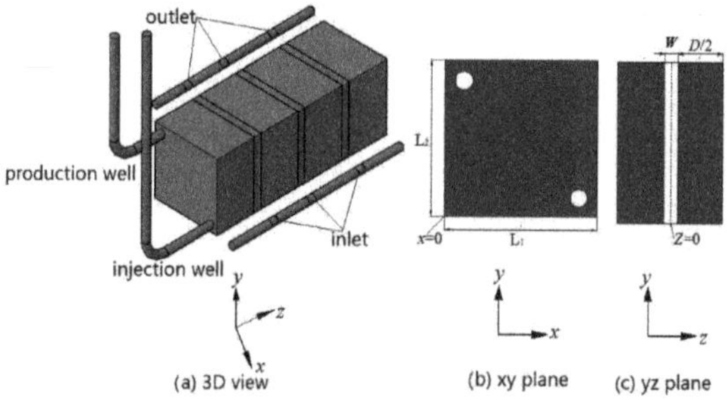

Figure 3. Conceptual model of a two-horizontal-well geothermal system.

2.3.2. Governing Equations

The continuity equation is described as follows:

$$\frac{\partial}{\partial t}(\gamma\rho_f) + \nabla \cdot (\rho_f K \cdot U) = 0 \tag{4}$$

where t denotes the time (s); γ is the porosity, which equals 1 when the fractures are completely open; ρ_f represents the fluid density (kg/m^3); K is the second-order area porosity tensor; U stands for the velocity vector; ∇ is the gradient operator.

Ignoring the convective acceleration and diffusion terms of the momentum equation, the momentum equation can be simplified to Darcy's law.

$$-\nabla P - \frac{\mu}{k}U + \rho_f g = 0 \tag{5}$$

where μ is the dynamic viscosity of the fluid (Pa·s); P is the pressure of the fluid (Pa); g is the gravitational acceleration (m/s^2).

The local energy governing equation is described as follows.

$$\frac{\partial(\rho_f\gamma C_f T)}{\partial t} - \frac{\partial P}{\partial t} + \nabla \cdot (\rho_f K \cdot U(C_f T)) = \nabla \cdot (\lambda_e K \cdot \nabla T) + \nabla(K \cdot U \cdot \tau) \tag{6}$$

where C_f is the specific heat capacity of the fluid, J/(kg·K); T is fluid Temperature (K); λ_e is the effective thermal conductivity of the crack, W/(m·K); τ is fluid shear stress, Pa.

2.3.3. Boundary and Hypotheses

(1) The model in Figure 3 has a repetitive characteristic of periodism in the z axis direction, the left and right walls are configured to be the periodic wall, while the interface between the fluid zone and the rock mass zone is set to the coupled wall.

(2) The matrix rock area is simplified to be homogeneous and isotropic impermeable blocks, namely the matrix permeability $\leq 9.87 \times 10^{-16}$ m^2. Chemical interactions between water and the rock matrix are also not taken into account.

(3) The rock's thermal diffusion coefficient is low so that the thermal conduction between the reservoir model in Figure 3 and the surrounding rock is ignored. Therefore, except for the above boundary, there is no heat flow over the other surfaces.

(4) Ignore the influence of temperature and pressure condition variety on the thermal property of the system. Thus, thermal conductivity coefficient, specific heat capacity and density of rock and water are constant during the whole mining stage.

(5) With a temperature gradient of 6 K/100 m (y axis), the initial average temperature of the rock mass is 473.15 K, but it may also range from 461.15 K to 485.15 K. There is no water loss during water circulation since the intake has been designated as a mass-flow inlet with an injection temperature of 60 °C. The outlet is used as a pressure-outlet.

2.3.4. Parameters Selection

Based on geothermal exploration findings and prior study experiences [24–26], a fundamental scenario of the Gonghe Basin in Qinghai was chosen as the research background to guide the selection of parameters. Table 1 contains the exact numerical model parameters for this instance. The model size is 400 m × 400 m × 400 m. The fracture aperture is settled as 0.002 m according to the literature and experience in hydrothermal geothermal systems [27]. The thermo-physical properties of matrix rock vary generally between 2–3.5 W/(mK), and 2.5 W/(mK) was employed. A geothermal examination of the Gonghe Basin in Qinghai found that the temperature can exceed 473.15 K at 3000 m below the surface and that the geothermal gradient is approximately 6–7 K/100 m [28]. As a result, the starting average temperature of the rock mass is set at 473.15 K and dispersed with a 6 K/100 m temperature gradient. According to Zeng's research, the water density in a broken reservoir ranges from 857 kg/m^3 to 992 kg/m^3 [29], and in this work, the average value of 925 kg/m^3 is used. The fracture fluid is regarded as a porous medium, and the modified Cubic Law [30] is used to estimate the permeability of the rough fracture while taking the influence of fracture roughness on heat transfer into account:

$$k = \frac{W^2}{12f} \tag{7}$$

$$f = 1 + 17\left(\frac{a}{2W}\right)^{1.5} \tag{8}$$

where k stands for the fracture permeability after reservoir stimulation such as hydraulic fracturing; W represents the fracture aperture; f denotes the Lomize roughness coefficient, which is calculated according to Equation (8); a stands for the asperity height, and is set to be 0.0005 m according to Wang's research [31].

Table 1. Parameters of the reservoir rock matrix in the numerical model.

Parameter	Value
Rock thermal conductivity	2.5 W/(m·K)
Rock specific heat	1000 J/(kg·K)
Rock density	2650 kg/m^3
Fracture permeability	1.9×10^{-7} m^2
Fracture height	400 m
Fracture length	400 m
Fracture aperture	0.002 m
Asperity height	0.0005 m
Injection temperature	333.15 K

By this method, the fracture permeability after reservoir reformation is calculated as 1.9×10^{-7} m^2, which was employed in the computer simulation.

It is also taken into account how temperature affects the mechanical and physical properties of water. The calculations below illustrate that the water's density, heat conductivity, heat capacity, and dynamic viscosity are all variables [6].

$$\rho_f = 838.5 + 1.4T - 3 \times 10^{-3}T^2 + 3.7 \times 10^{-7}T^3 \tag{9}$$

$$\begin{cases} \mu_f = 1.4 - 0.02T + 1.4 \times 10^{-4}T^2 - 4.6 \times 10^{-7}T^3 + 8.9 \times 10^{-10}T^4 \\ \quad -9.1 \times 10^{-13}T^5 + 3.8 \times 10^{-16}T^6 \; T \in [273.15,\ 413.15] \\ \mu_f = 0.004 - 2.1 \times 10^{-5}T + 3.9 \times 10^{-8}T^2 - 2.4 \times 10^{-11}T^3 \\ \quad T \in [413.15,\ 533.15] \end{cases} \tag{10}$$

$$C_f = 12010 - 80T + 0.3T^2 - 5.4 \times 10^{-4}T^3 + 3.6 \times 10^{-7}T^4 \tag{11}$$

$$\lambda_f = -0.9 + 8.9 \times 10^{-3}T - 1.6 \times 10^{-5}T^2 + 7.9 \times 10^{-9}T^3 \tag{12}$$

2.3.5. Domain Discretization

Figure 4 shows how the fracture, entry, and exit areas mesh. The region surrounding the well is split by an O-shape in the xy plane. The model is meshed into 10 layers with a minimum grid size of 0.1 m and a growth ratio of 1.1 inside a 1.6 m region close to the well. The grid size is raised by a factor of 1.2 in the region 1.6 m out from the well, with a minimum mesh size of 0.25 m. The fracture region is separated into 5 equal layers along the z-axis since the fracture aperture is only 2 mm wide. Each layer is equally 0.4 mm and permeability of each layer is 1.9×10^{-7} m^2. The model is partitioned into 10 boundary layers with a minimum grid size of 0.5 mm and a growth ratio of 1.1 in the rock mass region of 8 mm close to the fracture zone. Meshing occurs in the rock mass beyond the 8 mm fracture zone at a growth ratio of 1.3 and a minimum size of 13 mm.

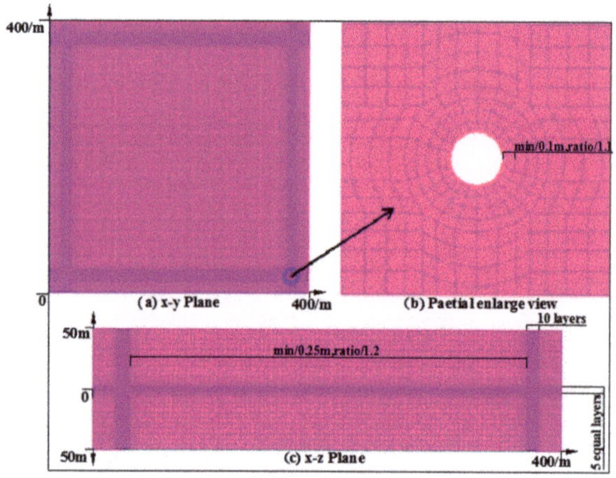

Figure 4. Meshing in the computational domain.

2.4. PLCB-Based Sequential Sampling Methods

The lower bounding function, which is a linear mix of approximation uncertainty and the anticipated value, is minimized in Cox and John's method to global optimization. The Lower Confidence Bounding (LCB) formulation is demonstrated as follows:

$$lcb = \hat{y}(x) - b\hat{\sigma}(x) \tag{13}$$

where $\hat{y}(x)$ is the predicted optimal value, $\hat{\sigma}(x)$ is the predicted uncertainty. The exploration/exploitation balance is regulated by a user-controlled variable b. Cox and John suggested $b = 2$. By incorporating a parameter a_i for the expected value, we proposed a cooling technique to address the challenge of selecting a well-working b. The original LCB function is therefore slightly updated to become as follows:

$$lcb = a_i\hat{y}(x) - b_i\hat{\sigma}(x) \tag{14}$$

In the above equation, a_i and b_i fluctuate during the whole optimization process and sampling procedure. The newly developed cooling method is regularly carried out by oscillating and monotone-decreasing control functions. When a larger value is assigned to b_i/a_i, the predicting uncertainty $\hat{\sigma}(x)$ has bigger weight, and the LCB minimization in Equation (11) focuses on the global search (exploration). Conversely, a smaller value of b_i/a_i assigns a bigger weight to local search (exploitation). Bigger b_i/a_i is suggested at an earlier stage, and then, smaller b_i/a_i is expected. As the sampling process proceeds, the value of b_i/a_i varies periodically to balance exploration and exploitation intelligently. The parameters a_i and b_i are formulated as follows:

$$\begin{gathered} a_i = 1 \\ b_i = (1 + \cos(\tfrac{i\pi}{n}))/\sin(\tfrac{i\pi}{n}) \end{gathered} \tag{15}$$

In the above equation, i stands for the iteration number (also known as the sequence sampling number), n controls the varying cycle and could be defined via users' preference.

An alternative method is to set the parameter a = 1, and then vary the parameter b according to the three control formulations as provided in the following:

$$b_i = \frac{b_0}{\text{In}(1+i)} \tag{16}$$

$$b_i = \frac{b_0}{i} \tag{17}$$

$$b_i = b_0 \exp(-0.8(i-1)^{\frac{1}{2}}) \tag{18}$$

where i stands for the iteration number, and b_0 represents the initial value of parameter b. Compared to Equation (12), formulations in Equations (13)–(15) are monotone-decreasing while Equation (12) is also an oscillating function.

3. Results and Discussion

3.1. Model Verification

Through a comparison of the dimensionless temperature, or T_D, between the works in this study and research in the literature [32], the suggested numerical simulation model is validated in this part. T_D is defined as $T_D = (T_{pro} - T_{inj})/(T_0 - T_{inj})$. Figure 5 demonstrates T_D of a 1D numerical model [32], 2D numerical model [33], and numerical model in this work. The evolution tendency of T_D in various stimulation models is essentially the same, as illustrated in Figure 5. The early stage's temperature is steady or increasing throughout the first year. However, the progress of T_D has steadily stagnated as mining time has increased. During the course of 20 years, T_D of the 2D numerical model decreases by about 42%, T_D of the 1D numerical model decreases by about 43.3%, and T_D of the numerical model in this work has a 45.6% drop. The comparison result shown in Figure 5 verify the rationality and effectiveness of the numerical model proposed in this paper.

3.2. Simulation Running

The numerical simulation method described in Section 2.3 is executed to obtain the outlet temperature of the circulation fluid. Figure 6 demonstrates the fluid field's simulated temperature distribution in the x-y plane (z = 0) after 20 years in a situation where the fracture space W is 100 m, injection flow q is 8 kg/s, and well spacing L is 990 m. Because circular zones of low temperature form and spread evenly around the entrance, it may be assumed that the outflow temperature is not altered in the early stages (Figure 6a–c).

The isotherm starts to bulge toward the outlet at 0.88 years (as shown in Figure 6d), which implies that the heat exchange between the flowing fluid and the high-temperature rock mass has altered the temperature in the vicinity of the outlet regions. Thus, the outlet temperature, T_{pro}, starts to drop (see Figure 7), and as the heat extraction time increases, the low temperature zones eventually expand to the complete areas (as shown in Figure 6e–i). Additionally, it shows that because the starting temperature of the rock mass is distributed in a gradient, the temperature at the lower left side of the entire fluid region is always higher than the top right side.

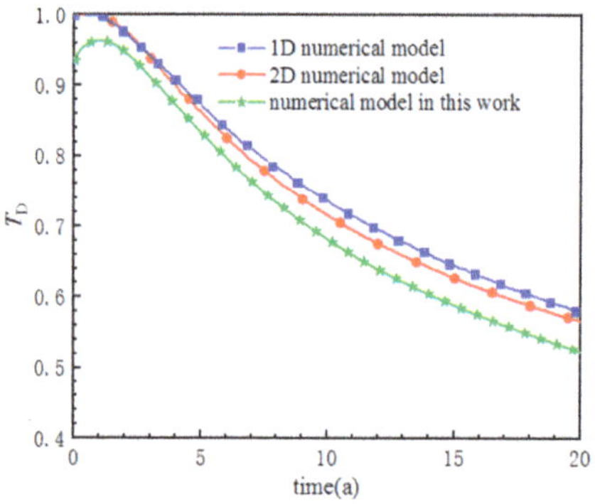

Figure 5. Comparison of dimensionless temperature of different numerical models.

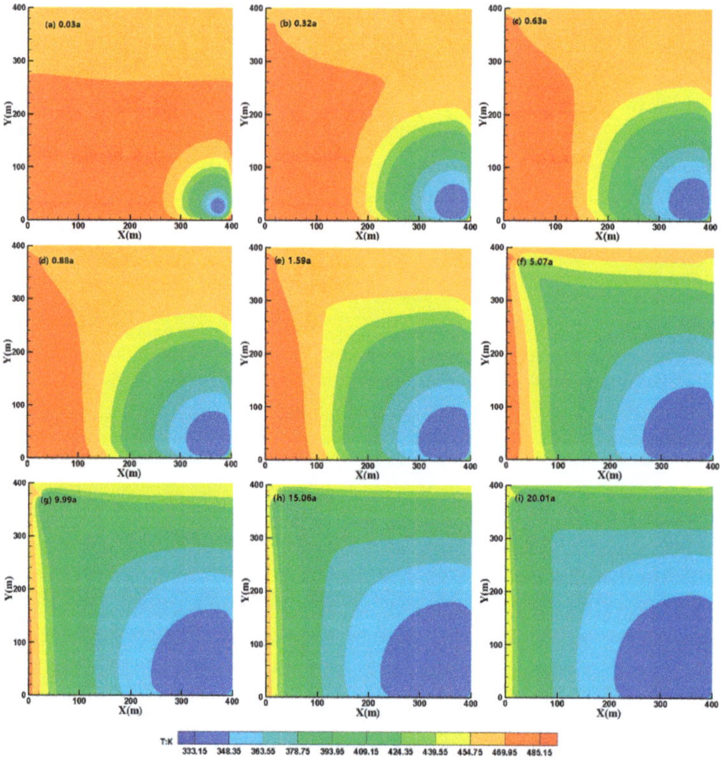

Figure 6. Simulated temperature distribution of the fluid field in the xy planes.

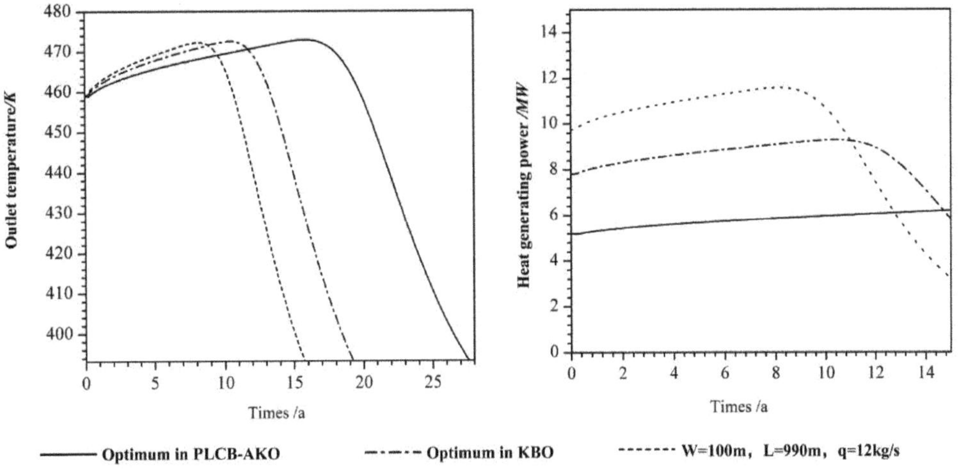

Figure 7. Outlet temperature and heat-generating power of different schemes.

Figure 7 depicts the development of the outlet temperature and heat-generating power of the optimum scheme through different approaches, in which the scheme (with w = 100 m, L = 990 m and q = 12 kg/s) is used as a reference group. It is obvious that the outlet temperature rose for a very brief period of time and then dropped to 393.15 K at 15.78 a in the reference scheme, while the optimum scheme of PLCB-AKO-1 dropped to 393.15 K at 27.16 a. Therefore, in the heat generation power curve, the optimum scheme of PLCB-AKO reached a stable output of 5.2 MW without a downward trend, while the heat-generation power curve drops at about 8 a and 14.5 a in the other two schemes.

3.3. Comparison with Kriging-Based Optimization

For comparison, the Kriging-based optimization approach is used to tackle the heat generation performance optimization problem (which is called KBO for short in this paper). Both the PLCB-based adaptive Kriging optimization method (PLCB-AKO for short) and KBO employ the fast LHD to generate the initial sample points.

When using the KBO technique, sample sizes of 30, 40, and 50 are set, and the starting sample size for Kriging in the suggested PLCB-AKO approach is set at 20. LHD is used to select sample points for Kriging modeling in KBO. To assess the precision of the Kriging approximation at the last iteration, three metrics are calculated: (1) relative root mean square error (RRMSE), (2) relative maximum absolute error (RMAE) and (3) relative average absolute error (RAAE). An estimate is more accurate if RRMSE and RAAE are smaller. On the other side, a big RMAE indicates that there is a localized sector of the design space where there is poor approximation. RMAE is therefore anticipated to be as minimal as is feasible. The three error measurements mentioned above are stated as follows:

$$
\begin{aligned}
RRMSE &= \frac{1}{STD}\sqrt{\frac{1}{N}\sum_{i=1}^{N}(y_i-\hat{y}_i)^2}, RMAE = \frac{1}{STD}\max|y_i-\hat{y}_i| \\
RAAE &= \frac{1}{N\times STD}\sum_{i=1}^{N}(y_i-\hat{y}_i)^2, STD = \sqrt{\frac{1}{N-1}\sum_{i=1}^{N}(y_i-\overline{y})^2}, i=1,\ldots,N
\end{aligned} \quad (19)
$$

in which N is the total number of validation points; $\overline{y}$ and y_i stand for the observed mean values and the actual responses, respectively, while $\hat{y}_i$ indicates the predicted value.

Table 2 shows the prediction accuracy of the Kriging approximation using various sample sizes in the KBO technique. It is obvious from Table 2 that the prediction accuracy of the Kriging approximation rises as the sample size grows.

Table 2. Prediction accuracy of Kriging in KBO approach.

Sample Size vs. Diff. Criteria	30	40	50
RRMSE	0.2175	0.1169	0.0682
RMAE	0.8012	0.4652	0.2013
RAAE	0.1435	0.0398	0.0732

Table 3 presents the PLCB-AKO approach's prediction accuracy during the most recent iteration, during which nine new samples were added using PLCB. It is proved that the proposed PLCB-AKO strategy used far fewer sample points, which results in far fewer computer simulations being run and a significant reduction in computing load.

Table 3. Prediction accuracy comparison of Kriging in PLCB-AKO and KBO.

Sample Size	Diff. Error Criteria	Kriging in KBO	Kriging in PLCB-AKO
50	RRMSE	0.0682	-
	RMAE	0.2013	-
	RAAE	0.0398	-
29(20+9)	RRMSE	-	0.0298
	RMAE	-	0.1632
	RAAE	-	0.0354

3.4. Survey of Different Infill Strategies

Different sequential sample infill strategies are compared in order to determine the most effective infill strategies for the suggested strategy. The original LCB [15] and the often used Expected improvement (EI) and extensions of EI (weighted EI, WEI), which have received a lot of attention over the last 10 years [34], are explored and compared with the suggested PLCB. The original LCB, EI, and weighted EI computations as well as the PLCB regulating functions as expressed in Equations (15)–(18) are also carried out in the Matlab environment. For each strategy, the same initial samples are located using the LHD method. PLCB, LCB, EI, and weighted EI are used to implement the sequential sampling technique. For simplicity, EI infill strategy within the proposed adaptive Kriging optimization is abbreviated to EI-AKO. Similar phrases include LCB-AKO for original LCB and WEI-AKO for weighted EI in adaptive Kriging optimization. Each technique was used 30 times separately to remove chance. Table 4 shows the mean and variance of normalized optimization outcomes obtained using various methods, where PLCB-1, PLCB-2, PLCB-3, and PLCB-4, respectively, stand for the controlling formulations in Equations (12)–(15). Figure 8 shows a box plot of the experimental data. The proposed PLCB-1-AKO approach clearly achieved the minimal mean of optimization.

Table 4. Mean and variance via different approaches.

	EI-AKO	WEI-AKO	PLCB-1-AKO	PLCB-2-AKO	PLCB-3-AKO	PLCB-4-AKO
Mean	0.6296	0.6399	0.4975	0.6144	0.6037	0.5852
Variance	0.0366	0.0359	0.0238	0.0326	0.0339	0.0304

Figure 9 depicts the lifetime power generation of different schemes when the cut-off temperature of current generation is 393.15 K. As the result of the adaptive Kriging-based method will not be exactly the same during each run, the average of total power generation obtained via EI-AKO, PLCB-AKO, WEI-AKO are demonstrated in Figure 9, which indicates little difference in total power generation, namely, 4.13 × 109 MJ, 4.23 × 109 MJ, 4.29 × 109 MJ, respectively.

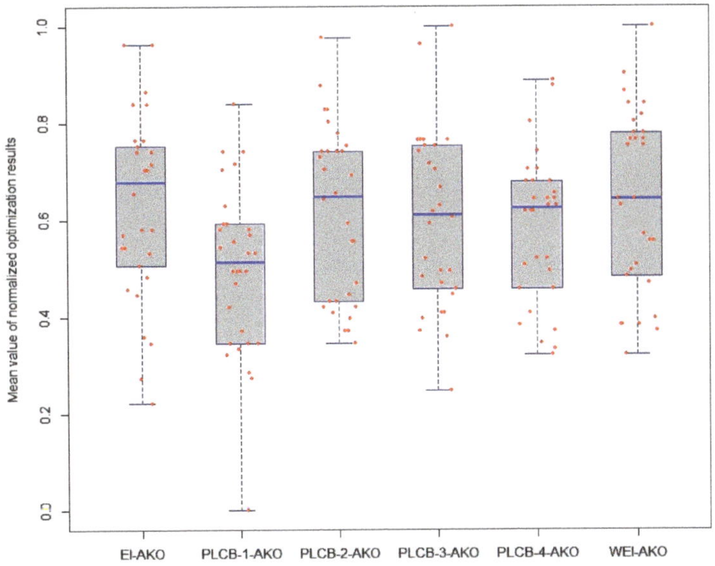

Figure 8. Box plot of normalized optimization results through different approaches.

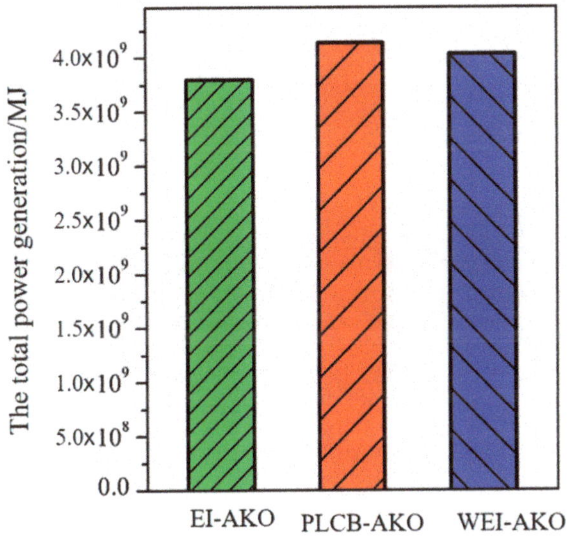

Figure 9. Lifetime power generation of different schemes.

The six aforementioned solutions are contrasted with one another in order to discover the optimal sample infill technique. The six sample infill procedures were subjected to the Friedman test [35,36], a well-known and often-employed non-parametric statistical test suited for multiple comparisons. Table 5 lists the results of the Friedman test's individual rankings and p-value calculations. Table 5 obviously demonstrates that the PLCB-1-AKO technique has a higher optimization statistical outcome among the evaluated options. The PLCB-1-AKO approach that has been suggested is suitable for the problem of optimizing heat generation performance.

Table 5. Results obtained through the Friedman test for different approaches.

Different Approaches	Rank	p-Value
EI-AKO	23.08	0.00053
WEI-AKO	21.92	
PLCB-1-AKO	10.38	
PLCB-2-AKO	19.13	
PLCB-3-AKO	18.98	
PLCB-4-AKO	17.5	

The Wilcoxon signed ranks tests [35,37], a non-parametric statistical test for pairwise comparisons, were adopted in this work to further investigate which of the five infill strategies is statistically different with respect to the suggested adaptive Kriging-based heat production performance optimization approach. The Wilcoxon signed ranks test findings for a pairwise comparison of the six alternative techniques are presented in Table 6 as a summary. The p-values in Table 6 show that, at the 0.05 level of significance, the suggested PLCB-1-AKO technique is statistically superior to alternative strategies.

Table 6. Results obtained through the Wilcoxon signed ranks for different approaches.

PLCB-1-AKO v.s.	R^+	R^-	p-Value
EI-AKO	387	78	2.76×10^{-5}
WEI-AKO	453.5	11.5	5.44×10^{-6}
PLCB-2-AKO	424.5	40.5	7.53×10^{-6}
PLCB-3-AKO	452.5	12.5	5.99×10^{-6}
PLCB-4-AKO	428	37	5.75×10^{-5}

4. Conclusions

In this research, a heat production performance optimization issue is addressed by combining an adaptive Kriging approximation with numerical simulations. An adaptive Kriging-based heat production performance optimization is developed for a two-horizontal-well geothermal system with rectangular multi-parallel fractures. The heat production performance optimization problem is formulated with Kriging approximation of heat generation power calculation. A PLCB-based adaptive strategy is introduced to update the Kriging approximation to balance the optimization accuracy and efficiency with newly located sampling points. The comparison with the Kriging-based optimization method shows that the suggested PLCB-based optimization method uses far fewer computer simulations and reduces the computational load significantly. The outlet temperature curve obtained with PLCB-AKO-1 rose for a longer time and dropped to 393.15 K at 27.16 a, while the heat generation power curve reached a stable output of 5.2 MW without a downward trend. Additionally, a comparison of other infill tactics shows that the suggested approach, which uses a control function that monotonically decreases and oscillates, achieves the best statistical optimization outcome. In the future, the heat losses and exchange in the tube segment from and to the ground surface would be taken into consideration in the heat production performance optimization problem for a two-horizontal-well geothermal system. A stochastic representation of the fracture distances and apertures considering the fracture surface roughness would be another interesting research direction.

Author Contributions: H.L. conceptualized this work and wrote the manuscript; J.Z. established the numerical model and completed the simulation; W.S. provided financial support for this study, B.D. participated in data analysis and manuscript writing. All authors have read and agreed to the published version of the manuscript.

Funding: This research was supported by The National Key Research and Development Programs of Henan Province, grant number 231111320800, The Open Fund of Shanghai Shallow Geothermal Energy Engineering Technology Center, grant number DRZX-202303, Geological Exploration Project of Qinghai Province Bureau of Geology and Mineral Exploration and Development, grant number Qingdikuangke [2023] No. 33, Qinghai Provincial Party Committee Talent Work Leading Group, grant number Qingrencaizi [2023] No. 1.

Institutional Review Board Statement: Not applicable.

Informed Consent Statement: Not applicable.

Data Availability Statement: All data, models, or code that support the findings of this study are available from the corresponding author upon reasonable request.

Conflicts of Interest: The authors declare that they have no competing interests.

References

1. Li, P.; Guan, P.; Zheng, J.; Dou, B.; Tian, H.; Duan, X.; Liu, H. Field Test and Numerical Simulation on Heat Transfer Performance of Coaxial Borehole Heat Exchanger. *Energies* **2020**, *13*, 5471. [CrossRef]
2. Goričanec, D.; Ivanovski, I.; Krope, J.; Urbancl, D. The Exploitation of Low-Temperature Hot Water Boiler Sources with High-Temperature Heat Pump Integration. *Energies* **2020**, *13*, 6311. [CrossRef]
3. Zheng, J.; Li, P.; Dou, B.; Fan, T.; Tian, H.; Lai, X. Impact research of well layout schemes and fracture parameters on heat production performance of enhanced geothermal system considering water cooling effect. *Energy* **2022**, *255*, 124496. [CrossRef]
4. Zhang, B.; Tian, H.; Dou, B.; Zheng, J.; Chen, J.; Zhu, Z.; Liu, H. Macroscopic and microscopic experimental research on granite properties after high-temperature and water-cooling cycles. *Geothermics* **2021**, *93*, 102079. [CrossRef]
5. Caf, A.; Urbancl, D.; Trop, P.; Goricanec, D. Exploitation of low-temperature energy sources from cogeneration gas engines. *Energy* **2016**, *108*, 86–92. [CrossRef]
6. Zheng, J.; Fan, T.; Dou, B.; Cui, G.; Tian, H. A Thermal-Hydraulic-Mechanical Coupling Simulation of Fluid Flow and Heat Transfer Specifically in Crossed-Rough Fractures in a Geothermal Reservoir. *J. Energy Eng.* **2022**, *148*, 04022032. [CrossRef]
7. Aliyu, M.D.; Chen, H.-P. Optimum control parameters and long-term productivity of geothermal reservoirs using coupled thermo-hydraulic process modelling. *Renew. Energy* **2017**, *112*, 151–165. [CrossRef]
8. Zhang, C.; Jiang, G.; Jia, X.; Li, S.; Zhang, S.; Hu, D.; Hu, S.; Wang, Y. Parametric study of the production performance of an enhanced geothermal system: A case study at the Qiabuqia geothermal area, northeast Tibetan plateau. *Renew. Energy* **2019**, *132*, 959–978. [CrossRef]
9. Wang, G.; Song, X.; Shi, Y.; Yang, R.; Yulong, F.; Zheng, R.; Li, J. Heat extraction analysis of a novel multilateral-well coaxial closed-loop geothermal system. *Renew. Energy* **2021**, *163*, 974–986. [CrossRef]
10. Wang, G.; Song, X.; Shi, Y.; Zheng, R.; Li, J.; Li, Z. Production performance of a novel open loop geothermal system in a horizontal well. *Energy Convers. Manag.* **2020**, *206*, 112478. [CrossRef]
11. Zheng, J.; Shao, X.; Gao, L.; Jiang, P.; Li, Z. A hybrid variable-fidelity global approximation modelling method combining tuned radial basis function base and kriging correction. *J. Eng. Des.* **2013**, *24*, 604–622. [CrossRef]
12. Schnoerr, D.; Sanguinetti, G.; Grima, R. Approximation and inference methods for stochastic biochemical kinetics—A tutorial review. *J. Phys. A Math. Theor.* **2017**, *50*, 093001. [CrossRef]
13. Shuai, W.; Li, E.; Wang, H.; Li, Y. Space mapping-assisted optimization of a thin-walled honeycomb structure for battery packaging. *Struct. Multidiscip. Optim.* **2020**, *62*, 937–955. [CrossRef]
14. Blatman, G.; Sudret, B. Efficient computation of global sensitivity indices using sparse polynomial chaos expansions. *Reliab. Eng. Syst. Saf.* **2010**, *95*, 1216–1229. [CrossRef]
15. Zheng, J.; Li, Z.; Gao, L.; Jiang, G. A parameterized lower confidence bounding scheme for adaptive metamodel-based design optimization. *Eng. Comput.* **2016**, *33*, 2165–2184. [CrossRef]
16. Zheng, J.; Lu, C.; Li, Z.; Jiang, G.; Zhao, H. Adaptive metamodel-based analytical target cascading for natural gas hydrates coring tool design optimization. *Eng. Optim.* **2021**, *54*, 1089–1109. [CrossRef]
17. Hu, W.; Yao, L.G.; Hua, Z.Z. Optimization of sheet metal forming processes by adaptive response surface based on intelligent sampling method. *J. Am. Acad. Dermatol.* **2008**, *197*, 77–88. [CrossRef]
18. Serani, A.; Pellegrini, R.; Wackers, J.; Jeanson, C.-E.; Queutey, P.; Visonneau, M.; Diez, M. Adaptive multi-fidelity sampling for CFD-based optimisation via radial basis function metamodels. *Int. J. Comput. Fluid Dyn.* **2019**, *33*, 237–255. [CrossRef]
19. Li, M.; Gou, Y.; Hou, Z.; Were, P. Investigation of a new HDR system with horizontal wells and multiple fractures using the coupled wellbore–reservoir simulator TOUGH2MP-WELL/EOS3. *Environ. Earth Sci.* **2015**, *73*, 6047–6058. [CrossRef]
20. Yu, P.; Dempsey, D.; Archer, R. A three-dimensional coupled thermo-hydro-mechanical numerical model with partially bridging multi-stage contact fractures in horizontal-well enhanced geothermal system. *Int. J. Rock Mech. Min. Sci. Géoméch. Abstr.* **2021**, *143*, 104787. [CrossRef]
21. Gong, F.; Guo, T.; Sun, W.; Li, Z.; Yang, B.; Chen, Y.; Qu, Z. Evaluation of geothermal energy extraction in Enhanced Geothermal System (EGS) with multiple fracturing horizontal wells. *Renew. Energy* **2020**, *151*, 1339–1351. [CrossRef]
22. Jiang, F.; Luo, L.; Chen, J. A novel three-dimensional transient model for subsurface heat exchange in enhanced geothermal systems. *Int. Commun. Heat Mass Transf.* **2013**, *41*, 57–62. [CrossRef]

23. Rybdylova, O.; Al Qubeissi, M.; Braun, M.; Crua, C.; Manin, J.; Pickett, L.M.; De Sercey, G.; Sazhina, E.M.; Sazhin, S.S.; Heikal, M. A model for droplet heating and its implementation into ANSYS Fluent. *Int. Commun. Heat Mass Transf.* **2016**, *76*, 265–270. [CrossRef]
24. Shaik, A.R.; Rahman, S.S.; Tran, N.H.; Tran, T. Numerical simulation of Fluid-Rock coupling heat transfer in naturally fractured geothermal system. *Appl. Therm. Eng.* **2011**, *31*, 1600–1606. [CrossRef]
25. Xia, Y.; Plummer, M.; Mattson, E.; Podgorney, R.; Ghassemi, A. Design, modeling, and evaluation of a doublet heat extraction model in enhanced geothermal systems. *Renew. Energy* **2017**, *105*, 232–247. [CrossRef]
26. Waples, D.W.; Tirsgaard, H. Changes in matrix thermal conductivity of clays and claystones as a function of compaction. *Pet. Geosci.* **2002**, *8*, 365–370. [CrossRef]
27. Zhang, J.; Standifird, W.B.; Roegiers, J.-C.; Zhang, Y. Stress-Dependent Fluid Flow and Permeability in Fractured Media: From Lab Experiments to Engineering Applications. *Rock Mech. Rock Eng.* **2007**, *40*, 3–21. [CrossRef]
28. Ma, Y.; Zhang, Y.; Huang, Y.; Zhang, Y.; Hu, Z. Experimental study on flow and heat transfer characteristics of water flowing through a rock fracture induced by hydraulic fracturing for an enhanced geothermal system. *Appl. Therm. Eng.* **2019**, *154*, 433–441. [CrossRef]
29. Zeng, Y.-C.; Su, Z.; Wu, N.-Y. Numerical simulation of heat production potential from hot dry rock by water circulating through two horizontal wells at Desert Peak geothermal field. *Energy* **2013**, *56*, 92–107. [CrossRef]
30. Cheng, Y.; Zhang, Y.; Yu, Z.; Hu, Z.; Ma, Y.; Yang, Y. Experimental and numerical studies on hydraulic fracturing characteristics with different injection flow rates in granite geothermal reservoir. *Energy Sci. Eng.* **2020**, *9*, 142–168. [CrossRef]
31. Huang, W.; Chen, J.; Cen, J.; Cao, W.; Li, Z.; Li, F.; Jiang, F. Heat extraction from hot dry rock by super-long gravity heat pipe: Effect of key parameters. *Energy* **2022**, *248*. [CrossRef]
32. Cheng, A.H.; Ghassemi, A.; Detournay, E. Integral equation solution of heat extraction from a fracture in hot dry rock. *Int. J. Numer. Anal. Methods Géoméch.* **2001**, *25*, 1327–1338. [CrossRef]
33. Zeng, Y.-C.; Wu, N.-Y.; Su, Z.; Wang, X.-X.; Hu, J. Numerical simulation of heat production potential from hot dry rock by water circulating through a novel single vertical fracture at Desert Peak geothermal field. *Energy* **2013**, *63*, 268–282. [CrossRef]
34. Kleijnen, J.P.C.; van Beers, W.; van Nieuwenhuyse, I. Expected improvement in efficient global optimization through bootstrapped kriging. *J. Glob. Optim.* **2011**, *54*, 59–73. [CrossRef]
35. Derrac, J.; García, S.; Molina, D.; Herrera, F. A practical tutorial on the use of nonparametric statistical tests as a methodology for comparing evolutionary and swarm intelligence algorithms. *Swarm Evol. Comput.* **2011**, *1*, 3–18. [CrossRef]
36. Zheng, J.; Lu, C.; Gao, L. Multi-objective cellular particle swarm optimization for wellbore trajectory design. *Appl. Soft Comput.* **2019**, *77*, 106–117. [CrossRef]
37. Zheng, J.; Li, Z.; Dou, B.; Lu, C. Multi-objective cellular particle swarm optimization and RBF for drilling parameters optimization. *Math. Biosci. Eng.* **2019**, *16*, 1258–1279. [CrossRef]

applied sciences

MDPI

Article

A Study on Three-Dimensional Multi-Cluster Fracturing Simulation under the Influence of Natural Fractures

Yuegang Li [1], Mingyang Wu [2], Haoyong Huang [3,4,*], Yintong Guo [2], Yujie Wang [5], Junchuan Gui [3,4] and Jun Lu [6]

1 Chongqing Shale Gas Exploration and Development Co., Ltd., Chongqing 401120, China; li_yg@petrochina.com.cn
2 State Key Laboratory of Geomechanics and Geotechnical Engineering, Wuhan Institute of Rock and Soil Mechanics, Chinese Academy of Sciences, Wuhan 430071, China; wmy@cqu.edu.cn (M.W.); ytguo@whrsm.ac.cn (Y.G.)
3 Shale Gas Research Institute of PetroChina Southwest Oil & Gas Field Company, Chengdu 610051, China; guijunchuan@petrochina.com.cn
4 Sichuan Key Laboratory of Shale Gas Evaluation and Exploitation, Chengdu 610051, China
5 PetroChina Southwest Oil and Gas Field Company, Chengdu 610051, China
6 Institute of Deep Earth Science and Green Energy, College of Civil and Transportation Engineering, Shenzhen University, Shenzhen 518060, China; junlu@szu.edu.cn
* Correspondence: huang_hy@petrochina.com.cn

Abstract: Multi-cluster fracturing has emerged as an effective technique for enhancing the productivity of deep shale reservoirs. The presence of natural bedding planes in these reservoirs plays a significant role in shaping the evolution and development of multi-cluster hydraulic fractures. Therefore, conducting detailed research on the propagation mechanisms of multi-cluster hydraulic fractures in deep shale formations is crucial for optimizing reservoir transformation efficiency and achieving effective development outcomes. This study employs the finite discrete element method (FDEM) to construct a comprehensive three-dimensional simulation model of multi-cluster fracturing, considering the number of natural fractures present and the geo-mechanical characteristics of a target block. The propagation of hydraulic fractures is investigated in response to the number of natural fractures and the design of the multi-cluster fracturing operations. The simulation results show that, consistent with previous research on fracturing in shale oil and gas reservoirs, an increase in the number of fracturing clusters and natural fractures leads to a larger total area covered by artificial fractures and the development of more intricate fracture patterns. Furthermore, the present study highlights that an escalation in the number of fracturing clusters results in a notable reduction in the balanced expansion of the double wings of the main fracture within the reservoir. Instead, the effects of natural fractures, geo-stress, and other factors contribute to enhanced phenomena such as single-wing expansion, bifurcation, and the bending of different main fractures, facilitating the creation of complex artificial fracture networks. It is important to note that the presence of natural fractures can also significantly alter the failure mode of artificial fractures, potentially resulting in the formation of small opening shear fractures that necessitate careful evaluation of the overall renovation impact. Moreover, this study demonstrates that even in comparison to single-cluster fracturing, the presence of 40 natural main fractures in the region can lead to the development of multiple branching main fractures. This finding underscores the importance of considering natural fractures in deep reservoir fracturing operations. In conclusion, the findings of this study offer valuable insights for optimizing deep reservoir fracturing processes in scenarios where natural fractures play a vital role in shaping fracture development.

Keywords: multi-cluster fracturing; deep reservoirs; FDEM; high geo-stress; fracture propagation

1. Introduction

Oil and gas resources continue to play a significant role in the global energy landscape. Therefore, there is a pressing need to ramp up the extraction and utilization of these resources [1]. Deep shale oil and gas resources at depths of over 3500 m account for a huge proportion of the current energy structure. Optimizing the construction and exploitation of deep oil and gas resources through construction techniques such as multi-cluster fracturing is a necessary means to achieve efficient exploitation of deep oil and gas resources [2]. It is noteworthy that deep oil and gas reservoirs are often found in high-stress environments with complex fracture characteristics (discontinuous structures such as bedding, joints, and fractures formed by the sedimentation and tectonic processes of deep reservoirs) [3], which significantly affects the effectiveness of reservoir transformation [4,5]. Therefore, an in-depth exploration of the impact of natural fractures on the multi-cluster fracturing construction technology of deep oil and gas reservoirs is of great significance for the development of deep oil and gas resources.

Both theoretical solutions and indoor tests can be used to study the propagation of fracturing fractures. Over the years, various theoretical solution models such as the KGD model, PKN model, and Penny model, among others, have been established through theoretical derivations [2,6–10]. Meanwhile, some scholars have found in their research that there may be multiple control mechanisms in the reservoir due to the influence of fracture toughness and the fracturing injection rate of reservoir rocks. Therefore, a theoretical model for artificial fracture propagation in reservoirs under the influence of comprehensive parameters such as viscosity control and toughness control has been derived. However, the reservoir rock is often assumed to be a homogenous and isotropic medium in theoretical derivation, meaning that the impact of naturally occurring fractures and the growth of numerous fracture groups cannot be considered. Of course, scholars have also conducted extensive physical simulation research using laboratory experimental methods [5,11–15]. In the initial stages, conventional triaxial loading systems were predominantly utilized, focusing on the influence of reservoir geo-stress by controlling surrounding rock and axial pressure. However, real reservoirs typically experience stress conditions under true triaxial stress, prompting some researchers to develop authentic triaxial experimental setups to simulate fracturing from single to multiple clusters. Through these experiments, they have been able to glean insights into the propagation patterns of artificial fractures. Nevertheless, replicating the impact of natural fractures in laboratory settings proves to be a challenge, and further examination is warranted for the transferability of specimen-scale results to reservoir-scale scenarios. Consequently, there is a growing interest in the simulation analysis of reservoir fracturing at the reservoir scale [16,17].

The principle of multi-cluster fracturing in tight reservoirs is illustrated in Figure 1. This method involves creating multiple hydraulic fracture networks in various blocks within the reservoir by strategically establishing multiple perforation positions and adjusting the fluid distribution across different clusters. The overarching objective is to enhance reservoir permeability and facilitate efficient development. Early reservoir fracturing simulations were mainly achieved through finite element methods. Subsequently, considering that reservoir rocks are a discontinuous medium, some new numerical simulation methods were proposed successively, such as the extended finite element method [18], particle flow dispersion element method [19,20], block discrete element method [2], discontinuous element method [21], finite discrete element method (FDEM) [22], etc. These methods make it possible to predict the fracturing construction effect of the target reservoir more accurately, but they have different advantages and limitations. For example, the extended finite element method uses virtual functions to construct a discontinuous displacement field, which can simulate the phenomenon of fracture deflection and propagation inside the element, but it is difficult to simulate the effects of natural fractures and other factors such as fracture bifurcation. The particle flow dispersion element method has an excellent performance in simulating reservoir fracturing at the micro-scale, but it usually assumes that the reservoir is composed of particles, which has great computational limitations and is no longer suitable for reservoir-scale simulation analysis [23,24]. The block discrete element method and finite discrete element method developed in recent years have effectively solved the scale limitations, thus enabling better simulation and evaluation of reservoir fracturing from micro- to macro-scales. Among them, the finite discrete element method [25] assumes that the reservoir is composed of rock blocks and their fractured interfaces (Figure 1). The rock block elements are used to simulate the deformation characteristics of the reservoir matrix during the fracturing process, and the interface elements are used to simulate the propagation of artificial fractures in the reservoir. Meanwhile, with years of development, the characteristics of natural fractures in reservoirs can be described by changing parameters such as the strength and stiffness of different interface elements, thereby achieving fracturing simulation on reservoirs containing natural fractures. Based on the FDEM method, Wu et al. [26] conducted a hydraulic fracturing inversion simulation study considering the natural bedding/fracture interfaces in rocks. Their findings confirmed the viability of utilizing the FDEM to simulate fracture formation in rocks containing such natural features. Additionally, they developed hydraulic fracturing models for reservoir areas considering various heterogeneous structures and conducted a detailed analysis on the impact of factors like natural discrete fractures and large embedded rock blocks on the propagation of hydraulic fractures [27]. Therefore, conducting a simulation study on fracturing in fractured reservoirs using the FDEM proves to be an effective method for unveiling the propagation behavior of hydraulic fractures in reservoir rocks containing natural fractures.

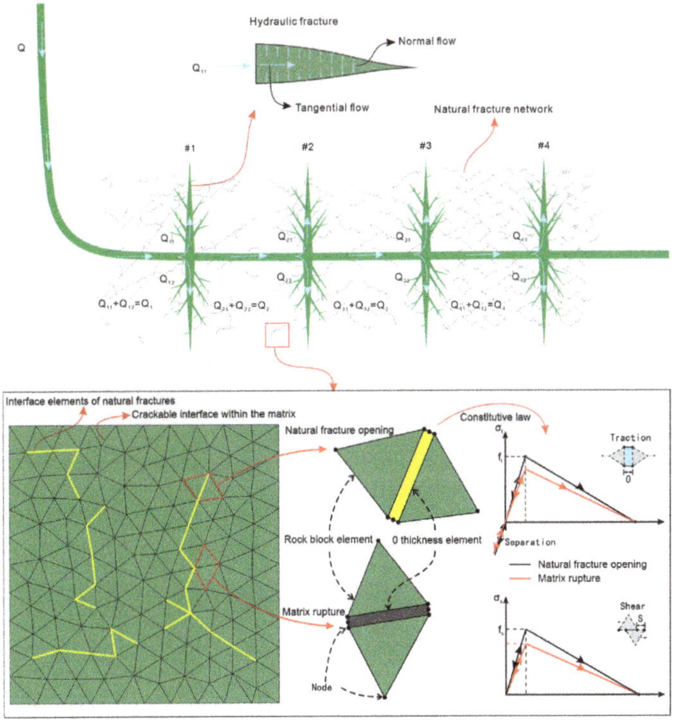

Figure 1. Schematic diagram of reservoir multi-cluster fracturing and FDEM principle.

Based on the finite discrete element method (FDEM), this paper establishes a three-dimensional multi-cluster fracturing simulation model considering the number of natural fractures. The conventional quantitative parameters such as fracture area, fracture length, and maximum fracture aperture were used to systematically analyze the expansion law of artificial fractures under the combined influence of natural fractures and multi-cluster fracturing. The numerical model, the comparative analysis results, and the conclusion are presented in Sections 2, 3, and 4, respectively.

2. Numerical Model

This paper uses cohesive elements and rock block elements to construct an FDEM model [28–30] and divides cohesive elements into cohesive elements within the matrix and cohesive elements representing natural fractures, thus achieving multi-cluster fracturing modeling of reservoirs containing natural fractures. The detailed modeling equations are as follows.

2.1. Simulation Method

Multi-cluster hydraulic fracturing in reservoirs is a typical fluid structure coupling process that involves the interaction between fluids and solids. When considering the interaction between fluids and solids during hydraulic fracturing, the stress balance equation of hydraulic fracturing can be expressed as [31]

$$\int_V (\overline{\sigma} - p_w \boldsymbol{I})\delta\varepsilon\delta dV = \int_S \boldsymbol{t}\cdot\delta\boldsymbol{v}dS + \int_V \boldsymbol{f}\delta\boldsymbol{v}dV \tag{1}$$

where $\overline{\sigma}$ represents the effective stress matrix, MPa; p_w denotes the pore pressure, MPa; $\delta\varepsilon/s^{-1}$ is the virtual strain rate matrix; $\boldsymbol{t}$ is the surface force matrix, N/m^2; δ_v (m/s) and $\boldsymbol{f}$ (N/m^3) are the virtual velocity matrix and the physical force matrix.

During the fracturing process, fluid density and matrix porosity influence the behavior of the fluid within the reservoir. The fluid continuity equation captures the conservation of mass during this process, providing a mathematical representation of the fluid's movement and distribution. It enables a more accurate description of the fluid flow dynamics within the reservoir [32].

$$\frac{d}{dt}\left(\int_V J\left(\frac{\rho_w}{\rho_w^0} n dV\right)\right) + \int_S \frac{\rho_w}{\rho_w^0} n\boldsymbol{n}\cdot v_w dS = 0 \tag{2}$$

where J represents the volume change ratio, dimensionless; ρ_w is the fluid density, kg/m^3; n_w is the void ratio, dimensionless; and v_w is the fluid seepage velocity, m/s.

It is important to mention that the fracturing process also involves the influence of fluid seepage within the matrix. The seepage behavior of reservoir rocks during fracturing can be calculated using Darcy's law, which determines the velocity and flow rate of fluids through factors such as fluid viscosity, permeability, and pressure gradient [33]:

$$v_w = -\frac{1}{n_w g \rho_w}\boldsymbol{k}\cdot\left(\frac{\partial p_w}{\partial x} - \rho_w \boldsymbol{g}\right) \tag{3}$$

where $\boldsymbol{k}$ represents a permeability matrix, m/s, and $\boldsymbol{g}$ is the gravity acceleration vector, m/s^2.

As fluid starts to seep into the reservoir rock mass, the influx of fluid will lead to an elevation in pore pressure, consequently facilitating the propagation of microfractures. It is generally believed that before hydraulic fractures open, the cohesive elements used to simulate hydraulic fractures follow a linear elastic relationship [6],

$$\sigma_{coh} = \begin{Bmatrix} \sigma_{coh\_n} \\ \sigma_{coh\_s} \\ \sigma_{coh\_t} \end{Bmatrix} = \boldsymbol{K}_{coh}\boldsymbol{\varepsilon}_{coh} = \begin{bmatrix} K_{coh\_nn} & K_{coh\_ns} & K_{coh\_nt} \\ K_{coh\_ns} & K_{coh\_ss} & K_{coh\_st} \\ K_{coh\_nt} & K_{coh\_st} & K_{coh\_tt} \end{bmatrix} \begin{Bmatrix} \varepsilon_{coh\_n} \\ \varepsilon_{coh\_s} \\ \varepsilon_{coh\_t} \end{Bmatrix}, \tag{4}$$

where $\boldsymbol{\sigma}_{coh}$ denotes the stress vector; σ_{coh_h}, σ_{coh_s}, σ_{coh_t} are the normal stress, the first tangent stress and the second tangent stress, respectively. Here, $\boldsymbol{K}_{coh}$ denotes the stiffness matrix; $\boldsymbol{\varepsilon}_{coh}$ is the strain matrix, and $\boldsymbol{\varepsilon}_{coh\_n}$, $\boldsymbol{\varepsilon}_{coh\_s}$, $\boldsymbol{\varepsilon}_{coh\_t}$ represent the normal strain, the first tangential strain, and the second tangential strain, respectively. They are defined as follows [25]:

$$\varepsilon_{coh\_n} = \frac{d_n}{T_0}, \varepsilon_{coh\_s} = \frac{d_s}{T_0}, \varepsilon_{coh\_t} = \frac{d_t}{T_0} \tag{5}$$

where d_n, d_s, d_t represent the normal displacement, the first displacement, and the second tangential displacement, respectively. Here, T_0 is the constitutive thickness.

With continued fluid injection, hydraulic fractures will gradually emerge within the reservoir rock mass, a phenomenon that can be simulated by modeling the damage and failure of interface elements. When the interface element is damaged, the anti-deformation ability of the interface element will significantly decrease, leading to the rapid opening of hydraulic fractures. The presence or absence of interface elements can be determined using the following formula [34]:

$$\left\{\frac{\langle\sigma_n\rangle}{\sigma_n^0}\right\}^2+\left\{\frac{\sigma_s}{\sigma_s^0}\right\}^2+\left\{\frac{\sigma_t}{\sigma_t^0}\right\}^2=\lambda \tag{6}$$

where σ_n is the normal stress; σ_s, σ_t represent the tangential stress, MPa; σ_s^0, σ_t^0 are the threshold stress, MPa; λ indicates that cohesive elements resist tension stress but not compression stress: $1 \leq \lambda \leq 1.05$.

After the hydraulic fractures are initiated, the fractures will not continue to open indefinitely due to factors such as interfacial forces. The deformation behavior after reaching the peak can be described by damage factors [28,35],

$$E=(1-d)\times E^0 \tag{7}$$

$$d=\frac{\delta_m^f\left(\delta_m^{max}-\delta_m^0\right)}{\delta_m^{max}\left(\delta_m^f-\delta_m^0\right)}, \tag{8}$$

where E_0, E are the initial elastic modulus and the elastic modulus after damage, Pa, respectively. Here, d is a damage factor, dimensionless. δ_m^{max}, δ_m^f, δ_m^0 signify the distinct displacements in the context of element behaviour. Specifically, δ_m^{max} refers to the maximum displacement experienced. δ_m^f and δ_m^0 denote the displacement.

When performing multi-cluster fracturing, the injected fluid will flow into the formation through the wellbore, resulting in flow distribution within the wellbore. Meanwhile, due to the long wellbore, there is also a corresponding pressure drop.

$$Q=\sum_{i=1}^{N} Q_i \tag{9}$$

$$p_0=p_{pf,i}+p_{cf,i}+p_{wf,i} \tag{10}$$

When fluid flows into the wellbore, flow resistance is easily generated due to the roughness of the wellbore's surface, and this behavior can be described by Bernoulli's equation [36]:

$$\begin{aligned}
&\nabla p-\rho g\Delta Z=(C_L+K_i)\tfrac{\rho v^2}{2},\\
&C_L=\tfrac{fL}{D_h},\\
&f=8\left[\left(\tfrac{8}{\mathrm{Re}}\right)^{12}+\tfrac{1}{(A+B)^{1.5}}\right]^{1/12},\\
&A=\left[-2.457\mathrm{In}\left(\left(\tfrac{7}{\mathrm{Re}}\right)^{0.9}+0.27\left(\tfrac{K_s}{D_h}\right)\right)\right]^{16},\\
&B=\left(\tfrac{37350}{\mathrm{Re}}\right)^{16}
\end{aligned} \tag{11}$$

$$p_f=0.087249\times\frac{\rho}{n^2D_p^4C^2}q_i^2 \tag{12}$$

where ΔP is the pressure difference at the node of the cluster, ΔZ is the elevation difference of the node, v is the fluid velocity in the wellbore, ρ is the fluid density, G is the acceleration of gravity, C_L is the loss coefficient, f is the friction force on the wellbore, L is the wellbore length, K_i is the loss term in a fixed direction, K_s is the roughness of the wellbore, D_h is the wellbore diameter, Re is the Reynolds number, n is the number of perforations, D_p is the perforation diameter, and C is a coefficient, generally 0.56~0.9 [6].

After the fracturing fluid is diverted from the wellbore, it will enter the fracturing fracture, and the fluid pressure drop inside can be described as

$$q = \frac{t^3}{12u} \nabla p, \tag{13}$$

where q is the volume flow vector per tangential element length; t is the thickness, m; μ is the viscous coefficient; p is the fluid pressure, MPa.

It is worth mentioning that there is a certain phenomenon of crossflow between the fluid used for fracturing and the existing fluid in the reservoir. Meanwhile, due to the high fluid pressure in the fractures during fracturing, more fluid flows into the rock matrix from the fractures. At this point, the fluid loss in the fracture is

$$\begin{cases} q_t = c_t(p_i - p_t) \\ q_b = c_b(p_i - p_b) \end{cases}, \tag{14}$$

where q_t, q_b represent the volumetric flow rate. The terms c_t, c_b are the filtration coefficient, $m/min^{0.5}$; p_t, p_b are the pore pressure, MPa; and p_i is the fluid pressure, MPa.

2.2. Model Setup

The computational complexity of a three-dimensional (3D) fracturing model usually varies in magnitude compared to a two-dimensional (2D) fracturing model. Therefore, compared to the 2D fracturing model, the 3D modeling process has appropriately simplified the model. The main parts that deal with changes include the following: ① reducing the geometric size of the model and increasing the layer thickness of the model; ② During mesh generation, the largest possible mesh was used, and after multiple geometric modeling and mesh generation, a model with suitable mesh quality and geometry was selected for subsequent numerical simulation modeling and fracturing analysis. Based on this, the specific conditions set for this 3D simulation model are as follows:

① The target reservoir has a significant tendency toward a natural bedding structure, so the construction of the bedding network is achieved through Python programming.

② Based on field data, the depth of the target block in the simulation typically falls within the range from 4018.5 m to 4263.5 m, with a fracturing construction interval usually set at 60 m. For ease of simulation calculations, it is assumed that the reservoir area size in the simulation is 60 m × 60 m × 3 m.

③ During fracturing, fluid will flow from the wellbore into the target reservoir. fp3d2 elements are used to simulate the fluid pressure drop and flow distribution in the wellbore, and the fpc3d2 element is used to simulate the frictional resistance when the wellbore fluid flows into the target area of the reservoir. By binding nodes, the simulation process of fluid flowing from the wellbore to the target reservoir can be achieved. In addition, considering the randomness of perforation spacing during the fracturing construction of the target reservoir, 38 random perforations are also set in the simulation (Figure 2b).

④ Based on the on-site data collected, the target block displays typical characteristics of carbonic basin subsidence. Therefore, by aligning the x, y, and z directions with the maximum horizontal principal stress, minimum horizontal principal stress, and vertical principal stress, respectively, the impermeable model boundaries are defined. The model is constrained by applying directional displacements, ensuring an accurate representation of the in situ stress conditions and preventing fluid flow across the boundaries.

⑤ Based on the on-site data from the target reservoir, representative model parameters are selected. The benchmark calculation example focuses on single-cluster fracturing, assuming a reservoir thickness of 3 m. The injection parameters include a displacement rate of 18 m^3/min and a total injection time of 3600 s. The in situ stress conditions are characterized by a minimum horizontal stress of 98 MPa, a maximum horizontal stress of 116 MPa, and a vertical stress of 112 MPa. The initial pore pressure is set to 90 MPa. Furthermore, the remaining rock material parameters are derived from the mechanical testing data of shale rocks obtained from the Lu214 block in the Sichuan Basin (as detailed in Table 1).

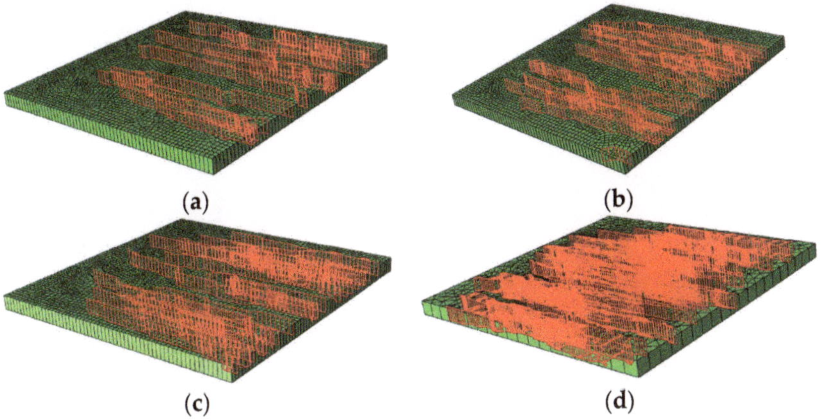

Figure 2. Schematic diagram of FDEM models with natural fractures: (**a**) 20 natural fractures; (**b**) 30 natural fractures; (**c**) 40 natural fractures; (**d**) 50 natural fractures.

Table 1. Main parameters used in simulation models.

Input Parameters	Value
Young's modulus (GPa)	40
Poisson's ratio	0.22
Permeability coefficient (m/s)	1×10^{-7}
Porosity	0.04
Tensile strength of natural fractures (MPa)	2
Critical damage displacement of natural fractures (m)	0.0001
Tensile strength of matrix interfaces (MPa)	6
Critical damage displacement of matrix interfaces (m)	0.001
Injection rate (m^3/min)	19–20
Fracturing fluid viscosity (mPa·s)	1
Pipe roughness (mm)	0.015×10^{-3}
Perforation diameter (m)	0.01

3. Results and Analysis

3.1. *Effect of Fracturing Cluster Number*

To investigate the behavior of multi-cluster fracturing under reservoir conditions in the target area, the fracture propagation results under different cluster numbers were first compared, and the results are shown in Figure 3. With inter-cluster distance increases, the fracturing simulation results of different numbers of clusters show different fracture propagation characteristics. As the number of clusters increases, the number of fracture elements and total area show an increasing trend. However, from single-cluster fracturing to triple-cluster fracturing, the increase in the number and area of fractures is relatively small. This indicates that as the number of clusters increases, the propagation of fractures is influenced by inter-cluster interactions. Concurrently, as the inter-cluster distance increases and the number of clusters is augmented, a reduction in fracture volume is observed. Intriguingly, the proportion of tensile failure within the fracture network shows a significant increase. This phenomenon suggests that single-cluster fracturing faces challenges in creating large-scale fractures. The injected fluid is primarily used to create and lengthen fractures. In contrast, multi-cluster fracturing facilitates the development of a more intricate fracture network and enhances the overall fracture volume. These observations underscore the influence of cluster configuration on fracture morphology and highlight the potential benefits of using multi-cluster fracturing strategies to optimize reservoir stimulation and enhance hydrocarbon recovery.

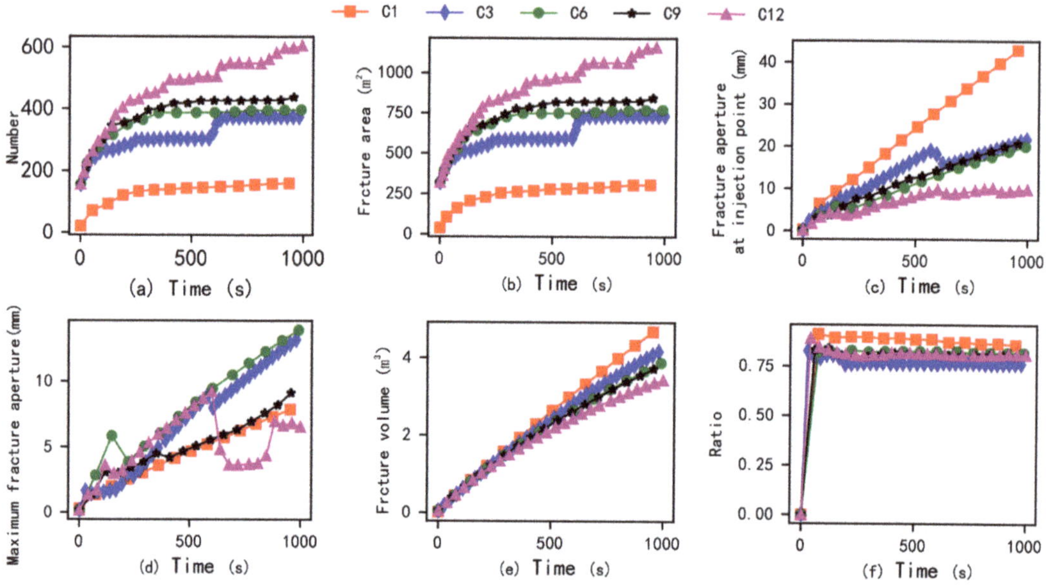

Figure 3. Results of conventional quantitative metrics under different cluster configurations ((**a**) number of damaged interface elements versus time; (**b**) fracture area versus time; (**c**) fracture aperture at injection point versus time; (**d**) maximum fracture aperture versus time; (**e**) fracture volume versus time; and (**f**) ratio of tensile failure versus time) (C1, C3, C6, C9, and C12 refer to fracturing scenarios involving 1 cluster, 3 clusters, 6 clusters, 9 clusters, and 12 clusters, respectively).

At a simulation time of 1000 s, the ultimate distribution of fractures is illustrated in Figure 4. It is noteworthy that as the number of clusters rises, the spacing between clusters diminishes within the model. Consequently, the total surface area covered by the induced fractures amplifies, whereas the aperture of fractures emanating from the injection point decreases. This finding indicates that with an increasing number of fracturing clusters, the area of medium-pressure fractures in the reservoir expands. However, as the total injection volume remains constant, there is a slight decreasing trend observed in the fracture aperture. Furthermore, the total volume of visible fractures decreases. These observations suggest that variations in the number of clusters and cluster spacing influence fracture morphology. The model tends to produce a higher quantity of smaller fractures influenced by a combination of factors including the model mesh, the presence of natural fractures, the number of fracturing clusters, and geo-stress conditions. This leads to an increase in the total fracture surface area and a reduction in the visible fracture volume. This behavior highlights the complex interplay between fracturing parameters and the resulting fracture network geometry.

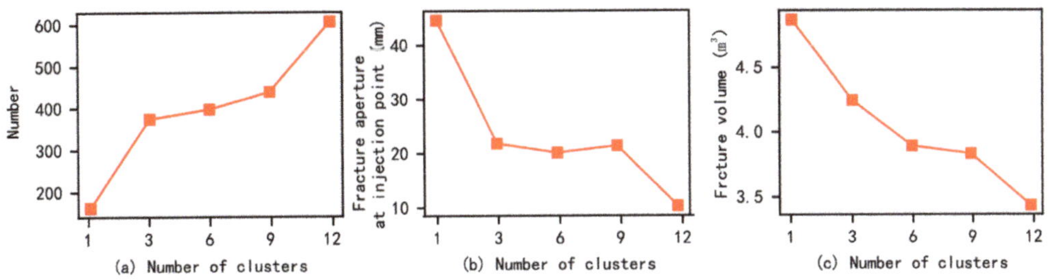

Figure 4. Results of conventional quantization parameters ((**a**) number of damaged interface elements versus number of fracturing clusters; (**b**) fracture aperture at injection point versus number of fracturing clusters; and (**c**) fracture volume versus number of fracturing clusters) (1000 s).

Figure 5 presents a comparative analysis of the final morphology of artificial fractures generated under varying cluster configurations, including different numbers of clusters and cluster spacing. In single-cluster fracturing, a single dominant fracture is mainly formed, with limited fracture initiation and propagation in the near-wellbore region. However, as the number of clusters is increased to three, three primary fractures are observed to develop, exhibiting both single-wing and double-wing propagation. This suggests that increasing the number of clusters can stimulate the formation of fractures near the wellbore and facilitate the development of multiple main fractures, enhancing the overall fracture complexity. When using six clusters of fracturing, multiple fractures appeared near the wellbore. When the number of clusters increases to nine, more fractures initiate simultaneously. It is worth noting that the fracture morphology was magnified 10 times in the simulation results to observe most of the artificial fractures. However, the main path of the fractures in cluster 6 fracturing is more obvious, while cluster 9 fracturing shows a discontinuous main path, indicating the presence of many locally small opening fractures, which is not conducive to the transportation of proppants and subsequent mining. More than 12 artificial fractures have sprouted in the near-wellbore area of 12-cluster fracturing. As the fracturing process progresses, competition expands, causing some of the newly formed fractures to not open further, ultimately leading to the expansion of 10 main fractures. The above results indicate that the more fracturing clusters there are, the smaller the cluster spacing, which usually significantly increases the volume of artificial fractures, thereby improving the effectiveness of reservoir transformation. Similar conclusions can also be observed in some studies on on-site earthquake monitoring methods [37–41]. For example, Chen et al. found through micro seismic monitoring of shale gas reservoirs in the southwestern region that an increase in the number of fracturing clusters can promote the occurrence of more micro seismic events, and micro seismic event points near the wellbore gradually gather to form multiple bands, indicating the possibility of producing more artificial fractures. At the same time, they also found that under the influence of natural fractures and other factors, many micro seismic events may occur in areas far from the wellbore, which is significant in the expansion results of the fracturing of a single cluster in Figure 5. It is worth mentioning that compared to the 3D view, only a portion of the main seam shape is observed in the top view. This is obviously due to some fractures opening too little. Therefore, increasing the number of clusters does not necessarily lead to a proportional increase in the number of main fractures. It is crucial to consider the influence of proppant size on the effective number of supported fractures. Proppant size plays a significant role in determining the aperture and conductivity of the fractures, which ultimately affects the overall effectiveness of the fracturing treatment. By optimizing the proppant size according to the reservoir and fracturing conditions, it is possible to maximize the number of fractures effectively supported and enhance hydrocarbon recovery.

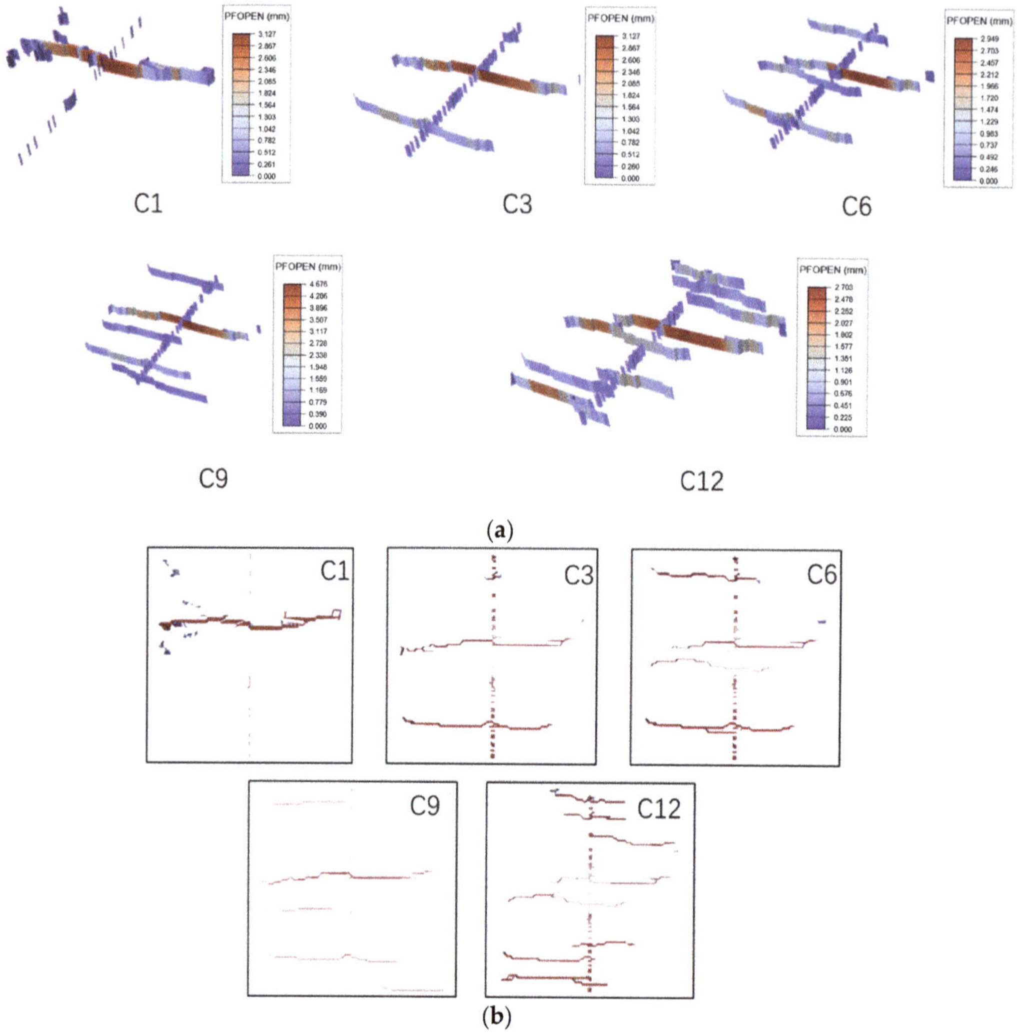

Figure 5. Morphology comparison results of artificial fractures under different cluster numbers ((**a**) 3D view and (**b**) top view) (C1, C3, C6, C9, and C12 refer to fracturing scenarios involving 1 cluster, 3 clusters, 6 clusters, 9 clusters, and 12 clusters, respectively).

3.2. *Effect of Number of Natural Fractures*

The distribution of natural fractures plays a crucial role in influencing fracturing effectiveness within reservoirs. Therefore, this section undertook a simulation analysis to investigate the impact of varying numbers of natural fractures in the simulated target block. Based on the field data obtained from the target block, the linear density of natural fractures within the block falls approximately within the range from 0.3 to 0.8. Considering a simulated area of 60 m × 60 m, the number of natural fractures in the model varies from 20 to 50. The simulation results obtained are shown in Figure 6.

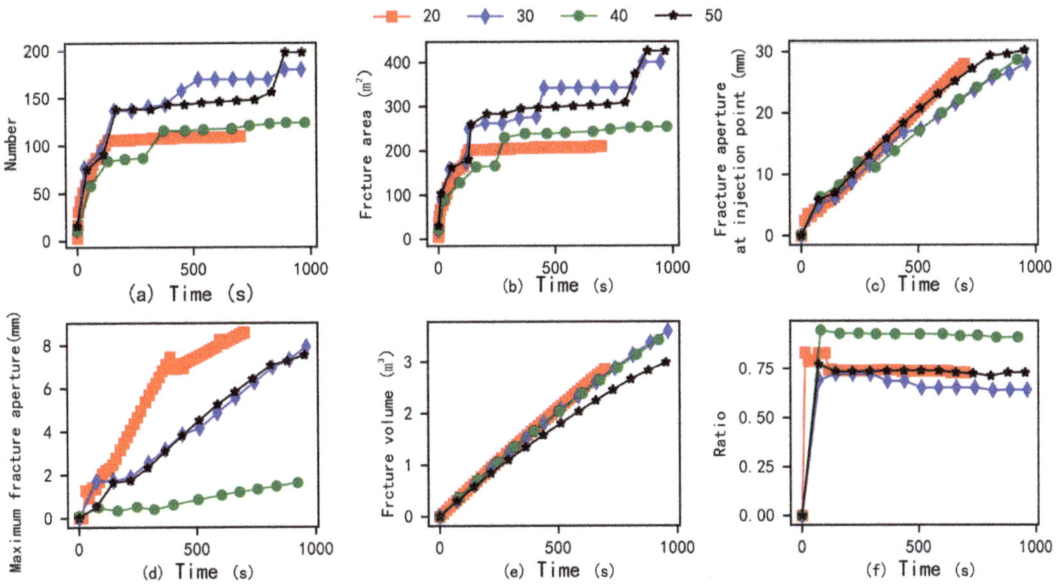

Figure 6. Results of conventional quantitative metrics under influence of number of natural fractures ((**a**) number of damaged interface elements versus time; (**b**) fracture area versus time; (**c**) fracture aperture at injection point versus time; (**d**) maximum fracture aperture versus time; (**e**) fracture volume versus time; and (**f**) ratio of tensile failure versus time) (numbers (20, 30, 40, 50) represent quantity of natural fractures in various simulation models).

Figure 6 shows that with the increase in fracture density, the number of fracture elements and the total area of fractures show an increasing trend. Meanwhile, the aperture and volume of fractures seem to show a certain decreasing trend. This may be due to the increased density of natural fractures, which makes reservoir fracturing more susceptible to the influence of fractures, resulting in the formation of many low-opening shear fractures. Therefore, with the increase in natural fractures, the shear fractures generated by shear stimulation significantly increase, and the fracture area increases. At the same time, since shear fractures are mostly low-opening fractures, the visible volume of artificial fractures shows a certain fluctuation and decreasing trend.

Figure 7 illustrates the opening of artificial fractures in reservoirs under the influence of varying natural fracture densities. As depicted in the figure, with the increase in natural fractures, the number of artificial fractures in the reservoir significantly increases, and more branches and extensions of artificial fractures occur. Simultaneously, an increase in natural fractures leads to a higher likelihood of large openings in artificial fractures near the reservoir wellbore. In addition, when the number of natural fractures increases to a certain extent, it may induce single-cluster fracturing to form a multi-cluster extended artificial fracture network. The above results indicate that natural fractures have a significant impact on the development of artificial fractures during reservoir fracturing, which may alter the position of the main fractures, the distribution of fractures near the wellbore, and the bending and branching morphology of the main fractures, thereby inducing the formation of complex fracture networks and improving the effectiveness of reservoir fracturing.

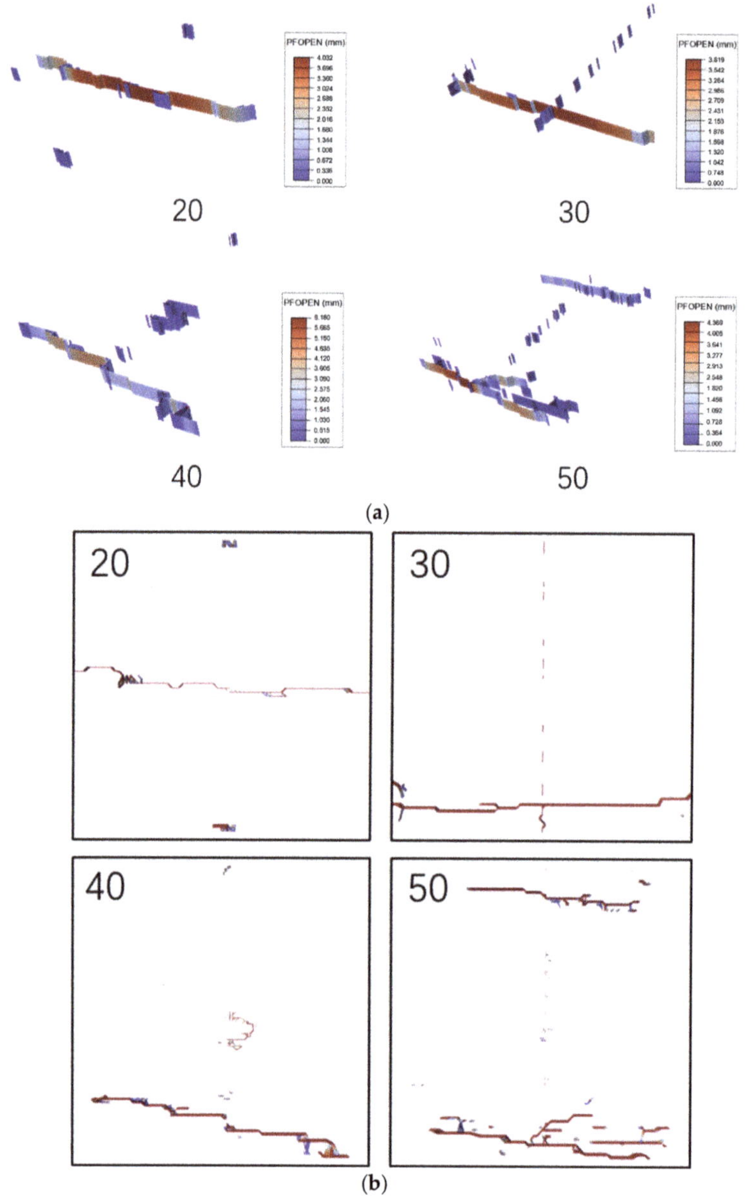

Figure 7. The results of the artificial fracture morphology in reservoirs under the influence of different natural fracture numbers ((**a**) is a 3D view and (**b**) is a top view) (the numbers (20, 30, 40, 50) represent the quantity of natural fractures in various simulation models).

3.3. Comprehensive Impact Analysis

The previous research results indicate that both the number of fracturing clusters and the number of natural fractures significantly affect the propagation pattern of artificial fractures in reservoirs [26,27]. Among them, the number of natural fractures is usually the internal condition of the reservoir, which is difficult to change through construction methods, while the number of fracturing clusters is a construction method that can be directly adjusted during on-site construction. If different cluster fracturing is used for areas with different natural fractures, it may further enhance the effectiveness of reservoir transformation. Therefore, to further discuss the comprehensive impact of multi-cluster fracturing construction and natural fractures, this section comprehensively compares and simulates the expansion of multi-cluster fracturing fractures in target reservoir areas with different numbers of natural fractures.

In Figure 8, there is a phenomenon of incomplete calculation in the fracturing results of six clusters. Therefore, the analysis is only based on the fracturing results of three and nine clusters. From the above figure, when the fracture density is 20, multi-cluster fracturing does not necessarily increase the area of reservoir fractures but has a significant impact on the volume of reservoir fractures. This phenomenon refers to the different behavior of artificial fractures within the reservoir caused by the presence of natural fractures. Notably, the presence of natural fractures often leads to a further increase in the stimulated reservoir volume (SRV). This observation highlights the importance of considering the interplay between natural fractures and hydraulic fracturing operations.

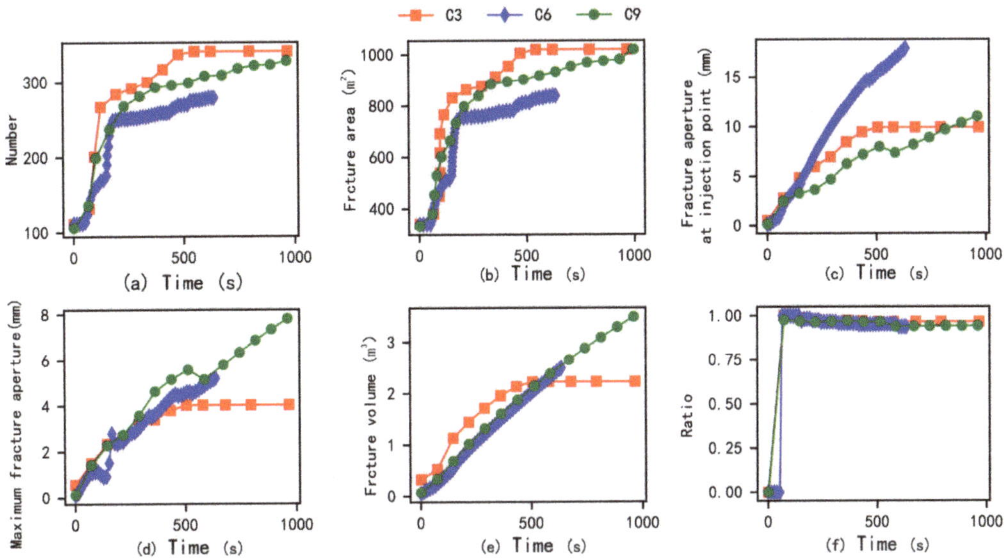

Figure 8. Results of conventional quantitative metrics under different cluster configurations when there are 20 natural fractures in region ((**a**) number of damaged interface elements versus time; (**b**) fracture area versus time; (**c**) fracture aperture at injection point versus time; (**d**) maximum fracture aperture versus time; (**e**) fracture volume versus time; and (**f**) ratio of tensile failure versus time) (C3, C6, and C9 correspond to fracturing scenarios involving 3 clusters, 6 clusters, and 9 clusters, respectively).

It can be significantly observed from Figure 9 that in this simulation model, there were three main fracture extensions in three clusters of fracturing, six main fracture extensions in six clusters of fracturing, and two main fractures and local complex fracture network extensions in nine clusters of fracturing. This result suggests that with an increasing number of fracturing clusters, complex fracture networks are more likely to form in the reservoir. Meanwhile, it is not difficult to observe from the results in the above figure that the two-dimensional view of the nine-cluster fracturing results is not significant, with many low-opening fractures. When considering fracture opening, the fracturing results of cluster 6 are significantly better than those of cluster 3 and cluster 9.

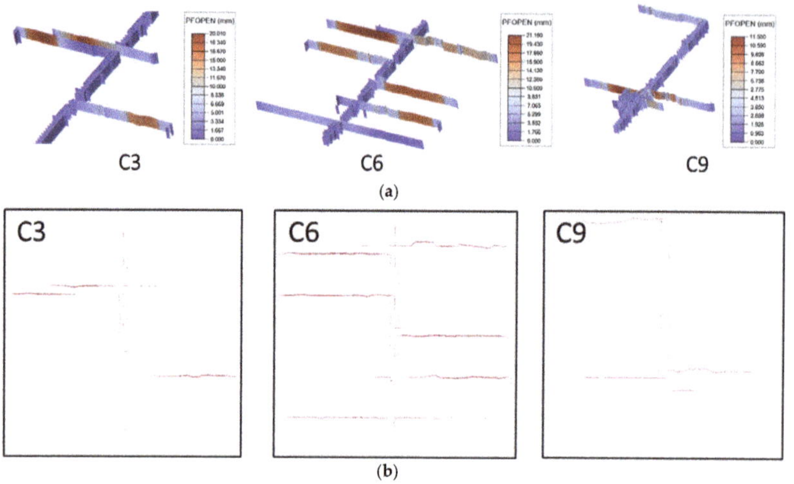

Figure 9. Fracture morphology under diverse cluster fracturing scenarios when there are 20 natural fractures in the region ((**a**) is a 3D view and (**b**) is a top view) (C3, C6, and C9 correspond to fracturing scenarios involving 3 clusters, 6 clusters, and 9 clusters, respectively).

Figure 10 presents comparative results of traditional quantitative metrics across various cluster configurations in the presence of 30 significant natural fractures within the region of interest. Compared to the simulation results with a fracture density of 20, the simulations of 1, 3, 6, 9, and 12 clusters were all fully calculated with a fracture density of 30. Therefore, the results for different clusters with a fracture density of 30 were further extracted. From the above figure, as the number of fracturing clusters increases, the artificial fracture area, fracture opening, and shear failure ratio of the reservoir all significantly increase.

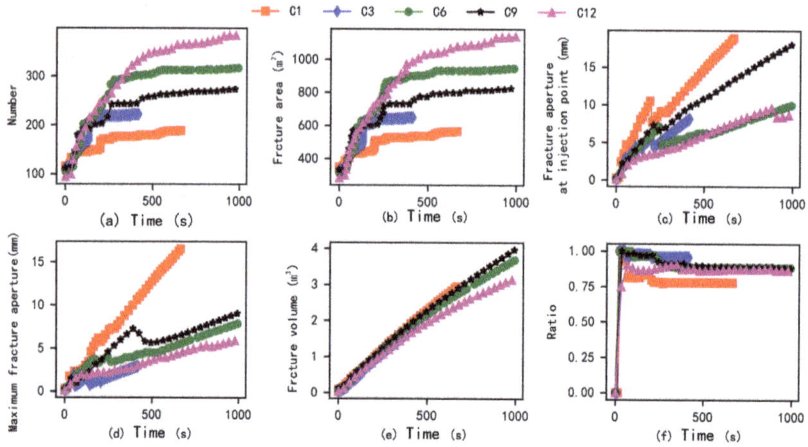

Figure 10. The comparison results of conventional quantitative parameters under the influence of different clusters when there are 30 natural fractures in the region ((**a**) number of damaged interface elements versus time; (**b**) fracture area versus time; (**c**) fracture aperture at injection point versus time; (**d**) maximum fracture aperture versus time; (**e**) fracture volume versus time; and (**f**) ratio of tensile failure versus time) (C1, C3, C6, C9, and C12 represent fracturing scenarios involving 1 cluster, 3 clusters, 6 clusters, 9 clusters, and 12 clusters, respectively).

The process of fracture morphology changes when the fracture number is 30 is shown in Figure 11. As the number of fracturing clusters increases, the main artificial fractures in the reservoir show an increasing trend. Unfortunately, when fracturing 1 cluster, 1 main fracture was formed, when fracturing 3 clusters, 3 main fractures were formed, when fracturing 6 clusters, 6 main fractures were formed, and when fracturing 12 clusters, 12 main fractures were formed. However, when fracturing nine clusters, only three main fractures were formed. We speculate that this is due to the large number of fractures near the wellbore generated in the simulation model of nine-cluster fracturing, indicating that implementing nine-cluster fracturing under the conditions of this model will make it difficult to achieve economic fracturing effects.

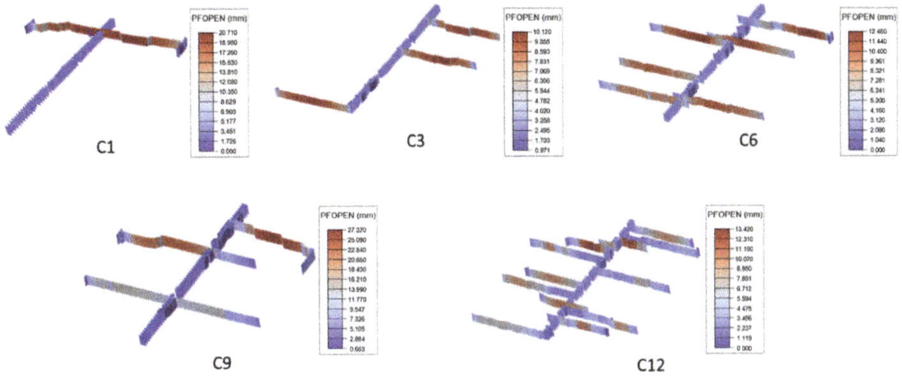

(a)

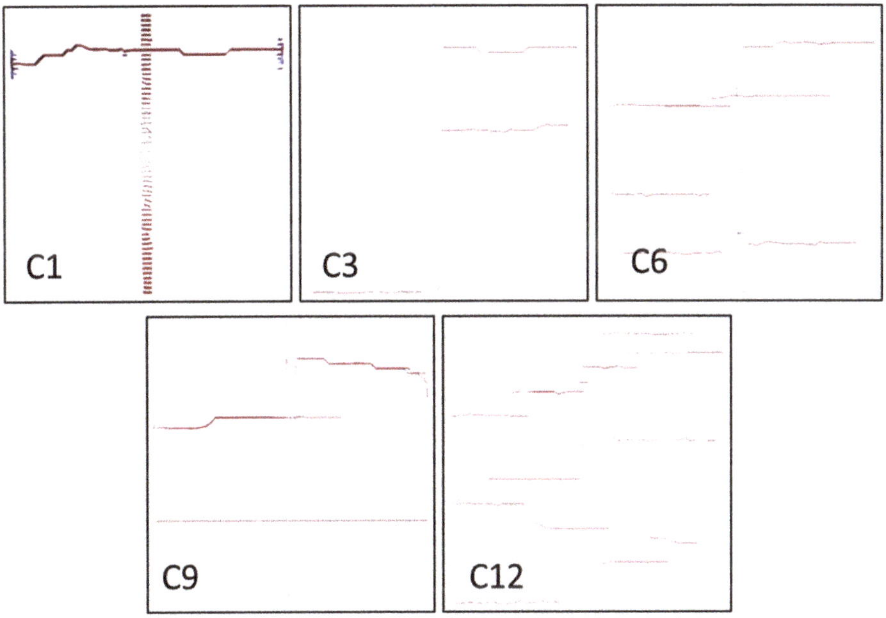

(b)

Figure 11. Fracture morphology under diverse cluster fracturing scenarios when there are 30 natural fractures in the region ((**a**) is a 3D view and (**b**) is a top view) (C1, C3, C6, C9, and C12 represent fracturing scenarios involving 1 cluster, 3 clusters, 6 clusters, 9 clusters, and 12 clusters, respectively).

Figure 12 presents a comparative analysis of conventional quantitative metrics across different cluster configurations in the presence of 40 natural fractures within the region of interest. In comparison to the simulation results obtained at a fracture density of 20, comprehensive simulations were conducted for 1 cluster, 3 clusters, 6 clusters, 9 clusters, and 12 clusters at a fracture density of 40. A thorough comparison was then performed on the conventional quantization parameters for various cluster configurations. As evident from the graphical representation of the results, the artificial fracture area, fracture aperture, and shear failure ratio within the reservoir all exhibit a noticeable upward trend with an increasing number of fracturing clusters. These observations highlight the significant influence of cluster configuration on fracture development and provide valuable insights for optimizing fracturing strategies to enhance reservoir stimulation and hydrocarbon recovery.

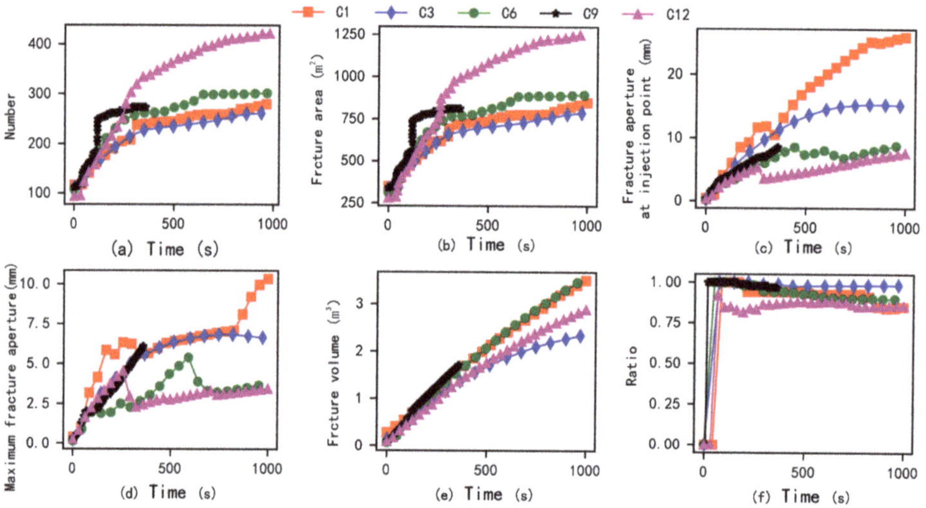

Figure 12. The comparative results of conventional quantitative metrics under varying cluster configurations in the presence of 40 natural fractures within the region ((**a**) number of damaged interface elements versus time; (**b**) fracture area versus time; (**c**) fracture aperture at injection point versus time; (**d**) maximum fracture aperture versus time; (**e**) fracture volume versus time; and (**f**) ratio of tensile failure versus time) (C1, C3, C6, C9, and C12 represent fracturing scenarios involving 1 cluster, 3 clusters, 6 clusters, 9 clusters, and 12 clusters, respectively).

Figure 13 illustrates the evolution of fracture morphology with varying fracturing cluster configurations at a fracture density of 40. As shown in the figure, the quantity of primary artificial fractures within the reservoir progressively increases as the number of fracturing clusters grows. This observation highlights the influence of cluster configuration on fracture development and the potential for optimizing fracturing strategies to enhance reservoir stimulation and hydrocarbon recovery. Among them, there was a phenomenon of multiple main fractures branching out during the fracturing of cluster 1, indicating that fracturing a single cluster may also create a local artificial network of main fractures. When fracturing three clusters, five main fractures were formed. When the expansion degree of the main fractures decreased, 5 main fractures were formed when fracturing 6 clusters, 3 main fractures and a local fracture network expansion area were formed when fracturing 9 clusters, and 12 main fractures were formed when fracturing 12 clusters. The results indicate that as the number of fracturing clusters increases, the number of artificial main fractures increases. It is worth mentioning that previous research results have shown that as the density of natural fractures increases, artificial fractures may exhibit more bending, branching, and other expansion phenomena, forming a complex network of fractures [42,43]. Comparing the simulation results under different numbers of natural fractures and clusters in this section, as the number of natural fractures increases, artificial fractures may become more complex. For example, when the simulated target block contains 40 natural fractures, even under single-cluster fracturing conditions, more complex fracture shapes are formed. Meanwhile, when the simulated target block contains 20 natural fractures, even under the conditions of fracturing of nine clusters, it may be difficult to form a complex artificial fracture network.

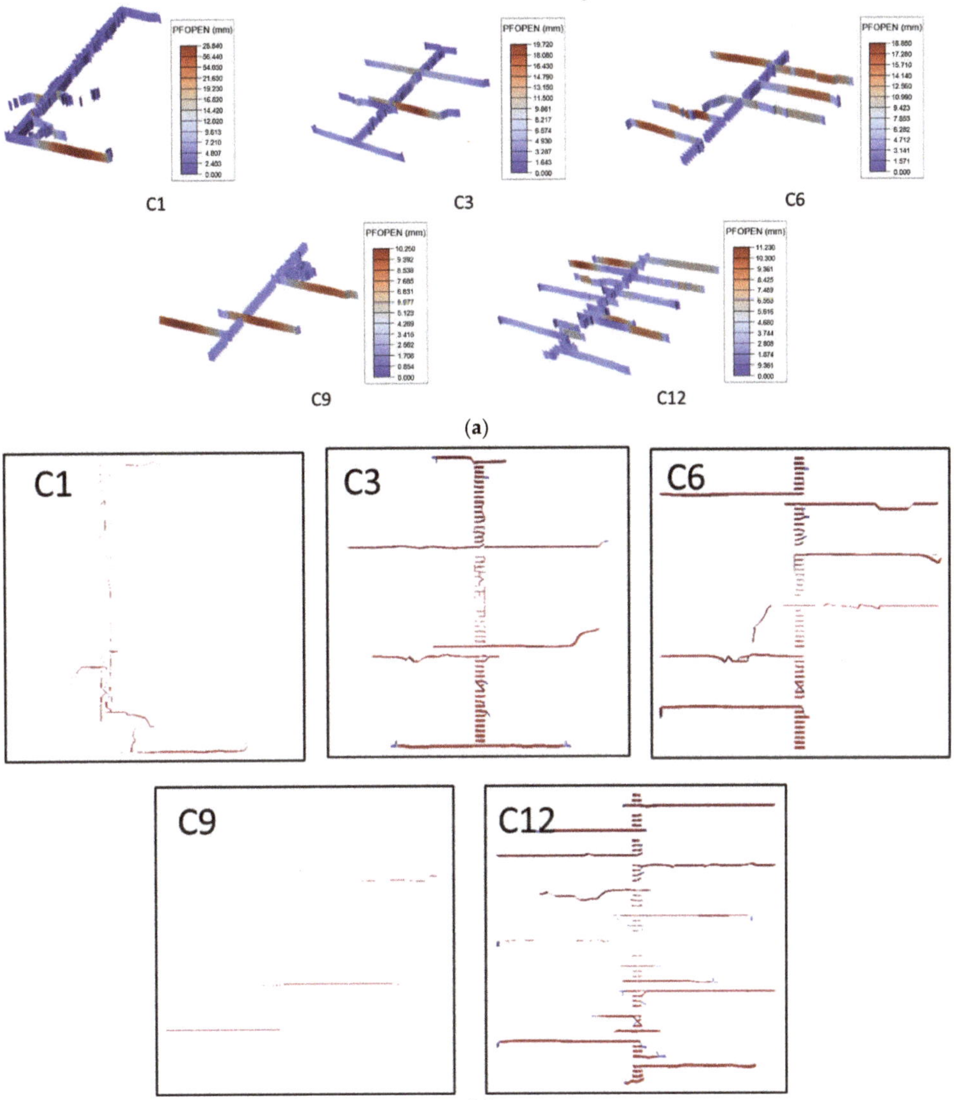

Figure 13. Fracture morphology under diverse cluster fracturing scenarios when there are 40 natural fractures in the region ((**a**) is a 3D view and (**b**) is a top view) (C1, C3, C6, C9, and C12 represent fracturing scenarios involving 1 cluster, 3 clusters, 6 clusters, 9 clusters, and 12 clusters, respectively).

4. Conclusions

Based on the FDEM and on-site data from a certain well block in Southwest China, multiple cluster fracturing simulation models were established considering the natural fractures and geological mechanical parameters of the target block. The changes in artificial fracture area, fracture aperture, fracture volume, fluid pressure, and fracture morphology in the simulation results are extracted, and the impact of the number of fracturing clusters and the number of natural fractures in the target block on the formation of artificial fractures was analyzed. The main conclusions are as follows:

(1) As the number of fracturing clusters increases, the number of artificial main fractures formed in the target block shows a significant increasing trend. However, when affected by the distribution of natural fractures, geo-stress, and other factors, it may also be difficult to form multiple main fractures through multi-cluster fracturing in the target block (Figure 9). Therefore, obtaining data on the spatial location and orientation of natural fractures may be more helpful in accurately estimating the fracturing effect of the target block.
(2) Due to the influence of the original random fracturing path and natural fractures of the reservoir, shear stimulation phenomena are prone to occur. Under these conditions, artificial fractures in the reservoir are prone to bending, branching, and other phenomena. When further affected by multiple construction methods, the artificial main fracture will be more prone to single-wing expansion rather than double-wing expansion.
(3) Multi-cluster fracturing construction may promote an increase in artificial fracture networks, but under the same injection amount, the aperture of artificial fractures will decrease. Therefore, increasing the injection rate appropriately during multi-cluster construction will be more conducive to the pumping of proppants and other materials.
(4) The increase in the number of natural fractures in the target block will help to obtain a more complex artificial fracture network. When the number of natural fractures reaches a certain threshold, even using a single-cluster fracturing construction process may form an artificial fracture network connected by multiple main fractures.

Author Contributions: Conceptualization, Y.L. and Y.G.; data curation, Y.G.; funding acquisition, J.L.; investigation, Y.L., M.W., H.H., Y.G., Y.W. and J.L.; methodology, M.W.; project administration, Y.L., H.H., and Y.G.; resources, H.H. and J.G.; supervision, Y.L., H.H., Y.W. and J.G.; writing—original draft, M.W.; writing—review and editing, Y.L. and H.H. All authors have read and agreed to the published version of the manuscript.

Funding: The authors gratefully acknowledge the financial support given by the Open Research Fund of State Key Laboratory of Geomechanics and Geotechnical Engineering, the Institute of Rock and Soil Mechanics, Chinese Academy of Sciences, Grant NO SKLGME022020. Meanwhile, this article is funded by the science and technology special project of PetroChina Co., Ltd., titled "Research on Large-scale Reservoir Increase, Production Increase, and Exploration & Development Technologies for Shale Gas" (NO. 2023ZZ14).

Institutional Review Board Statement: Not Applicable.

Informed Consent Statement: Not Applicable.

Data Availability Statement: The raw data and code supporting the conclusions of this article will be made available by the authors on request.

Conflicts of Interest: Author Yuegang Li was employed by the Chongqing Shale Gas Exploration and Development Co., Ltd., author Haoyong Huang and Junchuan Gui were employed by the Shale Gas Research Institute of PetroChina Southwest Oil & Gas Field Company, author Yujie Wang was employed by the PetroChina Southwest Oil and Gas Field Company. The remaining authors declare that the research was conducted in the absence of any commercial or financial relationships that could be construed as a potential conflict of interest.

References

1. Wu, M.; Liu, J.; Lv, X.; Shi, D.; Zhu, Z. A study on homogenization equations of fractal porous media. *J. Geophys. Eng.* **2018**, *15*, 2388–2398. [CrossRef]
2. Zheng, Y.; He, R.; Huang, L.; Bai, Y.; Wang, C.; Chen, W.; Wang, W. Exploring the effect of engineering parameters on the penetration of hydraulic fractures through bedding planes in different propagation regimes. *Comput. Geotech.* **2022**, *146*, 104736. [CrossRef]
3. Liu, Q.; Li, J.; Liang, B.; Liu, J.; Sun, W.; He, J.; Lei, Y. Complex wettability behavior triggering mechanism on imbibition: A model construction and comparative study based on analysis at multiple scales. *Energy* **2023**, *275*, 127434. [CrossRef]
4. Song, R.; Liu, J.; Cui, M. A new method to reconstruct structured mesh model from micro-computed tomography images of porous media and its application. *Int. J. Heat Mass Transf.* **2017**, *109*, 705–715. [CrossRef]
5. Wang, L.; Zhou, J.; Guo, Y.; Song, X.; Guo, W. Laboratory investigation and evaluation of the hydraulic fracturing of marine shale considering multiple geological and engineering factors. *Front. Earth Sci.* **2022**, *10*, 952655. [CrossRef]
6. Zhang, H.; Chen, J.; Li, Z.; Hu, H.; Mei, Y. Numerical Simulation of Multi-cluster Fracturing Using the Triaxiality Dependent Cohesive Zone Model in a Shale Reservoir with Mineral Heterogeneity. *Rock Mech. Rock Eng.* **2023**, *57*, 325–349. [CrossRef]
7. Nguyen, H.T.; Lee, J.H.; Elraies, K.A. A review of PKN-type modeling of hydraulic fractures. *J. Pet. Sci. Eng.* **2020**, *195*, 107607. [CrossRef]
8. Yew, C.H.; Weng, X. Chapter 1—Fracturing of a wellbore and 2D fracture models. In *Mechanics of Hydraulic Fracturing*, 2nd ed.; Gulf Professional Publishing: Boston, MA, USA, 2015; pp. 1–22.
9. Dontsov, E.V.; Peirce, A.P. Proppant transport in hydraulic fracturing: Crack tip screen-out in KGD and P3D models. *Int. J. Solids Struct.* **2015**, *63*, 206–218. [CrossRef]
10. Santillán, D.; Juanes, R.; Cueto-Felgueroso, L. Phase field model of fluid-driven fracture in elastic media: Immersed-fracture formulation and validation with analytical solutions. *J. Geophys. Res. Solid Earth* **2017**, *122*, 2565–2589. [CrossRef]
11. Song, R.; Liu, J.; Yang, C.; Sun, S. Study on the multiphase heat and mass transfer mechanism in the dissociation of methane hydrate in reconstructed real-shape porous sediments. *Energy* **2022**, *254*, 124421. [CrossRef]
12. Song, R.; Wang, Y.; Ishutov, S.; Zambrano-Narvaez, G.; Hodder, K.J.; Chalaturnyk, R.J.; Sun, S.; Liu, J.; Gamage, R.P. A Comprehensive Experimental Study on Mechanical Behavior, Microstructure and Transport Properties of 3D-printed Rock Analogs. *Rock Mech. Rock Eng.* **2020**, *53*, 5745–5765. [CrossRef]
13. Guo, C.; Xu, J.; Wei, M.; Jiang, R. Experimental study and numerical simulation of hydraulic fracturing tight sandstone reservoirs. *Fuel* **2015**, *159*, 334–344. [CrossRef]
14. Moghadasi, R.; Rostami, A.; Hemmati-Sarapardeh, A. Application of nanofluids for treating fines migration during hydraulic fracturing: Experimental study and mechanistic understanding. *Adv. Geo-Energy Res.* **2019**, *3*, 100–111. [CrossRef]
15. Yang, R.; Wang, Y.; Song, G.; Shi, Y. Fracturing and thermal extraction optimization methods in enhanced geothermal systems. *Adv. Geo-Energy Res.* **2023**, *8*, 136–140. [CrossRef]
16. Chen, B.; Barboza, B.R.; Sun, Y.; Bai, J.; Thomas, H.R.; Dutko, M.; Cottrell, M.; Li, C. A Review of Hydraulic Fracturing Simulation. *Arch. Comput. Methods Eng.* **2022**, *29*, 1–58. [CrossRef]
17. Yuan, J.; Jiang, R.; Zhang, W. The workflow to analyze hydraulic fracture effect on hydraulic fractured horizontal well production in composite formation system. *Adv. Geo-Energy Res.* **2018**, *2*, 319–342. [CrossRef]
18. Esfandiari, M.; Pak, A. XFEM modeling of the effect of in-situ stresses on hydraulic fracture characteristics and comparison with KGD and PKN models. *J. Pet. Explor. Prod. Technol.* **2023**, *13*, 185–201. [CrossRef]
19. Huang, L.; Liu, J.; Zhang, F.; Dontsov, E.; Damjanac, B. Exploring the influence of rock inherent heterogeneity and grain size on hydraulic fracturing using discrete element modeling. *Int. J. Solids Struct.* **2019**, *176–177*, 207–220. [CrossRef]
20. Huang, L.; Liu, J.; Zhang, F.; Fu, H.; Zhu, H.; Damjanac, B. 3D lattice modeling of hydraulic fracture initiation and near-wellbore propagation for different perforation models. *J. Pet. Sci. Eng.* **2020**, *191*, 107169. [CrossRef]
21. Zhu, X.; Feng, C.; Cheng, P.; Wang, X.; Li, S. A novel three-dimensional hydraulic fracturing model based on continuum–discontinuum element method. *Comput. Methods Appl. Mech. Eng.* **2021**, *383*, 113887. [CrossRef]
22. Wu, M.; Jiang, C.; Song, R.; Liu, J.; Li, M.; Liu, B.; Shi, D.; Zhu, Z.; Deng, B. Comparative study on hydraulic fracturing using different discrete fracture network modeling: Insight from homogeneous to heterogeneity reservoirs. *Eng. Fract. Mech.* **2023**, *284*, 109274. [CrossRef]
23. Huang, L.; Dontsov, E.; Fu, H.; Lei, Y.; Weng, D.; Zhang, F. Hydraulic fracture height growth in layered rocks: Perspective from DEM simulation of different propagation regimes. *Int. J. Solids Struct.* **2022**, *238*, 111395. [CrossRef]
24. Huang, L.; Tan, J.; Fu, H.; Liu, J.; Chen, X.; Liao, X.; Wang, X.; Wang, C. The non-plane initiation and propagation mechanism of multiple hydraulic fractures in tight reservoirs considering stress shadow effects. *Eng. Fract. Mech.* **2023**, *292*, 109570. [CrossRef]
25. Wu, M.; Zhang, D.M.; Wang, W.S.; Li, M.H.; Liu, S.M.; Lu, J.; Gao, H. Numerical simulation of hydraulic fracturing based on two-dimensional surface fracture morphology reconstruction and combined finite-discrete element method. *J. Nat. Gas Sci. Eng.* **2020**, *82*, 103479. [CrossRef]
26. Wu, M.; Wang, W.; Zhang, D.; Deng, B.; Liu, S.; Lu, J.; Luo, Y.; Zhao, W. The pixel crack reconstruction method: From fracture image to crack geological model for fracture evolution simulation. *Constr. Build. Mater.* **2020**, *273*, 121733. [CrossRef]
27. Wu, M.; Wang, W.; Song, Z.; Liu, B.; Feng, C. Exploring the influence of heterogeneity on hydraulic fracturing based on the combined finite–discrete method. *Eng. Fract. Mech.* **2021**, *252*, 107835. [CrossRef]

28. Liu, Q.; Sun, L.; Liu, P.; Chen, L. Modeling Simultaneous Multiple Fracturing Using the Combined Finite-Discrete Element Method. *Geofluids* **2018**, *2018*, 4252904. [CrossRef]
29. Rougier, E.; Munjiza, A.; Lei, Z.; Chau, V.T.; Knight, E.E.; Hunter, A.; Srinivasan, G. The combined plastic and discrete fracture deformation framework for FDEM. *Int. J. Numer. Methods Eng.* **2019**, *121*, 1020–1035. [CrossRef]
30. Yan, C.; Jiao, Y.-Y.; Zheng, H. A fully coupled three-dimensional hydro-mechanical finite discrete element approach with real porous seepage for simulating 3D hydraulic fracturing. *Comput. Geotech.* **2018**, *96*, 73–89. [CrossRef]
31. Wu, Z.; Sun, H.; Wong, L.N.Y. A Cohesive Element-Based Numerical Manifold Method for Hydraulic Fracturing Modelling with Voronoi Grains. *Rock Mech. Rock Eng.* **2019**, *52*, 2335–2359. [CrossRef]
32. Wu, M.; Gao, K.; Liu, J.; Song, Z.; Huang, X. Influence of rock heterogeneity on hydraulic fracturing: A parametric study using the combined finite-discrete element method. *Int. J. Solids Struct.* **2022**, *234–235*, 111293. [CrossRef]
33. Wu, Z.; Xu, X.; Liu, Q.; Yang, Y. A zero-thickness cohesive element-based numerical manifold method for rock mechanical behavior with micro-Voronoi grains. *Eng. Anal. Bound. Elem.* **2018**, *96*, 94–108. [CrossRef]
34. Sharafisafa, M.; Aliabadian, Z.; Sato, A.; Shen, L. Combined finite-discrete element modelling of hydraulic fracturing in reservoirs with filled joints. *Geoenergy Sci. Eng.* **2023**, *228*, 212025. [CrossRef]
35. Shi, F.; Wang, D.; Li, H. An XFEM-based approach for 3D hydraulic fracturing simulation considering crack front segmentation. *J. Pet. Sci. Eng.* **2022**, *214*, 110518. [CrossRef]
36. Cramer, D.; Friehauf, K.; Roberts, G.; Whittaker, J. Integrating DAS, Treatment Pressure Analysis and Video-Based Perforation Imaging to Evaluate Limited Entry Treatment Effectiveness. In Proceedings of the SPE Hydraulic Fracturing Technology Conference and Exhibition, The Woodlands, TX, USA, 5–7 February 2019; p. D031S007R001.
37. Chen, H.; Meng, X.; Niu, F.; Tang, Y.; Yin, C.; Wu, F. Microseismic Monitoring of Stimulating Shale Gas Reservoir in SW China: 2. Spatial Clustering Controlled by the Preexisting Faults and Fractures. *J. Geophys. Res. Solid Earth* **2018**, *123*, 1659–1672. [CrossRef]
38. Ou, C.; Liang, C.; Li, Z.; Luo, L.; Yang, X. 3D visualization of hydraulic fractures using micro-seismic monitoring: Methodology and application. *Petroleum* **2022**, *8*, 92–101. [CrossRef]
39. Lu, Z.; Jia, Y.; Cheng, L.; Pan, Z.; Xu, L.; He, P.; Guo, X.; Ouyang, L. Microseismic Monitoring of Hydraulic Fracture Propagation and Seismic Risks in Shale Reservoir with a Steep Dip Angle. *Nat. Resour. Res.* **2022**, *31*, 2973–2993.
40. Shang, X.; Long, S.; Duan, T. Fracture system in shale gas reservoir: Prospect of characterization and modeling techniques. *J. Nat. Gas Geosci.* **2021**, *6*, 157–172. [CrossRef]
41. Li, Y.; Cheng, C.H.; Toksöz, M.N. Seismic monitoring of the growth of a hydraulic fracture zone at Fenton Hill, New Mexico. *Geophysics* **1998**, *63*, 120–131. [CrossRef]
42. Dahi Taleghani, A.; Gonzalez-Chavez, M.; Yu, H.; Asala, H. Numerical simulation of hydraulic fracture propagation in naturally fractured formations using the cohesive zone model. *J. Pet. Sci. Eng.* **2018**, *165*, 42–57. [CrossRef]
43. Wang, H. Hydraulic fracture propagation in naturally fractured reservoirs: Complex fracture or fracture networks. *J. Nat. Gas Sci. Eng.* **2019**, *68*, 102911. [CrossRef]

applied sciences

MDPI

Article

An Experimental Study on the Physical and Mechanical Properties of Granite after High-Temperature Treatment Considering Anisotropy

Yan Qin [1,2,*], Linqing Wu [1], Qiong Wu [3], Nengxiong Xu [1,2,*], Guanjun Cai [3], Yuxi Guo [1] and Wenjing Zhou [1]

1 School of Engineering and Technology, China University of Geosciences (Beijing), Xueyuan Road 29, Beijing 100083, China; 2102200102@email.cugb.edu.cn (L.W.); 3002230011@email.cugb.edu.cn (Y.G.); 3002220002@email.cugb.edu.cn (W.Z.)
2 Engineering and Technology Innovation Center for Risk Prevention and Control of Major Project Geosafety, MNR, Xueyuan Road 29, Beijing 100083, China
3 Beijing Jingneng Geological Engineering Co., Ltd., 36 Mentougou Road, Mentougou District, Beijing 102300, China; wuqiongmiao@sina.com (Q.W.); caifeisoft@sina.com (G.C.)
* Correspondence: yanqin@cugb.edu.cn (Y.Q.); xunengxiong@cugb.edu.cn (N.X.); Tel.: +86-010-8232-3010(Y.Q.)

Abstract: The deep burial disposal of nuclear waste and dry hot rock mining relates to the effects of high temperatures on the physical and mechanical properties of granite. Previous studies have shown that due to the anisotropy of mineral arrangements during granite formation, the physical and mechanical properties of granite vary greatly with different temperatures. We conducted wave velocity tests, optical mirror tests, and uniaxial and conventional triaxial compression tests on granite in three orthogonal directions before and after high-temperature treatment. The main innovative conclusions are as follows: (1) High temperatures can cause the density of thermal cracks in the cross-section of granite, which varies with different sampling directions. Temperatures below 400 °C increase the anisotropy of granite, and there are obvious advantages in the development direction. (2) Under the same temperature conditions, granite samples taken parallel to the dominant direction of cracks exhibit the best mechanical properties. (3) In uniaxial compression tests, granite samples after high-temperature treatment are mostly subjected to tensile splitting failure. When the heating temperature is higher than 400 °C, a large number of transgranular cracks are generated during the thermal damage of granite at this temperature stage. Rock samples taken perpendicular to the dominant direction of the crack can generate radial cracks near the main failure surface, while rock samples taken parallel to the dominant direction of the crack can generate more axial cracks.

Keywords: granite; physical and mechanical properties; heat damage; anisotropy; microstructure

1. Introduction

In deep rock engineering, rocks often encounter high-temperature environments. For example, during the development of geothermal energy, the ambient temperature of dry hot rocks is above 200 °C [1,2]. In a deep burial disposal project involving high-level radioactive nuclear waste, the decay of radioactive elements in nuclear waste releases a large amount of heat, resulting in the surrounding rock being in a high-temperature environment for a long time [3,4]. In the process of engineering construction and operation, reconstruction after a building fire must also consider the impact of high temperatures on rocks [5–7]. A high temperature will significantly reduce the physical and mechanical properties of rocks, thus affecting engineering safety. Granite, a common magmatic rock, has the characteristics of low permeability, low density, and high strength and is widely used in buildings around the world. It is also the main reservoir rock for geothermal energy and the surrounding rock of deep burial disposal sites for high-level radioactive waste. Therefore, it is of great significance for engineering rock mechanics to study the physical and mechanical properties of granite after high-temperature treatment, such as in the deep burial disposal of nuclear waste and dry hot rock mining.

A high temperature causes the thermal expansion of mineral particles inside rocks and the extensive development of microcracks, leading to the loosening of the rock structure. On a macroscopic level, it is manifested that the physical and mechanical properties of rocks will deteriorate to varying degrees after different high-temperature treatment. In recent years, scholars have conducted both domestic and international research on the physical and mechanical properties of granite after high temperature treatment and have achieved significant results. Morrow et al. [8] and Chaki et al. [9] measured the porosity and permeability of granite after different temperature treatments, and the experimental results show that the porosity and permeability of granite increase with an increasing temperature. Additionally, the physical and mechanical properties of granite deteriorate. Sun et al. [10] measured the quality of granite treated at different temperatures ranging from room temperature to 800 °C and found that the quality of the rock sample decreased with an increasing heating temperature, which was related to the loss of free and bound water in the rock minerals. Yang et al. [11] reported that the density of granite decreases with an increasing temperature, and the magnitude of the decrease increases with an increasing heating temperature. Zhu et al. [12] found, through experiments, that the longitudinal wave velocity of granite decreases with an increasing temperature, and the higher the temperature is, the faster the decrease in the longitudinal wave velocity. Zhang et al. [13] conducted a comprehensive review of previous research on the variation in granite wave velocity after high-temperature treatment and divided the trend of the granite longitudinal wave velocity with temperature into three stages: at room temperature (~150 °C), there was no significant change in the longitudinal wave velocity; at 200–600 °C, the longitudinal wave velocity rapidly decreased with an increasing temperature; and at a heat-treatment temperature above 600 °C, the rate of decrease in the longitudinal wave velocity decreased. F. E. HEUZE [14] systematically summarized the main research results on high-temperature granite before 1983, compared the research results of different scholars through normalization, obtained the variation law of the main mechanical indicators of granite with temperature, and discussed future research directions. Liu et al. [15] studied the physical and mechanical properties of granite after high-temperature treatment and found that the compressive strength and elastic modulus of granite decreased with an increase in the temperature. Compared with the static elastic modulus, the dynamic elastic modulus of granite decreased faster. These authors believe that a high temperature has a greater impact on the physical properties of granite than on its mechanical properties. Zhao et al. [16] conducted Brazilian splitting tests on granite specimens with three different particle size distributions after different temperature treatments and reported that as the particle size and heterogeneity increased, the Brazilian tensile strength of untreated granite decreased. Moreover, when the heating temperature was less than 300 °C, the tensile strength of granite specimens with different particle sizes increased, remained unchanged, and decreased. When the temperature reached 400 °C, the temperature significantly decreased.

Through extensive literature research, numerous scholars have collected their research results in the field of high-temperature granite, and data on thermal loss related to the physical and mechanical properties of granite have been extracted and organized. By extracting and organizing these data, a series of physical and mechanical indicators, such as the wave velocity, peak strength, and elastic modulus, of granite are summarized as functions of temperature. To more intuitively demonstrate the impact of high temperatures on granite, the data were normalized, as shown in Figures 1 and 2. According to previous reports, the overall trends of the variations in the longitudinal wave velocity, uniaxial compressive strength, and elastic modulus of granite with respect to temperature after exposure to high temperatures are relatively similar, but the variation trends are slightly different. According to previous reports, the overall trends of the variations in the longitudinal wave velocity, uniaxial compressive strength, and elastic modulus of granite with respect to temperature after exposure to high temperatures are relatively similar, but the variation trends are slightly different. According to Chen et al. [17], the longitudinal wave velocity of granite does not change significantly when heated below 300 °C, while Ma et al.'s [18] experimental results suggest that the longitudinal wave velocity of granite significantly decreases when heated below 300 °C. In terms of the variation law of the uniaxial compressive strength, there are more significant differences among the achievements of different scholars. At heating temperatures below 200 °C, the uniaxial compressive strength shows three trends: an increase with the temperature [19], almost no change [20,21], and a decrease [15,22,23]. When the heating temperature exceeds 400 °C, there is also a significant difference in the rate of decrease in the uniaxial compressive strength with temperature. Alneasan [24] studied the effects of temperature on the fracture characteristics of granite under three loading modes and found that the fracture behavior is temperature-dependent and can transition from brittle to ductile. Wang [25,26] conducted cyclic loading tests on granite and used a specially designed acoustic emission testing system to monitor the thermal damage evolution of rocks in real time during heating and cooling processes. He established a thermal-damage-evolution model considering both heating and cooling processes. The numerical calculation results are in good agreement with the experimental results.

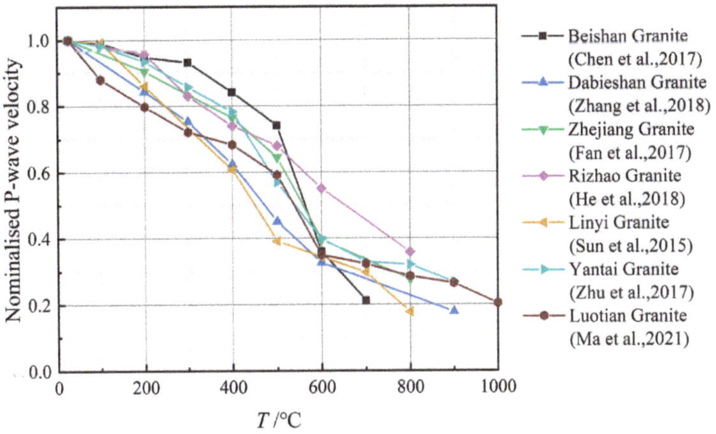

Figure 1. Normalized longitudinal wave velocity variation with different temperatures in previous studies [2,10,12,17,18,27,28].

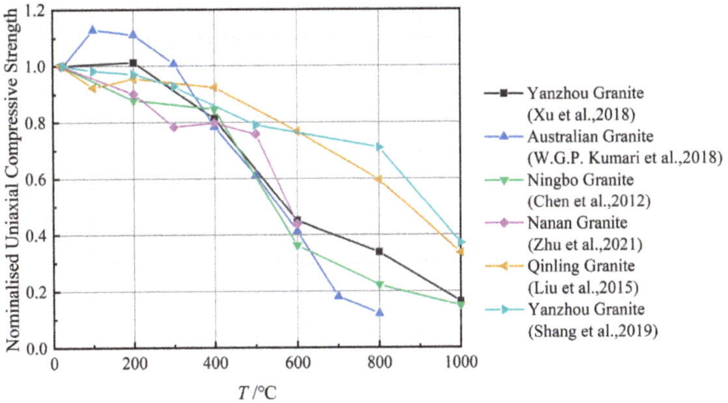

Figure 2. The variation law of the normalized uniaxial compressive strength with different temperatures in previous studies [15,19–23].

According to previous reviews, it is generally believed that the differences in the physical and mechanical properties of granite after exposure to high temperatures are due to the geological characteristics of the granite, that is, the different mineral compositions. However, this phenomenon also occurs when the mineral composition and content of the granite samples selected in the experiment are similar. Therefore, we speculate that anisotropy caused by different sampling directions is also an important influencing factor. Moreover, most scholars have conducted research on the anisotropy of granite under room-temperature conditions, with little consideration given to the effects of temperature on granite anisotropy, and there is a lack of comprehensive research on granite anisotropy.

2. Sample Preparation and Testing Steps

The granite used in the experiment was taken from Jining, Shandong Province, and had a gray-white color, was relatively fresh, and had a low degree of weathering. First, the mineral composition and microstructure of the samples were analyzed under normal temperature conditions to gain a preliminary understanding of the basic physical properties of the granite. An X-ray diffraction (XRD) analysis revealed that the main mineral components of the granite were quartz (24.4%), plagioclase (39.8%), potassium feldspar (30.6%), and biotite (5.2%).

To reduce errors and external influencing factors in the experiment, all the rock samples used in this experiment were taken from the same granite block. According to the International Society for Rock Mechanics Test Specification (ISRM 2007), all the samples were φ A 50 × 100 mm cylinders. To study the anisotropy of the physical and mechanical properties of granite, samples were taken along the three orthogonal directions of granite, according to the sampling scheme in Figure 3. Seventy-two samples were processed in each direction, totaling 216 granite samples. According to the wave speed test results, the three directions are named in a descending order of the wave speed: the Z direction, X direction, and Y direction.

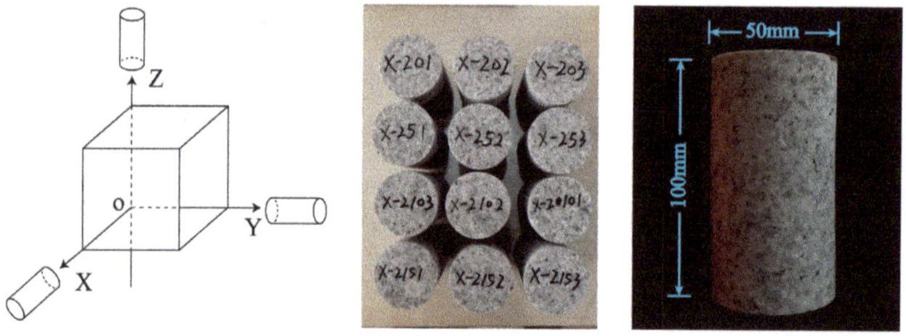

Figure 3. Sampling method and granite samples.

The heating temperature (T) was set at a temperature gradient of 200 °C and was set at 25 °C (room temperature), 200 °C, 400 °C, 600 °C, 800 °C, and 1000 °C. Uniaxial and triaxial compression tests with confining pressures of 5 MPa, 10 MPa, and 15 MPa were carried out on rock samples heated at different temperatures.

The instruments and equipment used in this experiment included electronic balances (with an accuracy of 0.001 g), vernier calipers (with an accuracy of 0.02 mm), a polarizing microscope (OLYMPUS BX53 comes from Beijing, China), an ultrasonic detector (Pundit-PL200 PE, 250 kHz comes from Beijing, China), a high-temperature furnace (SG-XL1200 comes from Beijing, China, with maximum temperature of 1200 °C and an accuracy of ±3 °C), and a mechanical testing machine (HSW-1000B comes from Beijing, China).

The experimental procedure was as follows: The rock samples were placed in a well-ventilated place at room temperature for 2 weeks to eliminate the influence of natural moisture content. As shown in Figure 4, the rock samples were placed into a high-temperature furnace according to temperature groups for heat treatment at a heating rate of 3 °C/min. After heating to the target temperature, the mixture was maintained at a constant temperature for 2 h. After the power was turned off, the samples were cooled naturally in the furnace to room temperature. Microscopic observation, wave velocity testing, and uniaxial and triaxial compression tests were carried out on the rock samples after high-temperature treatment, and the morphology of damaged samples was photographed.

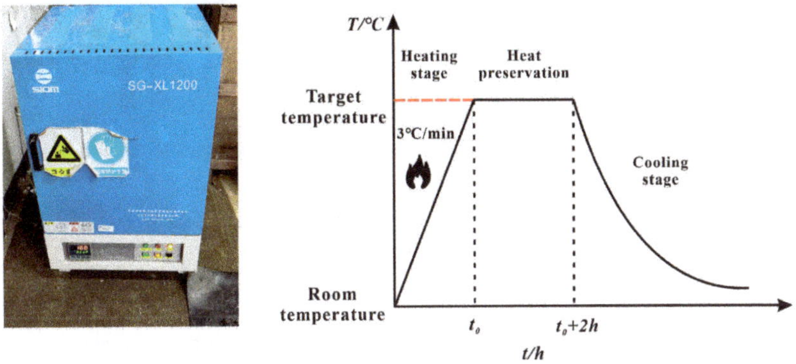

Figure 4. High-temperature furnace and rock sample heating process.

3. Test Results

The test results include the following: (1) changes in the appearance of the rock sample; (2) changes in microscopic properties; (3) changes in transverse and longitudinal wave velocities; (4) changes in the compressive strength and elastic modulus; (5) changes in cohesion and the internal friction angle; (6) anisotropic changes; and (7) destruction.

3.1. Changes in the Appearances of the Rock Samples

The surface color, tapping sound, and roughness of the granite rock samples after high-temperature treatment significantly changed compared to those of the untreated samples. Figure 5 shows the apparent color changes in some of the granite samples after being subjected to temperatures ranging from 20 °C to 1000 °C. The apparent color of the granite changed significantly with an increasing temperature. At room temperature, the granite grains appeared grayish white with a smooth surface. After heating at 200 °C, the color slightly shifted to white, and the knocking sound was crisp, indicating a smooth surface. After heating at 400 °C, the color changed to light flesh red with yellow spots, the knocking sound became turbid compared to before, and the surface was smoother. After heating at 600 °C, the overall color changed from gray to white, and the knocking sound became obviously cloudy, resulting in a rough surface. After heating at 800 °C, the white was lighter than that at 600 °C, and black substances appeared locally (which may appear on the surface where dark minerals melt and precipitate). The knocking sound was cloudy, the surface was rough, and the particles were difficult to erase. After heating at 1000 °C, the color of the granite became red compared with 800 °C, and the black material disappeared. The knock was very cloudy, and the surface was very rough and could be rubbed into particles.

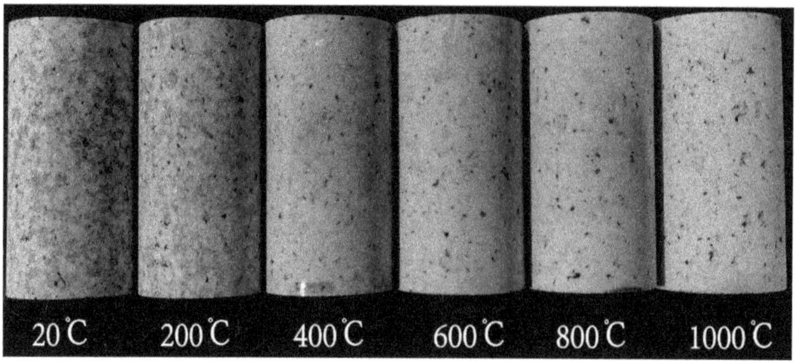

Figure 5. Comparison of granite appearance after different temperature treatments.

3.2. Changes in Microscopic Properties

As shown in Figure 6, the morphological changes in the mineral composition and microcrack-development characteristics of the granite samples after high-temperature treatment were observed using an optical microscope, and optical microscopic images of the samples treated at different temperatures were obtained. For the convenience of statistics and resolution, intergranular cracks are marked in red, intragranular cracks in the quartz are marked in yellow, intragranular cracks in the feldspar are marked in green, and intragranular cracks in the biotite are marked in orange. The rock was mainly composed of plagioclase, potassium feldspar, quartz, biotite, and a small amount of opaque minerals. In the original state, the contact edges between minerals in the rock were straight or smooth, a small number of primary cracks were observed in the quartz and feldspar crystals, and intergranular cracks were also rare. After heating at 200 °C, a small number of cracks could be observed inside the plagioclase and quartz particles, and intergranular cracks were distributed at the boundaries of the plagioclase and quartz. Compared with those in the initial state, the morphology of each mineral did not change much, but some plagioclase particles were slightly fragmented. At 400 °C, fragmentation occurred at the contact points between various minerals in the rock, with a small number of cracks appearing inside the potassium feldspar and many light brown microcracks appearing inside the quartz particles. Compared with those of the samples at room temperature, the morphology of each mineral did not change much, with biotite showing a slight reddish-brown color, indicating oxidation and obvious fragmentation within each mineral. As the heating temperature increased to 600 °C, the fragmentation phenomenon at the contact points between minerals in the rock became more pronounced than that in the sample at 400 °C, and the number of microcracks in the feldspar minerals significantly increased. Black mica exhibits a more reddish-brown color and extensive oxidation. Due to the occurrence of quartz at the 573 °C α–β phase transition, at high temperatures, (β) the volume of the phase quartz particles increased and recovered after cooling to room temperature. (α) The process of the expansion and contraction of quartz led to an increase in the number of broken minerals at the boundaries of the quartz minerals, an increase in the crack width, and an increase in the number of intergranular and intragranular cracks. After heating at 800 °C, the morphology of the various minerals changed significantly, with a large number of intergranular and intragranular cracks appearing. The fragmentation phenomenon of the internal and contact parts between minerals was more obvious. After high-temperature treatment at 1000 °C, the surfaces of the potassium feldspar, plagioclase feldspar, and quartz all became blurred, indicating strong weathering. Black mica became redder and had a greater degree of oxidation than 800 °C, while the degree of fracture of black mica was also greater. The various mineral components in the rock exhibited a large number of intragranular cracks, which were interconnected. The width of the intergranular cracks reached the maximum among all temperature levels.

Figure 6. Microscopic images of the microstructure of granite after different temperature treatments under a polarizing microscope.

Figure 7 shows the optical microscopic images of granite samples treated at different temperatures at the same ratio, which clearly reveal the development of cracks within the rock. In the initial state, the granite had fewer primary cracks, and the surface of the mineral particles was relatively smooth. When the temperature rose to 200 °C, the granite underwent thermal damage, leading to an increase in internal cracks. However, due to an increase in the mineral volume, the primary microcracks were close to a certain extent, and their width significantly decreased. At 400 °C, the width of the intergranular cracks significantly increased, and as the temperature increased, the cracks developed and extended. Compared to those at lower temperatures, the lengths of large cracks also increased, and the number of small cracks within the crystal increased. After heating at a high temperature of 600 °C, there was a significant difference in the microstructure of the granite under the light microscope compared to the morphology at temperatures below 400 °C, with a large increase in microcracks. Wide intergranular cracks developed and connected with each other, and broken particles appeared between the minerals. A large number of microcracks within each mineral particle were derived and developed, and transgranular cracks appeared for the first time. At 800 °C, the number of cracks significantly increased, the width widened, and the length increased. The internal cracks in the rock sample were interconnected and formed a crack network. Moreover, the small particles produced by the squeezing and crushing of minerals increased relative to those produced by lower temperature treatments. As the processing temperature increased to 1000 °C, the internal cracks in the rock sample became very dense, and the cracks became interconnected, forming a dense network of cracks. Mineral crystals were cut into small particles by these cracks. The phenomenon of particle fragmentation between minerals and within minerals was widely distributed, which can explain the significant decrease in the macroscopic mechanical properties of granite at 1000 °C.

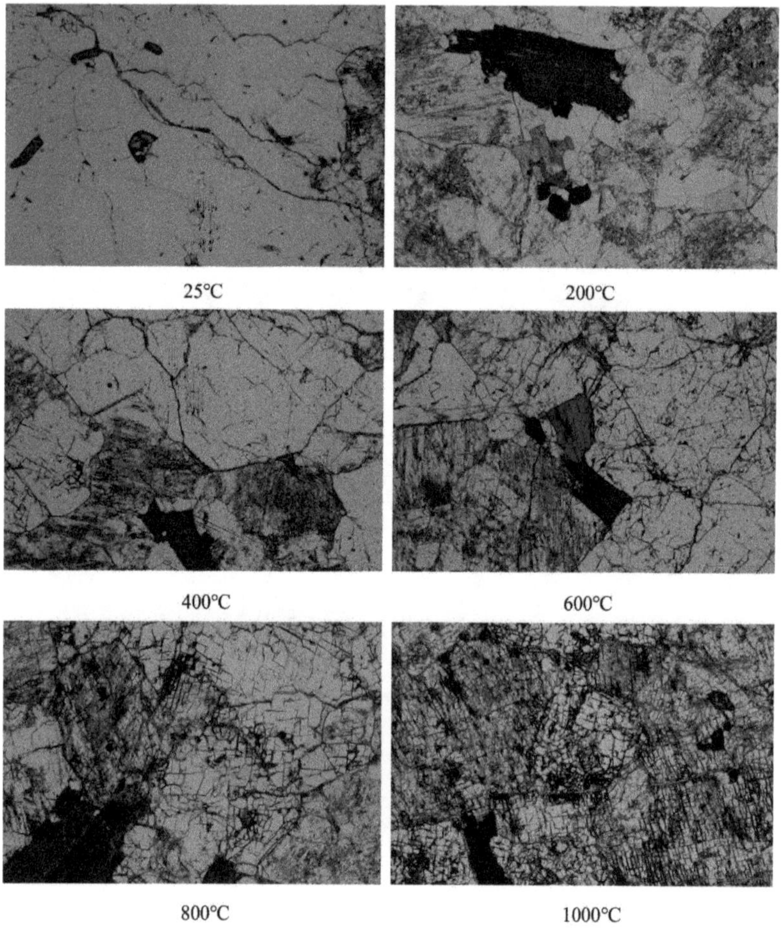

Figure 7. Crack photographs of granite sample slices after different temperature treatments.

3.3. *Changes in Longitudinal and Transverse Wave Velocities*

The effects of temperature on the longitudinal and transverse wave velocities of granite are shown in Figures 8 and 9, respectively. At temperatures ranging from 200 to 400 °C, the longitudinal wave velocity approximately linearly decreased, and the rate of decrease in the longitudinal wave velocity was significantly greater than that in the transverse wave velocity. At 400 °C, the longitudinal wave velocity of granite decreased to 2738 m/s, which was only 61% of that at room temperature. Although the transverse wave velocity decreased with an increasing temperature, it could still reach 76% of that at 400 °C. At 600 °C, the longitudinal wave velocity continued to decrease rapidly, with the average longitudinal wave velocity dropping to 1312 m/s, which was only 29% of that at room temperature. The transverse wave velocity decreased significantly faster than that at 400 °C, dropping to 42% at room temperature. As the temperature continued to increase, the rate of decrease in the wave velocity slowed. At 800 °C, the longitudinal wave velocity decreased to 1009 m/s, and the transverse wave velocity decreased to 846 m/s. When the heating temperature reached 1000 °C, the ultrasonic signal became extremely chaotic and unstable, and the wave velocity could not be read anymore. There was a significant change in the wave velocity of the granite before and after 400 °C, and 400 °C can be regarded as the threshold temperature for thermal damage to the granite wave velocity.

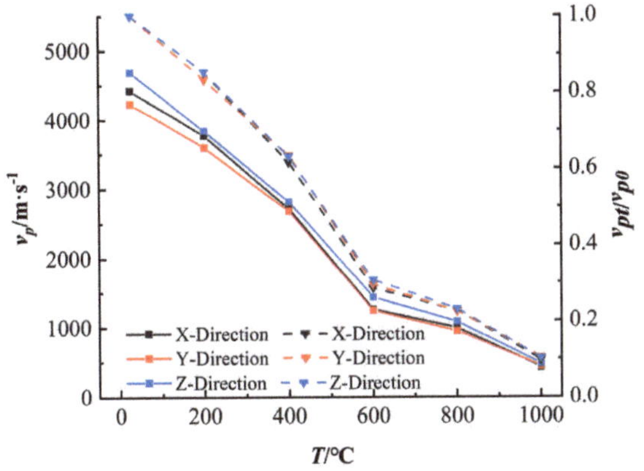

Figure 8. Temperature-dependent variation in the longitudinal wave velocity of granite in different sampling directions.

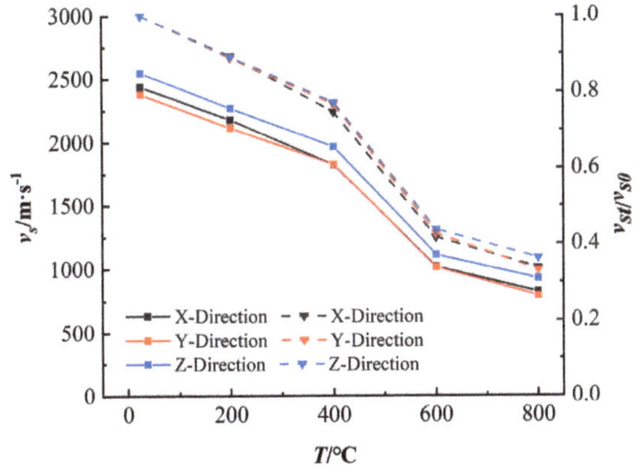

Figure 9. Variation in transverse wave velocity of granite with different temperatures in different sampling directions.

The law of the variation in wave velocity with different temperatures in the three orthogonal directions of granite can be summarized as follows: at room temperature, the Z-direction wave velocity was the highest, and the Y-direction wave velocity was the lowest. As the heating temperature increased, the longitudinal wave velocity of granite, in the first three sampling directions before 400 °C, decreased rapidly, and the Z direction had a faster deceleration rate, while the X direction and Y directions had a slower deceleration rate. The difference in the longitudinal wave velocity of granite in the three directions decreased rapidly, and the difference in the longitudinal wave velocity was the smallest at 400 °C. For the shear wave velocity, at 200 °C, the reduction rate of granite in the three directions was basically the same and still significantly different. At 400 °C, the reduction rate of the granite shear wave velocity in the X direction was significantly greater than that in the other two directions, rapidly decreasing to the same level as the shear wave velocity in the Y direction. When the temperature exceeded 400 °C, the variation patterns of the longitudinal and transverse wave velocities of granite in the three directions were basically the same. The wave velocities of granite in the X and Y directions tended to be consistent and significantly lower than those of granite in the Z direction. When the heating temperature reached 800 °C, the relationship between the longitudinal wave velocities of granite in the three directions was the same as that for normal temperatures, but the difference was not significant, while the transverse wave velocities of granite in the Z direction were still significantly greater than those in the other two directions.

In summary, there was a significant difference in the influence of temperature on the longitudinal and transverse wave velocities of granite. At the heating temperatures of 200–400 °C, the longitudinal wave velocity approximately linearly decreased, and the rate of decrease in the longitudinal wave velocity was significantly greater than that in the transverse wave velocity. At 600 °C, the longitudinal wave velocity continued to decrease rapidly, while the transverse wave velocity decreased significantly faster than at 400 °C. As the temperature continued to rise, the rate of the wave velocity decrease slowed down. There was a significant change in the wave velocity of granite before and after 400 °C.

3.4. *Changes in Uniaxial Compressive Strength and Elastic Modulus*

Figure 10 illustrates the relationship between the uniaxial compressive strength of granite and temperature variation. As shown in the Figure 10, the uniaxial compressive strength of granite decreased with an increasing temperature, but the uniaxial compressive strength of granite sampled in the different directions shows different trends: At room temperature, the Z-direction granite had the highest strength, followed by the X-direction granite, and the Y-direction granite had the lowest strength. As the temperature increased from room temperature to 400 °C, the strength of the granite decreased. The difference in the uniaxial compressive strength of granite in the three directions rapidly decreased within this temperature range, and the strength of granite in the Z direction was slightly greater than that in the X and Y directions. When the temperature increased to 600 °C, the strength of granite in the three directions decreased rapidly, and the strength of granite in the Y direction was relatively lower than that in the other two directions. As the temperature continued to rise, after the high temperature increased to 800 °C and 1000 °C, the strength differences in the three directions gradually disappeared and tended to be consistent.

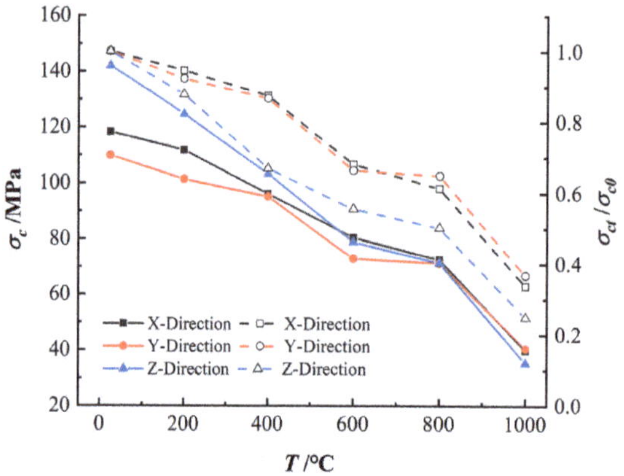

Figure 10. Temperature-dependent variation in the uniaxial compressive strength of granite under different sampling directions.

For samples obtained from different sampling directions, within the temperature range of room temperature to 600 °C, the changes in the strength of granite in the X and Y directions were relatively similar, both showing slight decreases in the uniaxial compressive strength after being subjected to temperatures below 400 °C. At 400 °C, the strength in the X and Y directions decreased by 12.62% and 13.32%, respectively, compared to that in the room-temperature state. As the temperature rose to 600 °C, the strength in the X and Y directions decreased by 31.89% and 33.57%, respectively, compared to that in the room-temperature state. However, the variation trend of granite sampled in the Z direction within this temperature range significantly differed. As the temperature increased, the uniaxial compressive strength of granite in the Z direction approximately decreased linearly, and the rate of decrease was much greater than that in the other two sampling directions. At 600 °C, the uniaxial compressive strength of granite in the Z direction decreased by 45.38% compared to that at room temperature. When the heating temperature was higher than 600 °C, the difference in the trend of the uniaxial compressive strength of granite in the three sampling directions gradually decreased, and the rate of strength reduction slowed down in the range of 600–800 °C. At 1000 °C, the strength of granite rapidly decayed to the minimum value, and the uniaxial compressive strength of granite in the X, Y, and Z directions decreased by 66.24%, 63.14%, and 75.08%, respectively, compared to that at room temperature.

Figure 11 shows the relationship between the elastic modulus of granite and temperature variation. The elastic modulus of granite in the X direction was 25.74 GPa at room temperature, which slightly decreased before 400 °C and decreased to 23.14 GPa at 400 °C for a decrease of 10.15% compared to room temperature. When the temperature was above 400 °C, the elastic modulus decreased almost linearly and dropped to 4.75 GPa at 1000 °C, a decrease of 81.54% compared to room temperature. The elastic modulus of granite in the Y direction at room temperature was 23.30 GPa, which was slightly lower than that in the X direction. As the temperature increased, the rate of decrease in the elastic modulus gradually accelerated. At 400 °C, the elastic modulus of granite in the Y direction was 20.03 GPa, which was 14.06% lower than that at room temperature. When the temperature increased to 800 °C, the rate of decrease in the elastic modulus slowed down compared to that at 600 °C. When the heating temperature exceeded 800 °C, the elastic modulus rapidly decreased again. At 1000 °C, the elastic modulus of the granite decreased to 4.62 GPa, which was 80.26% lower than that at room temperature. The elastic modulus of granite in the Z direction at room temperature was 25.88 GPa, which was the maximum value in the three sampling directions. As the temperature increased, the elastic modulus of the granite increased slightly at 200 °C. Within the temperature range of 200–600 °C, the elastic modulus decreased approximately linearly, with the fastest rate of decrease among the three directions. Compared to those at room temperature, the reduction amplitudes at 400 °C and 600 °C were 18.85% and 39.67%, respectively, which were significantly greater than those in the other two directions. As the temperature continued to increase, the variation in the elastic modulus with different temperatures was approximately in the Y direction. At 1000 °C, the elastic modulus decreased to 4.16 GPa, a decrease of 83.94% compared to room temperature.

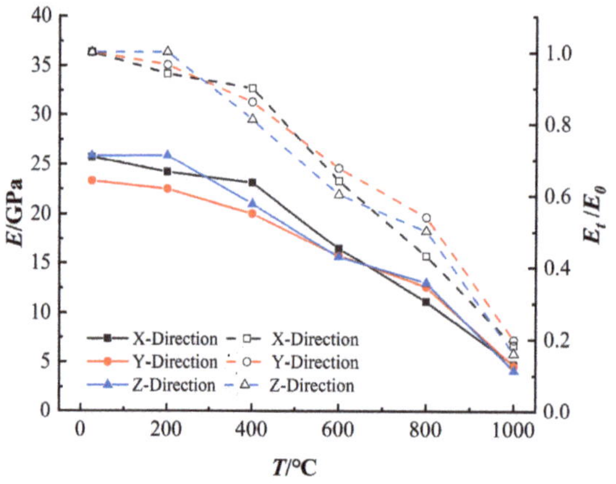

Figure 11. Temperature variation pattern of granite E in different sampling directions.

3.5. Changes in Cohesion and Internal Friction Angle

Cohesion and the internal friction angle are parameters of rock shear strength, where the internal friction angle reflects the friction characteristics of a rock. The main influencing factors include the material, mineral composition, porosity, etc., of the rock, while cohesion is an indicator of the internal bonding strength of a rock and is related to the mineral particle size, bonding composition, and bonding degree.

Figure 12 shows the variation trend of the internal friction angle of the rock samples in the different sampling directions with an increasing temperature, as well as the normalized internal friction angle with temperature variation. As shown in Figure 12, the internal friction angle of the rock samples in all three directions showed an overall upwards trend with an increasing temperature, indicating that a high temperature slightly increased the internal friction angle of the granite. Under normal temperature conditions, the internal friction angles of the granite in the three directions were arranged in descending order: Y direction > X direction > Z direction. As the temperature increased to 800 °C, the internal friction angle of the X-direction rock sample first increased, then decreased, and then increased again, while the internal friction angle of the Y-direction rock sample first slightly decreased. After 400 °C, the internal friction angle of the Z-direction rock sample increased at a relatively rapid rate, and it approximately linearly increased until it slightly decreased above 600 °C. When the temperature increased to 1000 °C, the internal friction angle of the rock samples in the three directions decreased to different degrees, but the values were still greater than those of the original state. Specifically, at 1000 °C, the feldspar melts formed a smooth glaze surface, resulting in a decrease in the friction coefficient and a decrease in friction between the mineral components. However, due to the high-temperature treatment, a large number of cracks were generated, minerals were broken, and small mineral particles filled the cracks. The friction characteristics of the particles were improved compared to those in the initial state. Therefore, compared with that at 800 °C, the internal friction angle at 1000 °C decreased but was still greater than that in the initial state.

Figure 13 shows the variation in cohesion with different temperatures in the rock samples from the different sampling directions, as well as the trend in cohesion with an increasing temperature after normalization. Although the cohesion of the rock samples from all three sampling directions decreased overall with an increasing temperature, there were differences in the variation patterns of the different temperature ranges. At 200 °C, the cohesion of the rock samples in all three sampling directions showed a decreasing trend, but the rate of decrease in the Y direction was significantly lower than that in the other two sampling directions. At 400 °C, the cohesion of the rock samples in the X direction slightly increased, while the cohesion of the rock samples in the other two directions continued to decrease. The decrease in cohesion of the rock samples in the Z direction was much greater in this temperature range than in the other two directions. As the temperature continued to increase, the cohesion of the rock samples in the three directions decreased steadily, and the rates of decrease were relatively close. The effects of temperature on the cohesion of the rock samples in the different sampling directions tended to be consistent. The normalized results reveal that the cohesion of the Z-direction rock sample was more sensitive to a high-temperature response than was that of the other two directions. As the temperature increased, the decrease in cohesion of the Z-direction rock sample relative to the initial state was significantly greater than that of the other two directions.

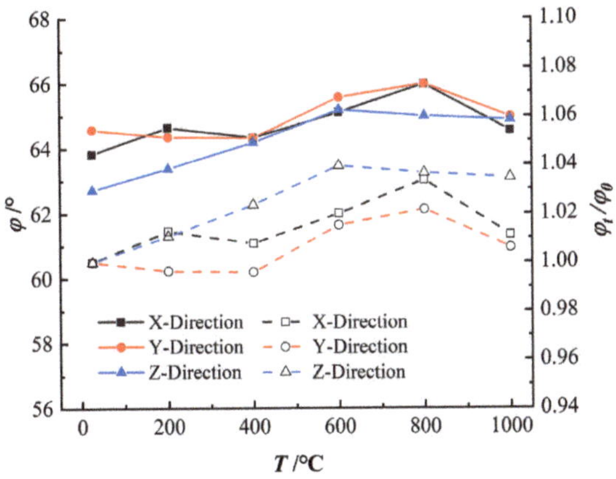

Figure 12. Temperature-dependent variation in the internal friction angle of granite in different sampling directions.

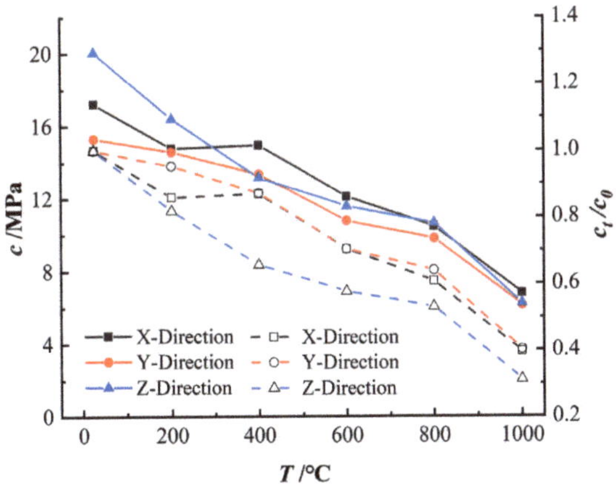

Figure 13. The variation law of granite cohesion with different temperatures in different sampling directions.

3.6. Anisotropic Variation

According to Formula (1), proposed by Birch et al. [29], the anisotropy indexes of rock physical and mechanical properties are obtained.

$$R_S = \frac{(S_{max} - S_{min})}{S_{mean}} \times 100\% \tag{1}$$

R_S is the anisotropy of a certain physical and mechanical index of granite, S_{max} is the maximum value of this index, S_{min} is the minimum value of this index, and S_{mean} is the average physical and mechanical index of rock samples in all sampling directions.

Figure 14 illustrates the variation of the granite wave velocity anisotropy with different temperatures. At a heating temperature of 200 °C, the granite's anisotropy decreased. This decrease in wave velocity was attributed to the removal of free water and weakly bound water from primary fractures within the granite. On the other hand, an increase in the temperature led to the expansion and mutual compression of mineral particles, which could cause some cracks to be compressed or even disappear. This phenomenon was random and irregular, resulting in a decrease in the anisotropy of the granite wave velocity at 200 °C. As the heating temperature increased, the anisotropy of the two types of rocks gradually increased. New microcracks were generated and propagated along the weakest surface of the mineral crystals and might have also connected to form large cracks. At 400 °C, the anisotropy of the granite wave velocity increased. At 600 °C, the anisotropy of the longitudinal wave velocity in the granite reached its maximum value, which was significantly greater than that at room temperature. After heating above 800 °C, the anisotropy of the longitudinal wave velocity in the granite decreased but was still greater than that in the normal temperature state. It can be inferred that temperatures above 400 °C can amplify the anisotropy of the longitudinal wave velocity in granite.

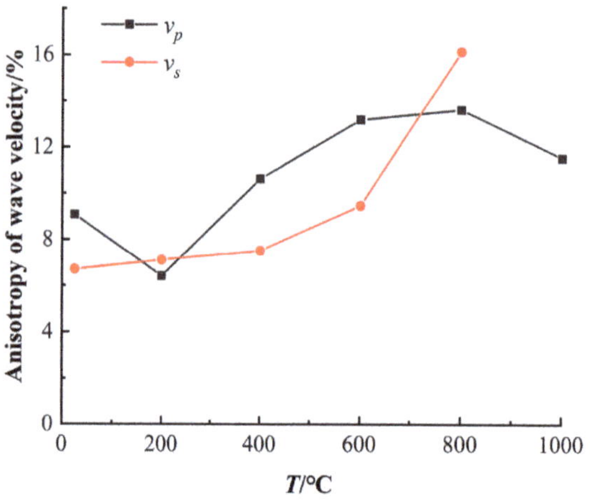

Figure 14. The variation law of wave velocity anisotropy with different temperatures.

The trend of the peak deviatoric stress anisotropy with respect to temperature under different confining pressure conditions is shown in Figure 15. Under uniaxial compression conditions, the peak deviatoric stress anisotropy first increased and then decreased with temperature. In the temperature range of 200~600 °C, the anisotropy of the peak deviatoric stress increased compared to that at room temperature, indicating that within this temperature range, thermal damage within the granite leads to a more pronounced anisotropy of the peak stress in the granite. When the heating temperature exceeded 600 °C, the anisotropy of the peak deviatoric stress in the granite rapidly deteriorated, and the difference in the peak deviatoric stress in each sampling direction decreased. As the confining pressure increased, it could be observed that the confining pressure can significantly reduce the anisotropy of the granite peak deviatoric stress. At various temperature levels, the anisotropy of the granite did not change much compared to that at room temperature, indicating that under the combined influence of a confining pressure and temperature, the peak deviatoric stress of granite can still show a certain degree of anisotropy.

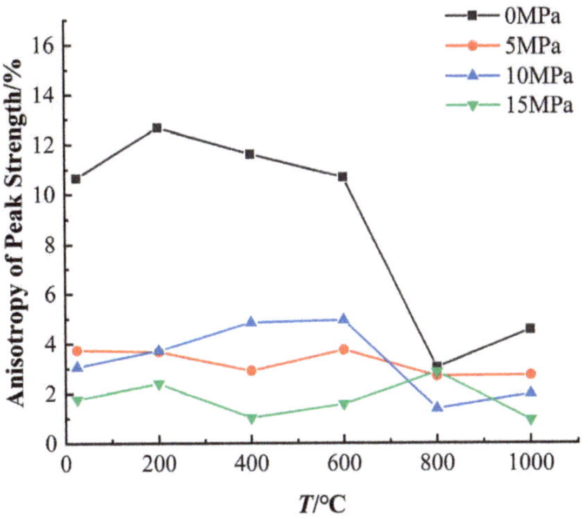

Figure 15. Changes in peak stress anisotropy with different temperatures.

Figure 16 shows the variation in the anisotropy of the elastic modulus with different temperatures. Under various confining pressure conditions, the anisotropy of the granite elastic modulus remained basically unchanged with different temperatures, and the anisotropy of the elastic modulus first increased and then decreased with an increasing temperature. The degree of anisotropy reached its maximum value at 400 °C, and as the temperature continued to rise, the degree of anisotropy began to decrease. As the confining pressure increased, the degree of anisotropy gradually decreased, indicating that a confining pressure can constrain the expression of anisotropy in the elastic modulus of granite. Under uniaxial conditions and a confining pressure of 5 MPa, there was little difference in the anisotropy of the elastic modulus between 200 °C and 400 °C. At the confining pressures of 10 MPa and 15 MPa, the anisotropy of the elastic modulus increased at 400 °C, compared to that at 200 °C. This phenomenon indicates that the confining pressure has a slight enhancing effect on the anisotropy in the temperature range of 200~400 °C.

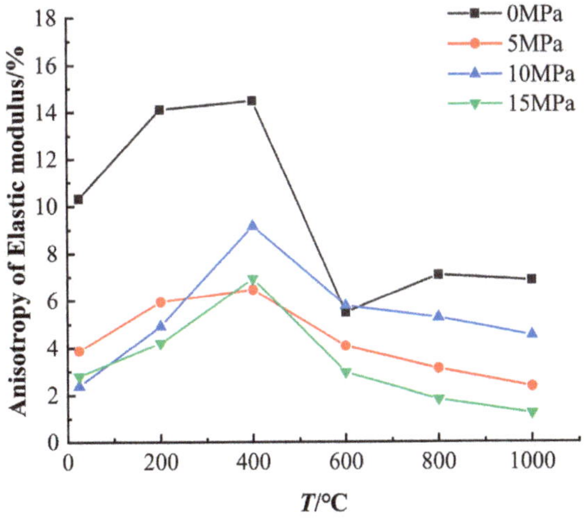

Figure 16. Temperature-dependent variation in the anisotropy of the elastic modulus.

3.7. Destructive Forms

The failure mode of granite is influenced by three factors: temperature, sampling direction, and confining pressure. Figure 17 shows photos of the failure modes of the granite in the three sampling directions under different temperatures and confining pressures. As the temperature and confining pressure increased, in the same sampling direction, the failure mode of the rock samples obviously changed from brittle failure to ductile failure. The failure mode of the granite under uniaxial conditions was mostly tensile splitting failure, while under triaxial compression conditions, the failure mode was shear failure. Although the failure modes of the rock samples in the different sampling directions were the same, there were certain differences in the morphology of the fracture surface.

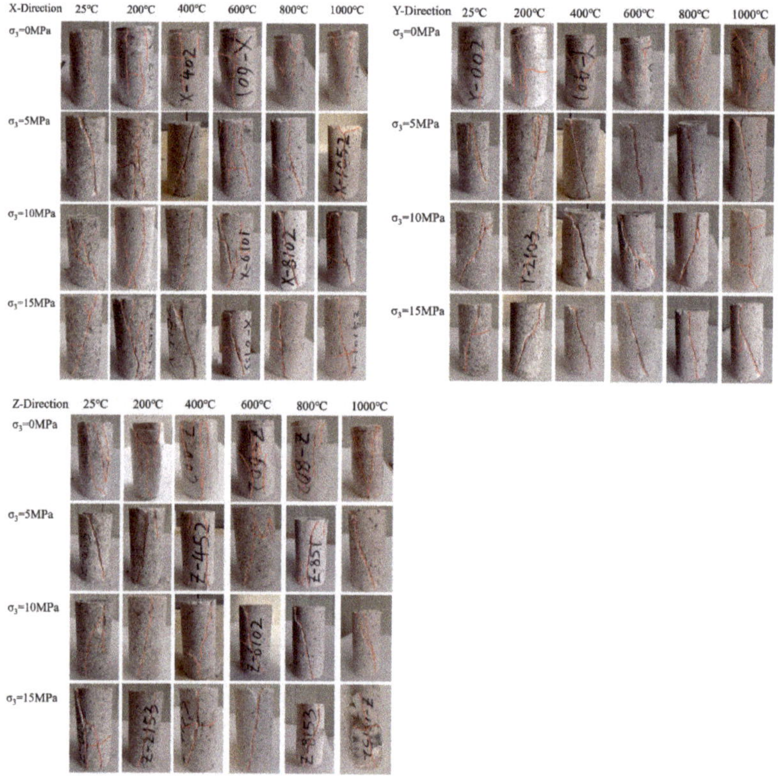

Figure 17. Failure morphology of granite in different sampling directions.

Under uniaxial compression conditions, the initial failure mode of the granite at room temperature was typical axial splitting failure.

The rock sample in the X direction underwent tensile failure along the axial direction, with the failure surface penetrating through the rock sample and no further cracks extending. As the heating temperature increased, the failure mode of the X-direction rock sample remained unchanged from 200 °C to 600 °C, but a high temperature caused the internal particles of the rock sample to become loose, and the number of local splitting surfaces that developed along the axis increased. At the same time, directional cracks that developed along the radial weak plane, observed in the photos of the rock samples heated at 200–600 °C, could also form radial cracks along the weak plane when the sample was damaged. When the heating temperature exceeded 600 °C, local shear failure was observed at the end of the granite in the X direction. Due to the relatively loose internal structure of the rock sample, lateral bulging was more obvious.

At 200 °C, local cracks along the axial direction appeared near the main fracture surface of the Y-direction rock sample, and there were horizontal cracks generated by the directional development of the original cracks. At 400 °C, due to the friction between the end face of the rock sample and the pressure plate, the failure mode exhibited a fractured cone at the upper end and axial splitting failure at the lower end of the cone. When the heating temperature was higher than 400 °C, the failure mode of rock samples in the Y direction changed to a combination of shear failure and tensile fracture failure. With an increasing heating temperature, additional small axial cracks were observed in the local area, and the lateral bulging phenomenon became more obvious.

The Z-direction rock sample also exhibited tensile splitting failure through the rock sample at heating temperatures of less than 400 °C. However, compared to those in the other two sampling directions, more macroscopic cracks along the axial direction appeared during the Z-direction rock sample failure, which was a manifestation of the directional development of cracks caused by thermal damage at the macroscopic level. As the temperature increased, local shear failure occurred in the Z direction of the rock samples at 600 °C and 800 °C but was still dominated by splitting failure, with a large number of axial cracks appearing. At 1000 °C, the failure mode of the rock samples in the Z direction completely changed to shear failure.

Under triaxial compression conditions, the failure mode of the granite in the different sampling directions was mainly shear failure, and the temperature and confining pressure were the main factors affecting the failure mode. At room temperature, as the confining pressure increased, the granite rock sample bore greater axial pressure, more cracks developed, and the failure surface morphology of the rock sample changed from simple to complex. At 5 MPa, the granite usually had a relatively obvious shear failure surface only, while at 10 MPa and 15 MPa, more derived cracks could be observed near the main failure surface. At 200 °C and 400 °C, as the confining pressure increased, the width of the main failure surface increased. This is because at this temperature stage, the cracks generated by granite thermal damage were mainly intergranular cracks. The confining pressure increased the frictional force between the mineral particles. When shear cracks appeared, the influence of the mineral particles on other surrounding mineral particles increased, leading to an increase in the crack width. When the heating temperature was higher than 400 °C, a large number of transgranular cracks were generated during the thermal damage of the granite at this temperature stage. The strength of the mineral particles decreased after experiencing a high temperature, and a large number of transgranular cracks were connected under high shear stress, forming a macroscopic shear failure surface. The mineral particles broke into a powder form, and the shear surface was smoother and more complete than that of the rock samples below 400 °C. As the confining pressure increased, the constraint force on the mineral particles increased, and the degree of crack closure in the rock sample increased. Therefore, at the same heating temperature, rock samples with higher confining pressures are more complete when damaged. Usually, only one main failure surface appears, and the development of local short cracks is significantly reduced.

4. Discussion

The relationships between the three sampling directions and the dominant development direction of cracks were analyzed through crack quantification. The quantitative statistical method for determining crack density refers to Gao et al. [30]. The specific steps are as follows: (1) To present the cracks in the sample after heat treatment more clearly, the cracks in the light microscopic images were extracted and reconstructed. (2) A circular analysis area was selected from the microscopic crack map, a reasonable reference line was used to determine the density, and the microscopic cracks were divided. (3) The microcrack density of the thermally cracked rocks was quantified using the crack density statistical method.

In this article, we selected a circular analysis area with a diameter of 2.04 mm to analyze the distribution characteristics of microcracks in the entire cross-section of the sample as much as possible. According to the microscopic crack density statistical method, we set 13 reference lines, numbered 1–13, with a spacing of 150, as shown in Figure 18 (μm). The length of each reference line is shown in Table 1.

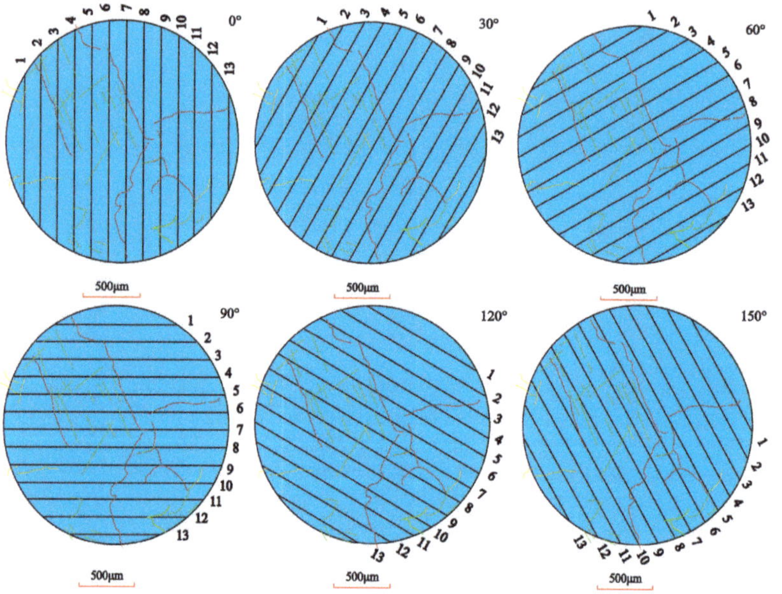

Figure 18. Schematic diagram of the reference line setting and statistical methods.

Table 1. Reference line length.

Reference Line Number	1	2	3	4	5	6	7	8	9	10	11	12	13
Length/mm	0.83	1.31	1.60	1.79	1.93	2.02	2.04	2.04	1.98	1.88	1.71	1.45	1.11

The number of intersection points between each reference line and the microscopic cracks were counted, and the points were divided by the length of each reference line to obtain the density of microscopic cracks on each reference line. The calculation formula for the microscopic crack density *LMD* is as follows:

$$LMD = \frac{N}{L} \tag{2}$$

In this formula, N is the number of microcracks on the reference line that intersect with the reference line, and L is the length of the reference line.

Using the vertical state of the reference line as the initial level, the line rotates clockwise, and the nonuniformity of the distribution of microcracks at 0°, 30°, 60°, 90°, 120°, and 150° is studied.

According to the above method, the crack distribution status of the rock samples from the three sampling directions was analyzed after different temperature treatments, and the average microcrack density of 13 reference lines was calculated at the 0°, 30°, 60°, 90°, 120°, and 150° reference line angles. When drawing, 30° is taken as the interval, and the reference line angle range is 0° to 180°, which is mapped to the range of 180° to 360°. A polar coordinate graph of the reference line angle microscopic crack density was drawn, as shown in Figure 18.

Figure 19 shows the distribution of the crack density on the granite cross-sections in the three sampling directions under room-temperature conditions. The rock samples were sampled along the X and Y directions. In the initial state, the distribution of primary cracks on the cross-section of the rock sample was significantly greater in one direction than in other directions. When turning from the angle with the lowest crack density to the angle with the highest crack density, the crack density showed an increasing trend from small to large; this also indicates that there was an advantageous direction of crack distribution on the cross-section of the rock samples in the X and Y directions. However, there was no obvious dominant distribution direction of the crack density on the cross-section of the rock samples sampled along the Z direction, and when rotating from the direction of minimum crack density to the direction of maximum crack density, the crack density did not follow the law of numerical increase. The distribution of the crack density was relatively chaotic, and no obvious trends were observed. Based on these findings, we can determine that the X and Y directions are perpendicular to the dominant crack development direction, while the Z direction is parallel to the dominant crack development direction.

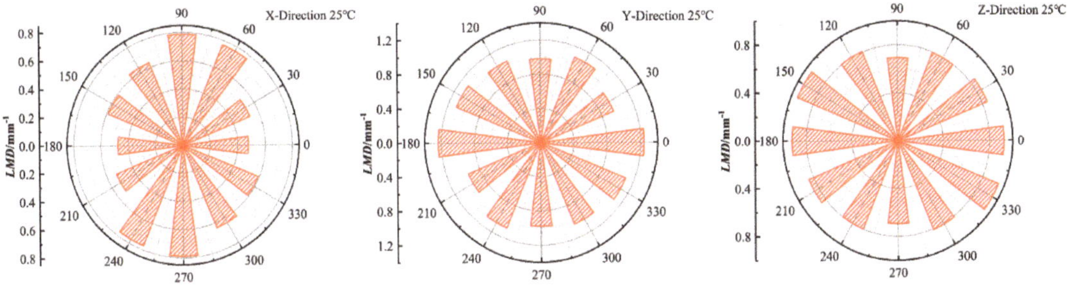

Figure 19. Distributional pattern of cracks on the cross-section of rock samples in different sampling directions at room temperature.

Figures 20 and 21 show the distributional patterns of cracks on the cross-sections of the granite samples taken in the three sampling directions after heating at 400 °C and 800 °C, respectively. At 400 °C, cracks in the cross-sections of the granite samples taken along the X and Y directions developed extensively along the preferred direction, and the density of cracks showed a monotonically increasing trend from small to large, indicating that hot cracks are derived and developed along a certain direction at this temperature. After heating to 800 °C, as a large number of microscopic cracks were generated, developed, and penetrated, the dominant direction of crack development became unclear, and the density of cracks in most directions was high. Thermal damage weakened the anisotropy of the crack distribution in the X and Y directions of the rock samples. For the rock samples sampled along the Z direction, as the heating temperature increased to 400 °C, the crack density increased. However, the pattern of crack-density increase with respect to the angle was relatively chaotic, with the highest crack density occurring in the 30° and 90° directions and the lowest crack density occurring in the 0° and 120° directions. There was no obvious pattern in the distribution of crack density. When the temperature increased to 600 °C and 800 °C, severe physical and chemical changes occurred inside the rock sample, and a large number of microscopic cracks were generated and interconnected to form a crack network. Therefore, the numerical value of the crack density changes relatively little with the reference line angle at lower temperatures, and there is no obvious direction for crack development.

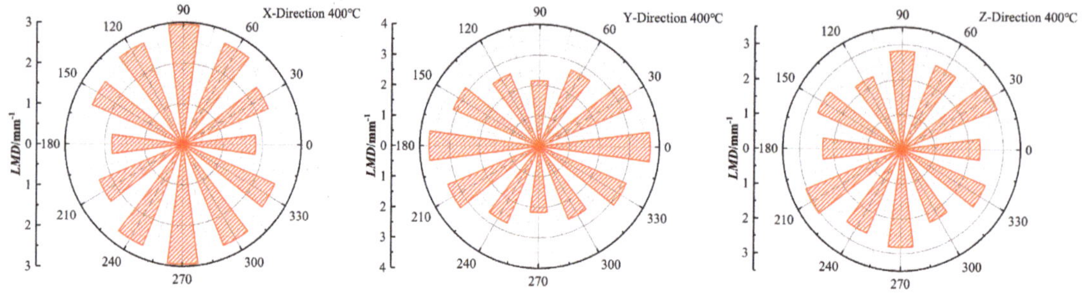

Figure 20. Distributional pattern of cracks on the cross-section of rock samples in different sampling directions above 400 °C.

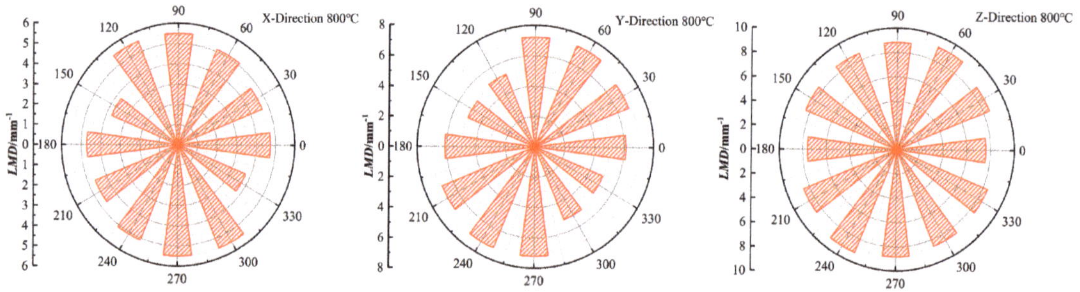

Figure 21. Distributional pattern of cracks on the cross-section of rock samples in different sampling directions above 800 °C.

Through the quantification of thermally induced microcracks and an analysis of the anisotropy of microcrack distribution, it can be seen that the changes in the macroscopic physical and mechanical properties of granite are closely related to the development of microcracks. The strength and deformation characteristics of granite exhibit obvious anisotropy, and the changes in the mechanical indicators of the rock samples with different sampling directions and temperatures reflect the directional development of microcracks at the macroscopic level. At room temperature, the rock samples parallel to the dominant direction of crack development had the lowest crack density in the cross-section, thus exhibiting the best mechanical properties. As the temperature increased to 400 °C, a large number of thermal cracks generated inside the granite developed in the dominant direction, leading to a significant increase in the anisotropy of the physical and mechanical properties of the granite. Under uniaxial compression conditions, the rock samples taken perpendicular to the dominant direction of the cracks produced radial cracks near the main failure surface, while the rock samples taken parallel to the dominant direction of the cracks produced more axial cracks. When the temperature increased to 800 °C and 1000 °C, due to the extensive development and interconnected connection of thermal cracks, microcracks in all directions were extensively developed, and the distribution of microcracks tended to be uniform, which greatly weakened the anisotropy of the physical and mechanical properties of the granite.

5. Conclusions

After subjecting granite samples to high-temperature treatment, a series of physical and mechanical tests were conducted along various sampling directions. This study focused on analyzing the impact of anisotropy on alterations in the physical and mechanical properties of the samples. The findings lead to the following conclusions:

(1) Temperature can cause significant changes in the appearance and morphology of granite, with a color transition from gray-white to white. Temperatures above 800 °C can cause granite to transform into flesh red. High temperatures above 400 °C can amplify the anisotropy of longitudinal and transverse wave velocities in granite. A temperature of 400 °C is the threshold temperature at which the physical properties of granite change with temperature. Quartz undergoes a phase transition at 573 °C, resulting in significant changes in mechanical properties between 400 °C and 600 °C and a decrease in the peak strength and elastic model. Under the same temperature conditions, granite sampled parallel to the dominant direction of cracks had the best mechanical properties and was more sensitive to temperature.
(2) There are differences in the crack density on granite cross-sections in different sampling directions. As the temperature increases, the difference in the crack density increases. The development of thermal damage cracks in granite is advantageous. High-temperature actions can cause a large number of mineral microcracks to develop in granite. Temperatures below 400 °C increase the anisotropy of granite. After exposure to higher temperatures, the cracks penetrate each other, the crack density tends to be uniform, the anisotropy weakens, and the confining pressure constrains the mineral particles and compact cracks. The confining pressure will constrain the mineral particles and compact cracks, and anisotropy weakens with the increase in the confining pressure.
(3) With an increasing temperature and confining pressure, the failure characteristics of granite gradually transition from brittle to plastic failure. When subjected to uniaxial compression, granite is mostly subjected to tensile splitting failure. Rock samples taken perpendicular to the dominant direction of the crack can generate radial cracks near the main failure surface, while rock samples taken parallel to the dominant direction of the crack can generate more axial cracks. As the confining pressure increases, the number of small cracks decreases, and the rock failure mode changes to shear failure.

Author Contributions: Conceptualization, Y.Q. and N.X.; methodology, Y.Q. and L.W.; software, Y.Q. and L.W.; validation, Y.Q. and Q.W.; formal analysis, Y.Q. and Y.G.; investigation, L.W. and G.C.; resources, Y.Q. and N.X.; data curation, Y.Q., L.W. and W.Z.; writing—original draft preparation, Y.Q., Q.W. and Y.G.; writing—review and editing, Y.Q., G.C. and Y.G.; visualization, Y.Q. and L.W.; supervision, Y.Q. and N.X.; project administration, Y.Q.; funding acquisition, N.X. All authors have read and agreed to the published version of the manuscript.

Funding: This research was funded by the National Natural Science Foundation of China (NSFC), under grant no. 42302315 and no. 42230709, And Central Guidance for Local Scientific and Technological Development Funds (YDZJSX20231B016).

Institutional Review Board Statement: Not applicable.

Informed Consent Statement: Not applicable.

Data Availability Statement: Data available on request due to restrictions privacy or ethical.

Acknowledgments: Thank you to the anonymous reviewers for their valuable feedback on the manuscript. The authors thank the National Natural Science Foundation of China (grant no. 42302315 and no. 42230709) and the Central Guidance for Local Scientific and Technological Development Funds (YDZJSX20231B016).

Conflicts of Interest: Author Guanjun Cai was employed by the Beijing Jingneng Geological Engineering Company. The remaining authors declare that the research was conducted in the absence of any commercial or financial relationships that could be construed as a potential conflict of interest.

References

1. Li, C.; Hu, Y.; Meng, T.; Meng, P.; Zhao, Z.; Zhang, C. Experimental study of the influence of temperature and cooling method on mechanical properties of granite: Implication for geothermal mining. *Energy Sci. Eng.* **2020**, *8*, 1716–1728. [CrossRef]
2. Zhang, F.; Zhao, J.; Hu, D.; Skoczylas, F.; Shao, J. Laboratory investigation on physical and mechanical properties of granite after heating and water-cooling treatment. *Rock Mech. Rock Eng.* **2018**, *51*, 677–694. [CrossRef]
3. Wang, Z.H.; Su, T.; Konietzky, H.; Tan, Y.L.; Zhou, G.L. Hydraulic properties of Beishan granite after different high temperature treatments. *Bull. Eng. Geol. Environ.* **2021**, *80*, 2911–2923. [CrossRef]
4. Zuo, J.P.; Wang, J.T.; Sun, Y.J.; Chen, Y.; Jiang, G.H.; Li, Y.H. Effects of thermal treatment on fracture characteristics of granite from Beishan, a possible high-level radioactive waste disposal site in China. *Eng. Fract. Mech.* **2017**, *182*, 425–437. [CrossRef]
5. Dionísio, A.; Martinho, E.; Pozo-António, J.S.; Braga, M.S.; Mendes, M. Evaluation of combined effects of real-fire and natural environment in a building granite. *Constr. Build. Mater.* **2021**, *277*, 122327. [CrossRef]
6. Ozguven, A.; Ozcelik, Y. Effects of high temperature on physico-mechanical properties of Turkish natural building stones. *Eng. Geol.* **2014**, *183*, 127–136. [CrossRef]
7. Bi, J.; Liu, P.; Gan, F. Effects of the cooling treatment on the dynamic behavior of ordinary concrete exposed to high temperatures. *Constr. Build. Mater.* **2020**, *248*, 118688. [CrossRef]
8. Morrow, C.; Lockner, D.; Moore, D.; Byerlee, J. Permeability of granite in a temperature gradient. *J. Geophys. Res. Solid Earth* **1981**, *86*, 3002–3008. [CrossRef]
9. Chaki, S.; Takarli, M.; Agbodjan, W.P. Influence of thermal damage on physical properties of a granite rock: Porosity, permeability and ultrasonic wave evolutions. *Constr. Build. Mater.* **2008**, *22*, 1456–1461. [CrossRef]
10. Sun, Q.; Zhang, W.; Xue, L.; Zhang, Z.; Su, T. Thermal damage pattern and thresholds of granite. *Environ. Earth Sci.* **2015**, *74*, 2341–2349. [CrossRef]
11. Yang, S.Q.; Ranjith, P.G.; Jing, H.W.; Tian, W.L.; Ju, Y. An experimental investigation on thermal damage and failure mechanical behavior of granite after exposure to different high temperature treatments. *Geothermics* **2017**, *65*, 180–197. [CrossRef]
12. Zhu, S.; Zhang, W.; Sun, Q.; Deng, S.; Geng, J.; Li, C. Thermally induced variation of primary wave velocity in granite from Yantai: Experimental and modeling results. *Int. J. Therm. Sci.* **2017**, *114*, 320–326. [CrossRef]
13. Zhang, W.; Sun, Q.; Zhang, Y.; Xue, L.; Kong, F. Porosity and wave velocity evolution of granite after high-temperature treatment: A review. *Environ. Earth Sci.* **2018**, *77*, 350. [CrossRef]
14. Heuze, F.E. High-temperature mechanical, physical and thermal properties of granitic rocks—A review. *Int. J. Rock Mech. Min. Sci. Geomech. Abstr.* **1983**, *20*, 3–10. [CrossRef]
15. Liu, S.; Xu, J. An experimental study on the physico-mechanical properties of two post-high-temperature rocks. *Eng. Geol.* **2015**, *185*, 63–70. [CrossRef]
16. Zhao, Z.; Liu, Z.; Pu, H.; Li, X. Effect of thermal treatment on Brazilian tensile strength of granites with different grain size distributions. *Rock Mech. Rock Eng.* **2018**, *51*, 1293–1303. [CrossRef]
17. Chen, S.; Yang, C.; Wang, G. Evolution of thermal damage and permeability of Beishan granite. *Appl. Therm. Eng.* **2017**, *110*, 1533–1542. [CrossRef]
18. Ma, T.; Zhu, G.; Peng, N.; Qiu, Y.; Liu, Y.; Zou, J. Physical-mechanical properties and thermal-induced damage of granite after high-temperature pretreatment. *Arab. J. Geosci.* **2021**, *14*, 1449. [CrossRef]
19. Kumari, W.G.P.; Ranjith, P.G.; Perera, M.S.A.; Chen, B.K. Experimental investigation of quenching effect on mechanical, microstructural and flow characteristics of reservoir rocks: Thermal stimulation method for geothermal energy extraction. *J. Pet. Sci. Eng.* **2018**, *162*, 419–433. [CrossRef]
20. Xu, X.L.; Karakus, M. A coupled thermo-mechanical damage model for granite. *Int. J. Rock Mech. Min. Sci.* **2018**, *103*, 195–204. [CrossRef]
21. Shang, X.; Zhang, Z.; Xu, X.; Liu, T.; Xing, Y. Mineral composition, pore structure, and mechanical characteristics of pyroxene granite exposed to heat treatments. *Minerals* **2019**, *9*, 553. [CrossRef]
22. Chen, Y.L.; Ni, J.; Shao, W.; Azzam, R. Experimental study on the influence of temperature on the mechanical properties of granite under uni-axial compression and fatigue loading. *Int. J. Rock Mech. Min. Sci.* **2012**, *56*, 62–66. [CrossRef]
23. Zhu, Z.; Tian, H.; Mei, G.; Jiang, G.; Dou, B.; Xiao, P. Experimental investigation on mechanical behaviors of Nanan granite after thermal treatment under conventional triaxial compression. *Environ. Earth Sci.* **2021**, *80*, 46. [CrossRef]
24. Alneasan, M.; Alzo'ubi, A.K. Temperature Effect on the Fracture Behavior of Granite Under Three Loading Modes (I, I/II, and II). Rock Mech. *Rock Eng.* **2023**, *56*, 2197–2211. [CrossRef]
25. Wang, Z.; Zhou, G.; Ge, X. Experimental study on damage characteristics of Beishan granite under single loading and multiple loading with AE techniques. *Sci. Rep.* **2023**, *13*, 8767. [CrossRef] [PubMed]
26. Zheng, Y.; Zhang, L.; Wu, P.; Guo, X.; Li, M.; Zhu, F. Physical and Mechanical Properties and Damage Mechanism of Sandstone at High Temperatures. *Appl. Sci.* **2024**, *14*, 444. [CrossRef]
27. Fan, L.F.; Wu, Z.J.; Wan, Z.; Gao, J.W. Experimental investigation of thermal effects on dynamic behavior of granite. *Appl. Therm. Eng.* **2017**, *125*, 94–103. [CrossRef]
28. He, L.; Yin, Q.; Jing, H. Laboratory Investigation of Granite Permeability after High-Temperature Exposure. *Processes* **2018**, *6*, 36. [CrossRef]

29. Birch, F. The velocity of compressional waves in rocks to 10 kilobars, part 2. *Elastic Prop. Equ. State* **1988**, *26*, 91–116.
30. Gao, J.; Xi, Y.; Fan, L.; Du, X. Real-time visual analysis of the microcracking behavior of thermally damaged granite under uniaxial loading. *Rock Mech. Rock Eng.* **2021**, *54*, 6549–6564. [CrossRef]

applied sciences

Article

Influence of Mixing Rubber Fibers on the Mechanical Properties of Expansive Clay under Freeze–Thaw Cycles

Zhongnian Yang [1,*], Zhenxing Sun [1], Xianzhang Ling [1,2], Guojun Cai [3], Rongchang Wang [1] and Xiang Meng [1]

1 School of Civil Engineering, Qingdao University of Technology, Qingdao 266520, China; sun.zhenxing@outlook.com (Z.S.); lingxianzhang@qut.edu.cn (X.L.); wang.rongchang@outlook.com (R.W.); mengxiangoan@outlook.com (X.M.)
2 School of Civil Engineering, Harbin Institute of Technology, Harbin 150006, China
3 College of Civil Engineering, Anhui Jianzhu University, Hefei 230009, China; focuscai@163.com
* Correspondence: yzhnqd@qut.edu.cn

Abstract: With the growth of the transportation industry, large volumes of waste tires are being generated, which necessitates the development of effective solutions for recycling waste tires. In this study, expansive clay was mixed with rubber fibers obtained from waste tires. Triaxial tests were conducted on the rubber fiber-reinforced expansive clay after freeze–thaw cycles. The experimental results of the unreinforced expansive clay from previous studies were used to evaluate the effect of mixing rubber fibers on the mechanical properties of rubber fiber-reinforced expansive clay under freeze–thaw cycles. The results demonstrate that the mixing of rubber fibers significantly reduces the effect of freeze–thaw cycles on the shear strength and elastic modulus of expansive clay. The shear strength and elastic modulus of the unreinforced expansive clay decrease markedly as the number of freeze–thaw cycles increases, while the shear strength and elastic modulus of the rubber fiber-reinforced expansive clay do not exhibit any remarkable change. A calculation model of the deviatoric stress–axial strain curves after freeze–thaw cycles was established. The model describes the deviatoric stress–axial strain behavior of rubber fiber-reinforced expansive clay and unreinforced expansive clay under different confining pressures and different numbers of freeze–thaw cycles.

Keywords: freeze–thaw cycles; mechanical properties; rubber fiber-reinforced soil; expansive clay; waste tires

1. Introduction

Automobile production has grown rapidly, owing to the expansion of the transportation industry [1]. Although automobiles have made our lives more convenient, they have also negatively affected the environment. One aspect of their environmental impact is the increasing amounts of waste tires, which is becoming an urgent environmental issue. According to statistical data, the annual global production of waste tires exceeds 10 million tons [2]. Currently, the primary techniques for disposing waste tires are underground landfills, centralized storage, heat recovery for energy production, and pyrolysis for other products [3–8]. Tire dumps require large amounts of land and are prone to catching fire, releasing significant levels of particulate matter (PM) into air during incineration [9–13]. Underground landfills are commonly used to dispose of waste tires. However, waste tires buried underground do not degrade even after many years and produce toxic substances that pollute underground water [14]. The European Union has enacted a law banning underground landfills for waste tires [15]. The pyrolysis of waste tires has been extensively studied in recent years, with various pyrolysis techniques being used to create useful products [16–19]. Rubber makes up approximately half of waste tire components; therefore, recycling rubber from waste tires has significant benefits, and an appropriate recycling method is urgently needed [14].

Numerous studies have been conducted to improve soil strength using rubber obtained from waste tires. The basic approach involves grinding waste tires into rubber granules and fibers, or cutting them into rubber strips and mixing them into the soil, which makes the soil valuable for engineering applications [14]. Studies on the influence of rubber granules on soil shear strength have demonstrated that mixing rubber granules with soil can increase the shear strength [20,21]. The addition of rubber granules mostly improved the shear strength by changing the internal friction angle of the soil [22,23]. According to Zhang et al. [24], mixing rubber strips with sand increases the internal friction angle by 68.1%. When the soil was subjected to dynamic loads, the addition of rubber enhanced its dynamic shear modulus, dynamic elastic modulus, and damping ratio [25]. In summary, mixing rubber with soil under certain conditions can change soil properties and enhance its value for engineering applications. However, approximately 23% of all land on Earth is covered with permafrost or seasonally frozen soil, where freeze–thaw cycling in seasonally frozen soil significantly affects the behavior of the soil and decreases its shear strength [26–28]. Studies have shown that freeze–thaw cycles reduce the strength of unreinforced soils, but there is no consensus on the effect of freeze–thaw cycles on the mechanical properties of rubber fiber-reinforced expansive clay [29–34]. Under dynamic loads, freeze–thaw cycles change the dynamic shear modulus of the soil and decrease the dynamic strength [35,36]. Although the characteristics of soils subjected to freeze–thaw cycles have been extensively studied, the impact of freeze–thaw cycles on rubber-reinforced soil has received relatively less attention.

The improvement of expansive soil has been a hot issue in geotechnical engineering. Expansive soil is a specialized clay rich in hydrophilic minerals such as montmorillonite, which is widely distributed worldwide. Expansive soil is considered problematic because it swells in volume when it absorbs water and shrinks and cracks when it dries. It is estimated that the cost of maintaining expansive soil already exceeds the combined economic losses caused by natural disasters worldwide [37]. Therefore, a feasible solution is needed to address the negative impacts of expansive soil on building structures.

This study aims to enhance expansive clay performance in freeze–thaw cycles and to find reliable methods for recycling large volumes of waste tires. Triaxial tests were performed on rubber fiber-reinforced expansive clay exposed to freeze–thaw cycles. To evaluate the effects of rubber fiber on the mechanical properties of expansive clay under freeze–thaw cycles, the triaxial test results of rubber fiber-reinforced expansive clay and unreinforced expansive clay were compared. An expansive clay stress–strain calculation model under freeze–thaw cycles was developed based on the test results.

2. Materials and Methods

2.1. Test Materials

The expansive clay was obtained from Henan Province, China. Based on the reference standards listed in Table 1, the maximum dry density of the test expansive clay was 1.858 g/cm^3 and the optimum moisture content was 15.71%. The expansive clay has a liquid limit of 57.9% and plastic limit of 22.3%, and is classified as high liquid limit clay (CH) according to the Unified Soil Classification System [38]. The particle size of the expansive clay is shown in Figure 1, and its uniformity coefficient (C_u) is 9.3.

One of the aims of this study was to find a sustainable solution for effective waste tire management. The rubber fibers were obtained by grinding waste tires.

Table 1. Characteristic parameters for the expansive clay.

Soil Parameters	Results	Reference Standards
Maximum dry density	1.858 g/cm^3	ASTM D698-12 [39]
Optimum moisture content	15.71%	
Natural moisture content	6.06%	ASTM D2216-19 [40]
Liquid limit	57.9%	ASTM D4318-17e1 [41]
Plastic limit	22.3%	
Free swell ratio	71%	Prakash et al. [42]
Specific gravity ds	2.72	ASTM D854-14 [43]
Uniformity coefficient (C_u)	9.3	ASTM D7928-17 [44]

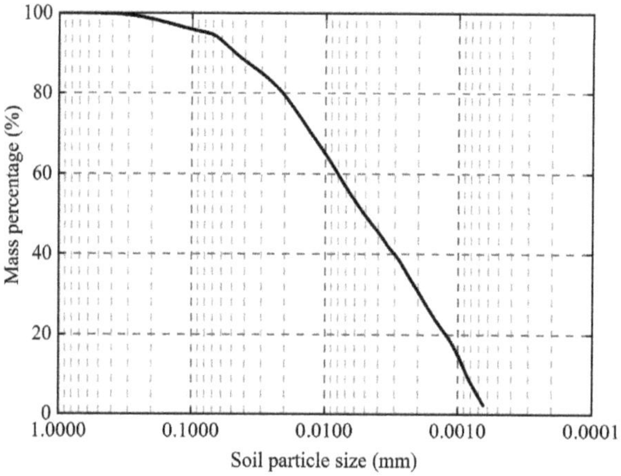

Figure 1. Particle size distribution curve of the expansive clay.

2.2. Sample Preparation

Akbarimehr et al. [45] measured the damping ratio of rubber-reinforced clay and suggested an optimum rubber content of 10%. When the fiber content exceeded 20% during the preparation of the specimens, the fibers clumped together and a uniform specimen could not be obtained. Similarly, Correia et al. [46] showed that specimens with high fiber content were not uniform. To avoid non-uniformity of the specimens, the rubber fiber content used in this study was 10% (ratio of fiber mass to dry expansive clay mass). Air-dried expansive clay was uniformly mixed with rubber fibers, and distilled water was added to increase the moisture content to 16% of the dry expansive clay mass. The mixture was stored in a sealed bag for 24 h to an even moisture distribution. Specimens with a diameter of 39.1 mm and height of 80 mm were obtained using the standard compaction effort [39]. Because it is more difficult to bring clay to saturation than sand or silt, the rubber fiber-reinforced expansive clay specimens were treated using the vacuum saturation method [47]. Some rubber fiber-reinforced expansive clay specimens are shown in Figure 2.

Figure 2. Rubber fiber-reinforced expansive clay specimen (**a**) before test and (**b**) after test.

2.3. Test Scheme

In its actual state, the expansive clay consolidation is completed after some time. Clays have limited permeability and do not release excess pore water pressure in time; hence, a consolidated undrained test was employed to simulate the actual drainage conditions of the clays. Eight strips of filter paper were uniformly applied around each specimen to accelerate the consolidation process. Confining pressures of 100, 200, and 300 kPa were applied to specimens with the same number of freeze–thaw cycles. In this study, triaxial tests were conducted on rubber fiber-reinforced expansive clay with different numbers of freeze–thaw cycles (N), which are 0, 1, 3, 6, 10, and 15. Wang et al. [48] suggested that temperatures approaching 0 °C may lead to incomplete freezing of the specimens. The specimens were frozen at −20 °C for 12 h and thawed at 20 °C for 12 h to ensure complete freezing and thawing of the specimens.

3. Results and Discussion

The stress–strain curves for some of the triaxial tests are shown in Figure 3 and the shear strength results are shown in Figure 4. With confining pressures of 100 kPa and 200 kPa, the shear strength changes more significantly during the first three freeze–thaw cycles. The shear strength declines after one freeze–thaw cycle. After three freeze–thaw cycles, the shear strength reaches the maximum value. After this, the shear strength gradually decreases with the increase in the number of freeze–thaw cycles. With the confining pressure of 300 kPa, the shear strength gradually decreases with the increase in the number of freeze–thaw cycles and reaches the turning point after six freeze–thaw cycles.

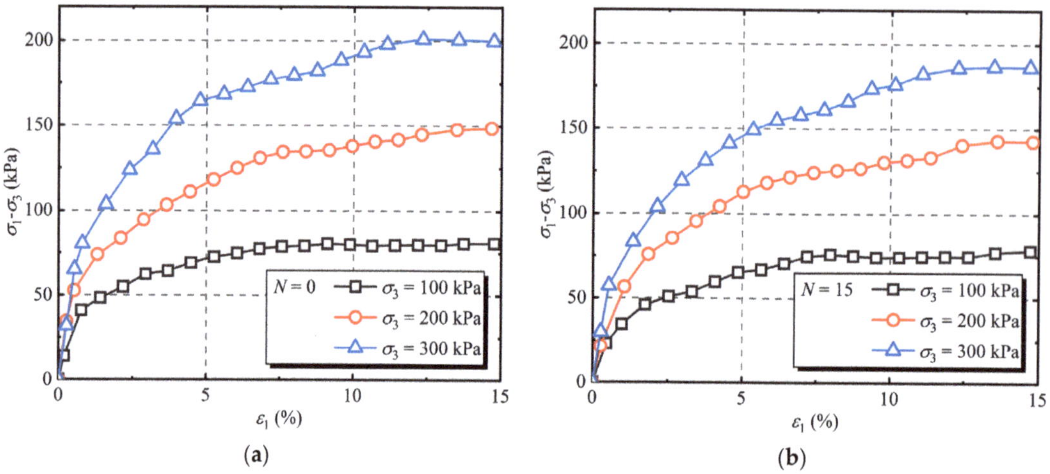

Figure 3. Part of the stress–strain curves. (**a**) $N = 0$; (**b**) $N = 15$.

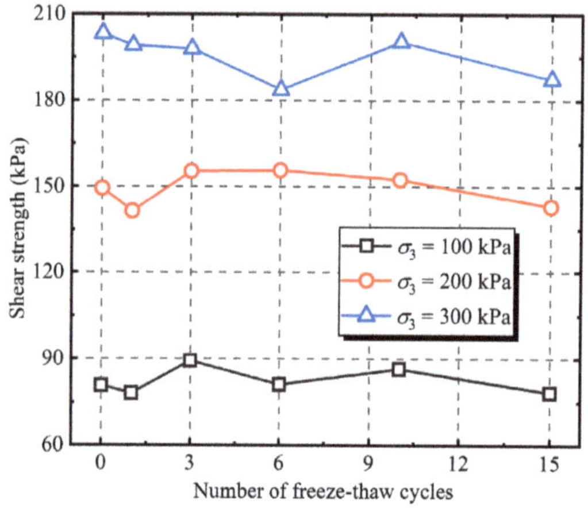

Figure 4. Shear strength with different number of freeze–thaw cycles.

3.1. Effect of Mixing Rubber Fibers on Shear Strength of Expansive Clay

Numerous studies have shown that the freeze–thaw cycling process significantly reduces the shear strength of unreinforced soils [29–34]. Figure 4 shows the relationship between the shear strength of rubber fiber-reinforced expansive clay and the number of freeze–thaw cycles. The shear strength decreases after 15 freeze–thaw cycles. However, the maximum decrease in shear strength was 9.6% for all confining pressures, which was significantly different from the performance of unreinforced expansive clay under freeze–thaw cycles. The changes in shear strength of unreinforced expansive clay under freeze–thaw cycles from previous studies are referenced to show how mixing rubber fibers affects the shear strength, and the test conditions are summarized in Table 2 [49,50]. Liu et al. [51] indicated that the freezing or thawing temperature had almost no effect on shear strength; therefore, the magnitude of freezing or thawing temperatures was not differentiated in this study.

The normalized shear strength was obtained by dividing the shear strength after freeze–thaw cycles by the shear strength without freeze–thaw cycles. The effect of mixing rubber fibers on the shear strength under freeze–thaw cycles can be determined by comparing the normalized shear strength of the rubber fiber-reinforced expansive clay with that of the unreinforced expansive clay. The results are shown in Figure 5.

Table 2. Test conditions.

References	Soil Types	Test Method	Number of Freeze–Thaw Cycles	Confining Pressures/kPa
Tang et al. [49]	Expensive clay	CU	0, 1, 3, 5, 9	100, 200, 300
Zhao et al. [50]	Expensive clay	CU	0, 1, 4, 6, 10	50, 100, 200, 300

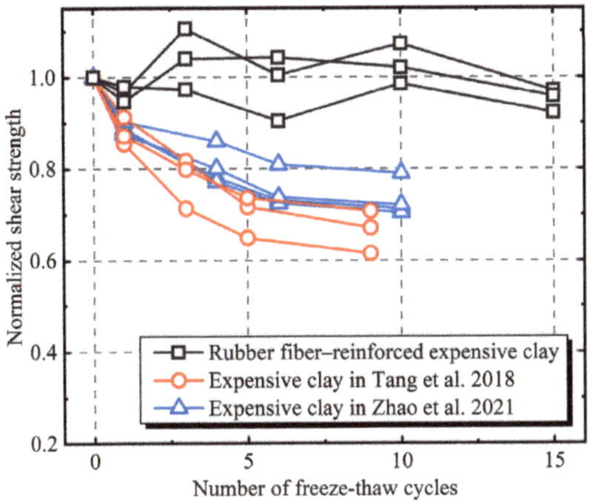

Figure 5. Variation in normalized shear strength with freeze–thaw cycles [49,50].

Figure 5 shows the variation in the normalized shear strength with the number of freeze–thaw cycles, and the different curves in the same color represent different confining pressures. The normalized shear strength of rubber fiber-reinforced expansive clay changed the least after 15 freeze–thaw cycles. However, the shear strength of unreinforced expansive clay decreased by approximately 34% (average) after 9 freeze–thaw cycles in Tang et al. [49], and by approximately 27% (average) after 10 freeze–thaw cycles in Zhao et al. [50]. Unreinforced expansive clay generates cracks under freeze–thaw cycles and the cracks develop with the increasing number of freeze–thaw cycles [52–54]. The cracks reduce the shear strength of the expansive clay [53]. When cracks are generated in the expansive clay, the rubber fibers produce a tensile force to weaken the development of the cracks, thus reducing the effect of freeze–thaw cycles on the shear strength of the expansive clay [55,56]. It can be observed from the above discussion that mixing rubber fibers enhances the ability of the expansive clay to resist the effects of freeze–thaw cycles and thus the service life of expansive clay engineering in seasonal frozen regions is prolonged.

3.2. Effect of Mixing Rubber Fibers on Elastic Modulus of Expansive Clay

Elastic modulus is an important mechanical parameter for soil materials. The stress–strain curves shown in Figure 3 are not straight lines, indicating that the secant elastic modulus is different at different strain levels. Some scholars have defined the elastic modulus of soil as the secant elastic modulus at 1% axial strain [34,49]. In this study, the elastic modulus was calculated from the secant elastic modulus at 2% axial strain based on the stress–strain curves of the specimens, as expressed in Equation (1), and the results are shown in Figure 6.

$$E_{2\%} = \frac{\Delta\sigma}{\Delta\varepsilon} = \frac{(\sigma_1 - \sigma_3)_{2\%}}{\varepsilon_{1,\,2\%}} \quad (1)$$

where $E_{2\%}$ is the elastic modulus, $(\sigma_1 - \sigma_3)_{2\%}$ is the deviatoric stress at 2% axial strain, and $\varepsilon_{1,\,2\%}$ is 2% axial strain.

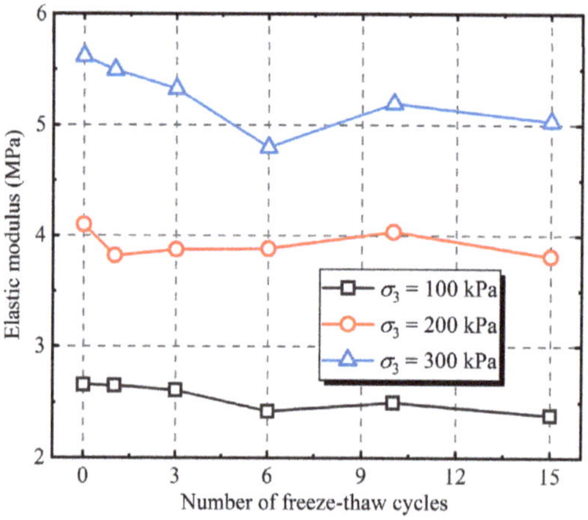

Figure 6. Elastic modulus with different number of freeze–thaw cycles.

The results shown in Figure 6 indicate that the elastic modulus of rubber fiber-reinforced expansive clay decreases slowly from an overall point of view at the confining pressures of 100 kPa and 200 kPa. However, at the confining pressure of 300 kPa, the decrease in the elastic modulus of the rubber fiber-reinforced expansive soil is more significant. The results of the elastic modulus of unreinforced expansive clay under freeze–thaw conditions from previous studies were referred to compare the effect of mixing rubber fibers on the elastic modulus of expansive soil [49,50]. The results of the normalized elastic modulus are shown in Figure 7 and the test conditions are presented in Table 2.

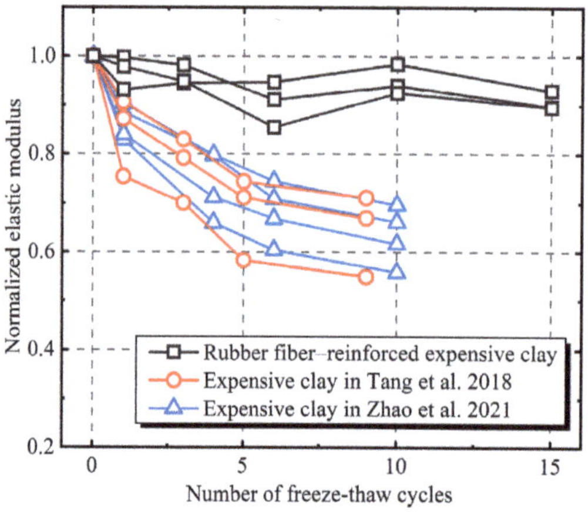

Figure 7. Variation in normalized elastic modulus with freeze–thaw cycles [49,50].

Figure 7 shows the variation in the normalized elastic modulus with the number of freeze–thaw cycles, and the different curves in the same color represent different confining pressures. The normalized elastic modulus of rubber fiber-reinforced expansive clay showed a decreasing trend within 15 freeze–thaw cycles. The elastic modulus of rubber fiber-reinforced expansive clay decreased by approximately 10% (average) after 15 freeze–thaw cycles. However, the elastic modulus of unreinforced expansive clay decreased by approximately 36% (average) after 9 freeze–thaw cycles in Tang et al. [49], and by approximately 37% (average) after 10 freeze–thaw cycles in Zhao et al. [50]. The rubber fibers produce tensile forces that weaken the development of cracks (caused by freeze–thaw cycles) in the expansive clay, thus reducing the effect of freeze–thaw cycles on the elastic modulus of the expansive clay [53,55,56]. Based on the above discussion, the mixing of rubber fibers lowers the effect of freeze–thaw cycles on the elastic modulus and thus improves the engineering performance of expansive clay in seasonal frozen regions.

4. Stress–Strain Curves after Freeze–Thaw Cycles of Expansive Clay

The shape of the stress–strain curves shown in Figure 3 is very similar to that of the nonlinear elastic model proposed by Duncan and Chang [57]. The relationship between the deviatoric stress and axial strain is expressed as Equation (2):

$$(\sigma_1 - \sigma_3) = \frac{\varepsilon_1}{a + b\varepsilon_1} \tag{2}$$

where $\sigma_1 - \sigma_3$ is the deviatoric stress, ε_1 is the axial strain, and a and b are the two model parameters.

The a and b in Equation (2) correspond to the initial elastic modulus and ultimate strength in the deviatoric stress–axial strain curve, respectively, and this correspondence is expressed in Equation (3):

$$\left.\begin{aligned} E_i &= \frac{1}{a} \\ (\sigma_1 - \sigma_3)_{\text{ult}} &= \frac{1}{b} \end{aligned}\right\} \tag{3}$$

where E_{i} is the initial elastic modulus, and $(\sigma_1 - \sigma_3)_{\text{ult}}$ is the ultimate strength.

Equation (2) converted to Equation (4) is expressed as follows:

$$\frac{\varepsilon_1}{(\sigma_1 - \sigma_3)} = a + b\varepsilon_1 \tag{4}$$

Equation (4) shows that there is a linear relationship between $\varepsilon_1/(\sigma_1 - \sigma_3)$ and ε_1. The $\varepsilon_1/(\sigma_1 - \sigma_3)$-$\varepsilon_1$ curves are drawn according to the test results of rubber fiber-reinforced expansive clay and the results of unreinforced expansive clay conducted by Tang et al. [49] and Zhao et al. [50]. The partial results are displayed in Figure 8. The intercept a and slope b were obtained by linear fitting of the $\varepsilon_1/(\sigma_1 - \sigma_3)$-$\varepsilon_1$ curves, and the initial elastic modulus and ultimate strength after different numbers of freeze–thaw cycles at different confining pressures were obtained by substituting the correspondence expressed in Equation (3). The results are summarized in Table 3.

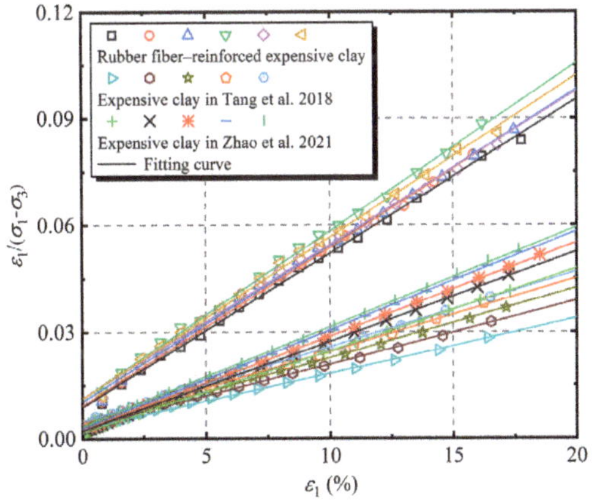

Figure 8. Relationship between $\varepsilon_1/(\sigma_1$-$\sigma_3)$ and ε_1 [49,50].

Table 3. Results for initial elastic modulus and ultimate strength.

References	σ_3/kPa	N	a	b	R^2	E_i/MPa	$(\sigma_1-\sigma_3)_{ult}$/kPa
This study	100	0	0.01157	0.01161	0.998	8.64	86.13
		1	0.00947	0.01246	0.996	10.56	80.26
		3	0.01794	0.01008	0.996	5.57	99.21
		6	0.01728	0.01124	0.998	5.79	88.97
		10	0.01741	0.01067	0.997	5.74	93.72
		15	0.01787	0.01168	0.997	5.60	85.62
	200	0	0.01219	0.00596	0.997	8.20	167.79
		1	0.01240	0.00630	0.996	8.06	158.73
		3	0.01479	0.00545	0.997	6.76	183.49
		6	0.01507	0.00559	0.995	6.64	178.89
		10	0.01352	0.00569	0.997	7.40	175.75
		15	0.01440	0.00608	0.997	6.94	164.47
	300	0	0.00876	0.00433	0.997	11.42	230.95
		1	0.00919	0.00445	0.997	10.88	224.72
		3	0.01038	0.00439	0.997	9.63	227.79
		6	0.01119	0.00472	0.997	8.94	211.86
		10	0.01141	0.00431	0.995	8.76	232.02
		15	0.01124	0.00455	0.997	8.90	219.78
Tang et al. [49]	100	0	0.00305	0.00543	0.997	32.79	184.16
		1	0.00546	0.00580	0.997	18.32	172.41
		3	0.00774	0.00637	0.998	12.92	156.99
		5	0.00862	0.00724	0.998	11.60	138.12
		9	0.00958	0.00772	0.997	10.44	129.53
	200	0	0.00381	0.00236	0.995	26.25	423.73
		1	0.00465	0.00277	0.994	21.51	361.01
		3	0.00436	0.00333	0.997	22.94	300.30
		5	0.00524	0.00359	0.998	19.08	278.55
		9	0.00592	0.00376	0.998	16.89	265.96
	300	0	0.00276	0.00156	0.994	36.23	641.03
		1	0.00312	0.00178	0.996	32.05	561.80
		3	0.00331	0.00195	0.995	30.21	512.82
		5	0.00405	0.00205	0.998	24.69	487.80
		9	0.00483	0.00211	0.999	20.70	473.93
Zhao et al. [50]	50	0	0.00344	0.00712	0.999	29.07	140.45
		1	0.00603	0.00806	0.999	16.58	124.07
		4	0.00894	0.00891	0.999	11.19	112.23
		6	0.01032	0.00951	0.999	9.69	105.15
		10	0.01293	0.00961	0.999	7.73	104.06
	100	0	0.00223	0.00479	0.999	44.84	208.77
		1	0.00383	0.00527	0.990	26.11	189.75
		4	0.00448	0.00606	0.999	22.32	165.02
		6	0.00499	0.00647	0.999	20.04	154.56
		10	0.00626	0.00654	0.999	15.97	152.91
	200	0	0.00244	0.00309	0.999	40.98	323.62
		1	0.00278	0.00349	0.999	35.97	286.53
		4	0.00313	0.00381	0.999	31.95	262.47
		6	0.00422	0.00406	0.999	23.70	246.31
		10	0.00475	0.00414	0.999	21.05	241.55
	300	0	0.00159	0.00231	0.999	62.89	432.90
		1	0.00212	0.00252	0.999	47.17	396.83
		4	0.00258	0.00263	0.999	38.76	380.23
		6	0.00288	0.00277	0.999	34.72	361.01
		10	0.00352	0.00280	0.999	28.41	357.14

Table 3 shows the results for ultimate strength and initial elastic modulus. R^2 are all above 0.99, which indicates that $\varepsilon_1/(\sigma_1-\sigma_3)$ and ε_1 exhibit a highly linear relationship. Figure 9 shows the relationship between the ultimate strength and initial elastic modulus.

Analysis of the data shows that there is a proportional relationship between the initial elastic modulus and ultimate strength, and the correspondence is expressed as Equation (5):

$$\frac{(\sigma_1 - \sigma_3)_{\text{ult}}}{E_{\text{i}}} = M \tag{5}$$

where M is the coefficient between the initial elastic modulus and ultimate strength.

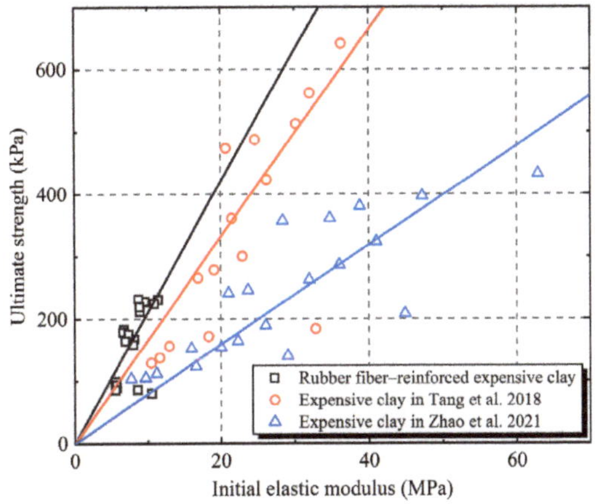

Figure 9. Relationship between ultimate strength and initial elastic modulus [49,50].

Analysis of the data reveals a certain regularity in the variation in the ultimate strength. The normalized ultimate strength is defined as the ratio of the ultimate strength after freeze–thaw cycles over the ultimate strength without freeze–thaw cycles, expressed as Equation (6):

$$R = \frac{(\sigma_1 - \sigma_3)_{\text{ult}, N}}{(\sigma_1 - \sigma_3)_{\text{ult}, 0}} \tag{6}$$

where R is the normalized ultimate strength, $(\sigma_1 - \sigma_3)_{\text{ult}, N}$ is the ultimate strength after N freeze–thaw cycles, and $(\sigma_1 - \sigma_3)_{\text{ult}, 0}$ is the ultimate strength without freeze–thaw cycles.

The normalized ultimate strengths were averaged (average values for different confining pressures) and are shown in Figure 10. The relationship between the average value of the normalized ultimate strength and the number of freeze–thaw cycles was established and shown as Equation (7), where the parameter λ is a deterioration parameter.

$$R_{\text{ave}} = \lambda\left(e^{-\lambda N} - 1\right) + 1 \tag{7}$$

where R_{ave} is the average normalized ultimate strength, λ is the deterioration parameter of the ultimate strength under freeze–thaw cycles, and N is the number of freeze–thaw cycles.

In Equation (7), λ reflects the soil ultimate strength as affected by freeze–thaw cycles. The value of λ is between 0 and 1. The ultimate strength is less affected by freeze–thaw cycles when the value of λ is close to 0. The ultimate strength is more affected by freeze–thaw cycles when the value of L is close to 1. The ultimate strength is $(1 - \lambda)$ times the ultimate strength without freeze–thaw cycles if the expansive clay is subjected to an infinite number of freeze–thaw cycles, according to Equation (7). Figure 10 shows that $\lambda = 0.336$ for unconsolidated expansive clay in Tang et al. [49] and $\lambda = 0.275$ in Zhao et al. [50]. However, the ultimate strength of rubber fiber-reinforced expansive clay is almost unaffected by freeze–thaw cycles, with $\lambda = 0$.

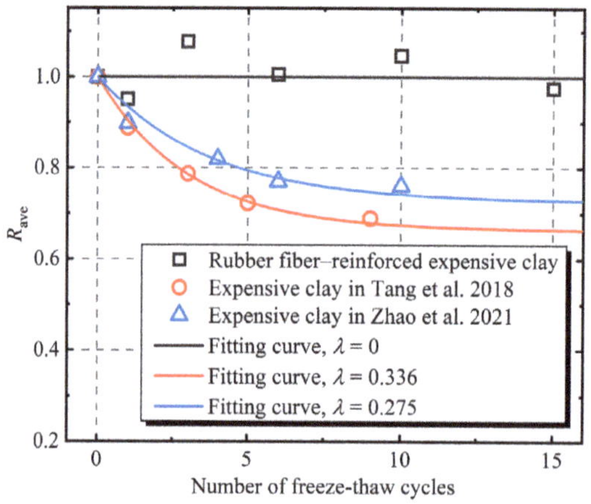

Figure 10. Relationship between R_{ave} and N [49,50].

The ultimate strength after freeze–thaw cycling is obtained through the deterioration parameters and the ultimate strength without freeze–thaw cycles, expressed as Equation (8):

$$(\sigma_1 - \sigma_3)_{\text{ult},\,N} = R_{\text{ave}} \cdot (\sigma_1 - \sigma_3)_{\text{ult},\,0} \quad (8)$$

Equation (9) shows the Mohr–Coulomb strength theory for soil, in which the shear strength of the soil can be divided into cohesive and frictional components.

$$\tau_{\text{f}} = c + \sigma \cdot \tan\varphi \quad (9)$$

where τ_{f} denotes the shear strength, c denotes the cohesive force, φ denotes the internal friction angle, and σ denotes the normal stress on the shear surface.

The ultimate strength can be obtained from the cohesion force and internal friction angle according to the geometrical relationship of Mohr's circle, as expressed in Equation (10):

$$(\sigma_1 - \sigma_3)_{\text{ult},\,0} = 2\frac{c + \sigma_3 \cdot \tan\varphi}{\sqrt{1 + \tan^2\varphi} - \tan\varphi} \quad (10)$$

By substituting Equations (3), (5), (7), (8), and (10) into Equation (2), the formula for the deviatoric stress after different numbers of freeze–thaw cycles at different confining pressures can be obtained, and is expressed as Equation (11):

$$(\sigma_1 - \sigma_3) = \frac{2(c + \sigma_3 \cdot \tan\varphi) \cdot \left[\lambda\left(\mathrm{e}^{-\lambda N} - 1\right) + 1\right]}{\left(\sqrt{1 + \tan^2\varphi} - \tan\varphi\right) \cdot (M + \varepsilon_1)} \cdot \varepsilon_1 \quad (11)$$

Equation (11) demonstrates that the deviatoric stress is related to the number of freeze–thaw cycles, confining pressure, and axial strain. The deviatoric stress–axial strain curve after any number of freeze–thaw cycles at any confining pressure can be obtained according to Equation (11). The cohesive force and internal friction angle were determined using conventional triaxial tests. The parameters needed for Equation (11) are listed in Table 4. The deviatoric stress–axial strain curves were calculated for different confining pressures after different numbers of freeze–thaw cycles, and the partial results shown in Figure 11 indicate that Equation (11) yielded satisfactory results.

Table 4. Parameters.

References	c/kPa	$\tan\varphi$	M	σ_3/kPa	N
This study	7	0.275	0.0210	100, 200, 300	0, 1, 3, 6, 10, 15
Tang et al. [49]	6	0.575	0.0166	100, 200, 300	0, 1, 3, 5, 9
Zhao et al. [50]	30	0.395	0.0080	50, 100, 200, 300	0, 1, 4, 6, 10

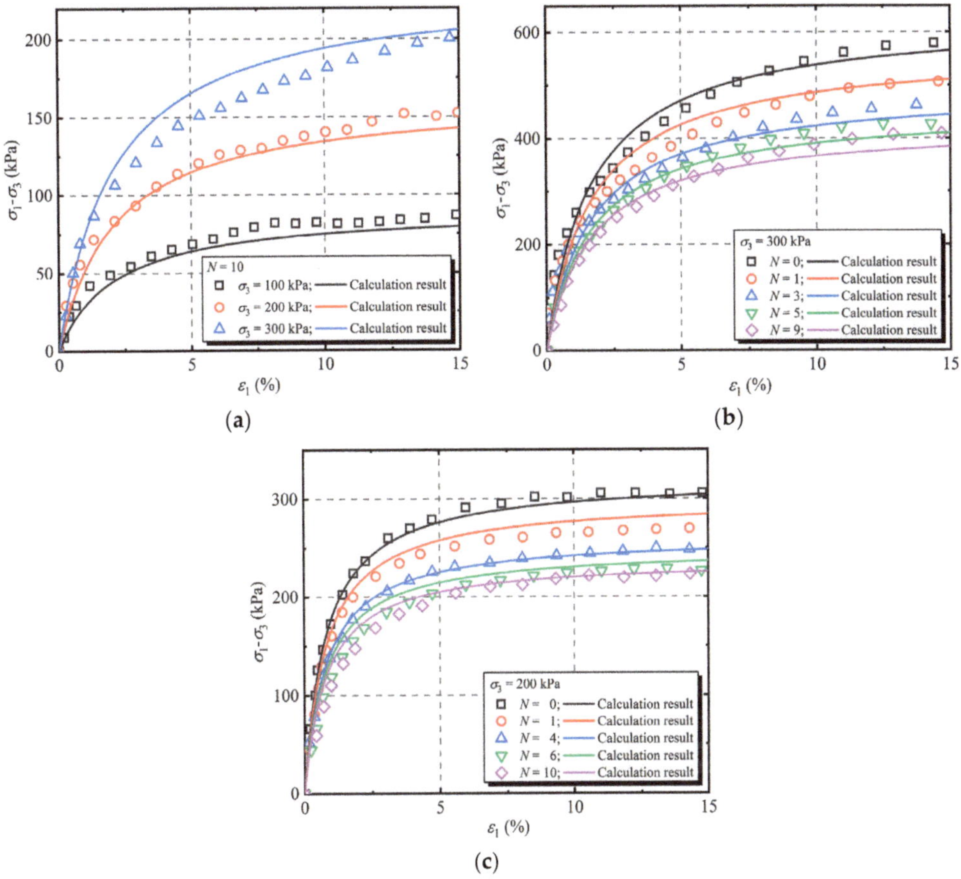

Figure 11. Test results and calculation results with different test conditions. (**a**) N = 10 in this study; (**b**) σ_3 = 300 kPa in Tang et al. [49]; (**c**) σ_3 = 200 kPa in Zhao et al. [50].

5. Conclusions

In this study, the mechanical properties of rubber fiber-reinforced expansive clay under freeze–thaw cycles were investigated and compared with those of unreinforced expansive clay. The main conclusions are as follows.

(1) The mixing of rubber fibers reduced the effect of freeze–thaw cycles on the shear strength and elastic modulus of expansive clay. The shear strength and elastic modulus of rubber fiber-reinforced expansive clay decreased slightly with an increase in the number of freeze–thaw cycles. However, the shear strength and elastic modulus of the unreinforced expansive clay decreased rapidly with an increase in the number of freeze–thaw cycles.

(2) The relationship between the number of freeze–thaw cycles and the ultimate strength was established, which can describe the change in ultimate strength under freeze–thaw cycles. The ultimate strength of rubber fiber-reinforced expansive clay does not change markedly as the number of freeze–thaw cycles increases, and the deterioration parameter λ is 0. In contrast, the ultimate strength of the unreinforced expansive clay significantly decreased after freeze–thaw cycles, with the deterioration parameters as 0.336 and 0.275.
(3) A calculation model of the deviatoric stress–axial strain curves after freeze–thaw cycles was established, which can describe the deviatoric stress–axial strain behavior of rubber fiber-reinforced expansive clay and unreinforced expansive clay under different confining pressures and numbers of freeze–thaw cycles.

This paper investigated the static response of rubber fiber-reinforced expansive soil under freeze–thaw cycles. The subgrade soils are subjected to dynamic loads, and future research on the dynamic response of rubber fiber-reinforced expansive soil under freeze–thaw cycles can be carried out.

Author Contributions: Conceptualization, Z.Y.; data curation, writing—original draft, Z.S.; methodology, X.L.; supervision, G.C.; project administration, R.W. and X.M. All authors have read and agreed to the published version of the manuscript.

Funding: This research was funded by the National Natural Science Foundation of China, grant number 42177153, and the National Major Scientific Research Instrument Development Project, grant number 41627801.

Institutional Review Board Statement: Not applicable.

Informed Consent Statement: Not applicable.

Data Availability Statement: The data presented in this study are available on request from the corresponding author.

Conflicts of Interest: The authors declare no conflicts of interest.

References

1. Yang, J.T. The Impact of Urbanization Process on Civil Car Ownership in China. In Proceedings of the 5th Annual International Conference on Social Science and Contemporary Humanity Development (SSCHD), Wuhan, China, 15–16 November 2019; Volume 376, pp. 612–617. [CrossRef]
2. Ferdous, W.; Manalo, A.; Siddique, R.; Mendis, P.; Yan, Z.G.; Wong, H.S.; Lokuge, W.; Aravinthan, T.; Schubel, P. Recycling of landfill wastes (tyres, plastics and glass) in construction—A review on global waste generation, performance, application and future opportunities. *Resour. Conserv. Recycl.* **2021**, *173*, 105745. [CrossRef]
3. Dobrota, D.; Dobrota, G.; Dobrescu, T. Improvement of waste tyre recycling technology based on a new tyre markings. *J. Clean Prod.* **2020**, *260*, 121141. [CrossRef]
4. Tian, X.L.; Zhuang, Q.X.; Han, S.; Li, S.M.; Liu, H.C.; Li, L.; Zhang, J.X.; Wang, C.S.; Bian, H.G. A novel approach of reapplication of carbon black recovered from waste tyre pyrolysis to rubber composites. *J. Clean Prod.* **2021**, *280*, 124460. [CrossRef]
5. Han, W.W.; Han, D.S.; Chen, H.B. Pyrolysis of Waste Tires: A Review. *Polymers* **2023**, *15*, 1604. [CrossRef] [PubMed]
6. Mavukwana, A.E.; Fox, J.A.; Sempuga, B.C. Waste tyre to electricity: Thermodynamics analysis. *J. Environ. Chem. Eng.* **2020**, *8*, 103831. [CrossRef]
7. Samesová, D.; Ponist, J.; Hybská, H.; Pochyba, A.; Schwarz, M.; Salva, J. Determination of aerobic and anaerobic biological degradability of waste tyres. *Environ. Sci. Eur.* **2023**, *35*. [CrossRef]
8. Qu, B.Y.; Li, A.M.; Qu, Y.; Wang, T.; Zhang, Y.; Wang, X.; Gao, Y.; Fu, W.; Ji, G.Z. Kinetic analysis of waste tire pyrolysis with metal oxide and zeolitic catalysts. *J. Anal. Appl. Pyrolysis* **2020**, *152*, 104949. [CrossRef]
9. Zhou, Y.; Zhu, G.Q.; Yu, X.F.; Pan, R.L.; Chu, T.W.; Liang, Z.H. The flame development process and thermal radiation of tire stacks in a large-scale experiment. *J. Therm. Anal. Calorim.* **2022**, *147*, 2481–2494. [CrossRef]
10. Raudonyte-Svirbutaviciene, E.; Stakeniene, R.; Joksas, K.; Valiulis, D.; Bycenkiene, S.; Zarkov, A. Distribution of polycyclic aromatic hydrocarbons and heavy metals in soil following a large tire fire incident: A case study. *Chemosphere* **2022**, *286*, 131556. [CrossRef]
11. Cuevas, J.; Gonzalez-Santamaria, D.E.; Garcia-Delgado, C.; Ruiz, A.; Garralon, A.; Ruiz, A.I.; Fernandez, R.; Eymar, E.; Jimenez-Ballesta, R. Impact of a tire fire accident on soil pollution and the use of clay minerals as natural geo-indicators. *Environ. Geochem. Health* **2020**, *42*, 2147–2161. [CrossRef]

12. Kumari, S.; Elumalai, S.P.; Jain, M.K. Study on effect of tire burning on particulate matter concentration and respiratory deposition doses to the workers and inhabitants during road pavement activity. *Air Qual. Atmos. Health* **2022**, *15*, 1413–1426. [CrossRef]
13. Mentes, D.; Toth, C.E.; Nagy, G.; Muranszky, G.; Poliska, C. Investigation of gaseous and solid pollutants emitted from waste tire combustion at different temperatures. *Waste Manag.* **2022**, *149*, 302–312. [CrossRef] [PubMed]
14. Bulei, C.; Todor, M.P.; Heput, T.; Kiss, I. Directions for material recovery of used tires and their use in the production of new products intended for the industry of civil construction and pavements. *IOP Conf. Ser. Mater. Sci. Eng.* **2018**, *294*, 012064. [CrossRef]
15. Mmereki, D.; Machola, B.; Mokokwe, K. Status of waste tires and management practice in Botswana. *J. Air Waste Manag. Assoc.* **2019**, *69*, 1230–1246. [CrossRef] [PubMed]
16. Jiang, G.S.; Pan, J.Q.; Deng, W.Z.; Sun, Y.Z.; Guo, J.; Che, K.; Yang, Y.; Lin, Z.Y.; Sun, Y.C.; Huang, C.J.; et al. Recovery of high pure pyrolytic carbon black from waste tires by dual acid treatment. *J. Clean Prod.* **2022**, *374*, 133893. [CrossRef]
17. Zou, C.; Ren, Y.; Li, S.G.; Hu, H.Y.; Cao, C.Y.; Tang, H.; Li, X.; Yao, H. Effects of molten salt thermal treatment on the properties improvement of waste tire pyrolytic char. *Waste Manag.* **2022**, *149*, 53–59. [CrossRef]
18. Pan, Y.H.; Yang, D.C.; Sun, K.; Wang, X.W.; Zhou, Y.G.; Huang, Q.X. Pyrolytic transformation behavior of hydrocarbons and heteroatom compounds of scrap tire volatiles. *Fuel* **2020**, *276*, 118095. [CrossRef]
19. Zheng, D.H.; Cheng, J.; Wang, X.L.; Yu, G.Q.; Xu, R.N.; Dai, C.N.; Liu, N.; Wang, N.; Chen, B.H. Influences and mechanisms of pyrolytic conditions on recycling BTX products from passenger car waste tires. *Waste Manag.* **2023**, *169*, 196–207. [CrossRef] [PubMed]
20. Irani, N.; Ghasemi, M. Effect of scrap tyre on strength properties of untreated and lime-treated clayey sand. *Eur. J. Environ. Civ. Eng.* **2021**, *25*, 1609–1626. [CrossRef]
21. Zhang, T.; Cai, G.J.; Duan, W.H. Strength and microstructure characteristics of the recycled rubber tire-sand mixtures as lightweight backfill. *Environ. Sci. Pollut. Res.* **2018**, *25*, 3872–3883. [CrossRef]
22. Daud, K.A. Soil improvement using waste tire chips. *Int. J. Civ. Eng. Technol.* **2018**, *9*, 1338–1345. [CrossRef]
23. Anbazhagan, P.; Manohar, D. Energy absorption capacity and shear strength characteristics of waste tire crumbs and sand mixtures. *Int. J. Geotech. Earthq.* **2015**, *6*, 28–49. [CrossRef]
24. Zhang, H.B.; Yuan, X.F.; Liu, Y.; Wu, J.Q.; Song, X.G.; He, F.D. Experimental study on the pullout behavior of scrap tire strips and their application as soil reinforcement. *Constr. Build. Mater.* **2020**, *254*, 119288. [CrossRef]
25. Lv, J.H.; Yang, Z.N.; Shi, W.; Lu, Z.C.; Zhang, Q.; Ling, X.Z. Dynamic Characteristics of Rubber Reinforced Expansive Soil (ESR) at Positive and Negative Ambient Temperatures. *Polymers* **2022**, *14*, 3985. [CrossRef] [PubMed]
26. Zhou, Z.W.; Ma, W.; Zhang, S.J.; Mu, Y.H.; Li, G.Y. Effect of freeze-thaw cycles in mechanical behaviors of frozen loess. *Cold Reg. Sci. Techol.* **2018**, *146*, 9–18. [CrossRef]
27. Zhang, Y.G.; Liu, S.H.; Deng, G.; Fang, B.X.; Zhang, Y.Y.; Lu, Y. Effect of freeze-thaw cycles on mechanical behavior of clay-gravel mixtures. *J. Mt. Sci.* **2022**, *19*, 3615–3626. [CrossRef]
28. Yang, A.W.; Yang, S.P.; Zhong, X.K.; Zhang, X.W. Mechanical Properties and Structural Evolution of Sludge-cured Lightweight Soils Subjected to Freeze-thaw Cycles. *KSCE J. Civ. Eng.* **2022**, *26*, 4478–4488. [CrossRef]
29. Orakoglu, M.E.; Liu, J.K.; Lin, R.; Tian, Y.H. Performance of Clay Soil Reinforced with Fly Ash and Lignin Fiber Subjected to Freeze-Thaw Cycles. *J. Cold Reg. Eng.* **2017**, *31*, 04017013. [CrossRef]
30. Xu, J.; Li, Y.F.; Lan, W.; Wang, S.H. Shear strength and damage mechanism of saline intact loess after freeze-thaw cycling. *Cold Reg. Sci. Technol.* **2019**, *164*, 102779. [CrossRef]
31. Zaimoglu, A.S.; Calik, Y.; Akbulut, R.K.; Yetimoglu, T. A Study on Freeze-Thaw Behavior of Randomly Distributed Fiber-Reinforced Soil. *Period. Polytech.-Civ. Eng.* **2016**, *60*, 3–9. [CrossRef]
32. Wang, M.; Meng, S.J.; Sun, Y.Q.; Fu, H.Q. Shear strength of frozen clay under freezing-thawing cycles using triaxial tests. *Earthq. Eng. Eng. Vib.* **2018**, *17*, 761–769. [CrossRef]
33. Ghazavi, M.; Roustaei, M. Freeze-thaw performance of clayey soil reinforced with geotextile layer. *Cold Reg. Sci. Technol.* **2013**, *89*, 22–29. [CrossRef]
34. Orakoglu, M.E.; Liu, J.K. Effect of Freeze-Thaw Cycles on Triaxial Strength Properties of Fiber-Reinforced Clayey Soil. *KSCE J. Civ. Eng.* **2017**, *21*, 2128–2140. [CrossRef]
35. Cui, G.H.; Cheng, Z.; Zhang, D.L.; Ma, S.X.; Liu, Z.Q. Effect of Freeze-Thaw Cycles on Dynamic Characteristics of Undisturbed Silty Clay. *KSCE J. Civ. Eng.* **2022**, *26*, 3831–3846. [CrossRef]
36. Yang, Z.N.; Lu, Z.C.; Shi, W.; Wang, C.; Ling, X.Z.; Liu, X.; Guan, D.; Cheng, Z.J. Dynamic Properties of Expansive Soil-Rubber under Freeze-Thaw Cycles. *J. Mater. Civ. Eng.* **2023**, *35*, 04023026. [CrossRef]
37. Selvakumar, S.; Soundara, B. 24—Swelling behavior of expansive soils stabilized with expanded polystyrene geofoam inclusion. In *New Materials in Civil Engineering*; Samui, P., Kim, D., Iyer, N.R., Chaudhary, S., Eds.; Butterworth-Heinemann: Oxford, UK, 2020; pp. 745–776.
38. *D2487-17, A*; Standard Practice for Classification of Soils for Engineering Purposes (Unified Soil Classification System). ASTM International: West Conshohocken, PA, USA, 2017. [CrossRef]
39. *D698-12, A*; Standard Test Method for Laboratory Compaction Characteristics of Soil Using Standard Effort. ASTM International: West Conshohocken, PA, USA, 2021. [CrossRef]

40. *D2216-19, A;* Standard Test Methods for Laboratory Determination of Water (Moisture) Content of Soil and Rock by Mass. ASTM International: West Conshohocken, PA, USA, 2019. [CrossRef]
41. *D4318-17e1, A;* Standard Test Methods for Liquid Limit, Plastic Limit, and Plasticity Index of Soils. ASTM International: West Conshohocken, PA, USA, 2017. [CrossRef]
42. Prakash, K.; Sridharan, A. Free swell ratio and clay mineralogy of fine-grained soils. *Geotech. Test. J.* **2004**, *27*, 220–225. [CrossRef]
43. *D854-14, A;* Standard Test Methods for Specific Gravity of Soil Solids by Water Pycnometer. ASTM International: West Conshohocken, PA, USA, 2014. [CrossRef]
44. *D7928-17, A;* Standard Test Method for Particle-Size Distribution (Gradation) of Fine-Grained Soils Using the Sedimentation (Hydrometer) Analysis. ASTM International: West Conshohocken, PA, USA, 2017. [CrossRef]
45. Akbarimehr, D.; Eslami, A.; Esmail, A. Geotechnical behaviour of clay soil mixed with rubber waste. *J. Clean Prod.* **2020**, *271*, 122632. [CrossRef]
46. Correia, N.S.; Rocha, S.A.; Lodi, P.C.; McCartney, J.S. Shear strength behavior of clayey soil reinforced with polypropylene fibers under drained and undrained conditions. *Geotext. Geomembr.* **2021**, *49*, 1419–1426. [CrossRef]
47. Estabragh, A.R.; Bordbar, A.T.; Javadi, A.A. Mechanical Behavior of a Clay Soil Reinforced with Nylon Fibers. *Geotech. Geol. Eng.* **2011**, *29*, 899–908. [CrossRef]
48. Wang, D.-Y.; Ma, W.; Niu, Y.-H.; Chang, X.-X.; Wen, Z. Effects of cyclic freezing and thawing on mechanical properties of Qinghai–Tibet clay. *Cold Reg. Sci. Technol.* **2007**, *48*, 34–43. [CrossRef]
49. Tang, L.; Cong, S.Y.; Geng, L.; Ling, X.Z.; Gan, F.D. The effect of freeze-thaw cycling on the mechanical properties of expansive soils. *Cold Reg. Sci. Technol.* **2018**, *145*, 197–207. [CrossRef]
50. Zhao, G.T.; Han, Z.; Zou, W.L.; Wang, X.Q. Evolution of mechanical behaviours of an expansive soil during drying-wetting, freeze-thaw, and drying-wetting-freeze-thaw cycles. *Bull. Eng. Geol. Environ.* **2021**, *80*, 8109–8121. [CrossRef]
51. Liu, J.K.; Chang, D.; Yu, Q.M. Influence of freeze-thaw cycles on mechanical properties of a silty sand. *Eng. Geol.* **2016**, *210*, 23–32. [CrossRef]
52. Lu, Y.; Liu, S.H.; Weng, L.P.; Wang, L.J.; Li, Z.; Xu, L. Fractal analysis of cracking in a clayey soil under freeze-thaw cycles. *Eng. Geol.* **2016**, *208*, 93–99. [CrossRef]
53. Luo, Z.L.; Zheng, Y.J.; Zhang, X.; Wang, L.H.; Gao, Y.; Liu, K.X.; Han, S.J. Effect of Phase Change Materials on Mechanical Properties of Stabilized Loess Subgrade Subjected to Freeze-Thaw Cycle. *J. Mater. Civ. Eng.* **2023**, *35*, 04023217. [CrossRef]
54. Li, T.G.; Kong, L.W.; Guo, A.G. The deformation and microstructure characteristics of expansive soil under freeze-thaw cycles with loads. *Cold Reg. Sci. Technol.* **2021**, *192*, 103393. [CrossRef]
55. Ahmadi, S.; Ghasemzadeh, H.; Changizi, F. Effects of A low-carbon emission additive on mechanical properties of fine-grained soil under freeze-thaw cycles. *J. Clean Prod.* **2021**, *304*, 127157. [CrossRef]
56. Fan, K.; Pei, Q.; Liu, L.; Han, Z.; Zou, W. Strength and microstructure of a lignin fiber-reinforced expansive soil in cold regions. *Geosynth. Int.* **2022**, *29*, 622–629. [CrossRef]
57. Duncan, J.M.; Chang, C.-Y. Nonlinear analysis of stress and strain in soils. *J. Soil Mech. Found. Div.* **1970**, *96*, 1629–1653. [CrossRef]

applied sciences

MDPI

Article

Constitutive Characteristics of Rock Damage under Freeze–Thaw Cycles

Yaoxin Li [1], Zhibin Wang [1], Haiqing Cao [2] and Tingyao Wu [2,*]

1 School of Civil Engineering, Hunan University of Science and Technology, Changsha 411100, China; 21020201063@mail.hnust.edu.cn (Y.L.); wangzhibin123465@163.com (Z.W.)
2 College of Civil Engineering, Chongqing University, Chongqing 400045, China; caohaiqing12342024@163.com
* Correspondence: wutingyao@cug.edu.cn

Abstract: Freeze–thaw effect is one of the most important environmental conditions that rocks may be subjected to. Through laboratory model tests, the damage characteristics of rocks under the FTC were studied. Based on assuming that the strength of rocks subject to the FTC follows the Weibull distribution, the cumulative damage variable of the number of FTCs was introduced. A cumulative damage constitutive model of shear strength attenuation of rock that meets the Mohr–Coulomb criterion is established. The rationality and applicability of the proposed damage constitutive model are verified by comparing the results of rock shear strength parameters under cyclic freeze–thaw loads.

Keywords: FTC; rock; damage; constitutive

1. Introduction

In many engineering applications, rocks are subject to freeze–thaw cycles (FTC). Under the action of freeze–thaw cycle and low temperature frostbite, rock and soil will have irreversible damage. This is an important safety hazard and disaster inducement for geotechnical engineering in cold areas [1,2]. For example, frost heave rupture of highway tunnels in cold regions leads to instability. Freeze–thaw collapse leads to landslide, multi-slope collapse of frozen rock in mining areas, dynamic disturbance of frozen rock caused by earthquakes, etc. The compressive strength of rock plays an important role in the construction of rock engineering and is a key parameter in the construction process [3,4]. Therefore, the influence of FTC on the compressive strength of rock is discussed, and the intrinsic relationship between rock strength and FTC is understood. This is of great significance to the research and practice of rock engineering [5].

The compressive strength of rock before and after different FTCs has been obtained, and it was found that the compressive strength of rock will decrease more along with the increase in the number of FTCs [6]. In addition, it was found that, after the freezing temperature of the rock sample was reduced, the compressive strength decreased more significantly [7]. Berea rock was selected for FTC treatment, and the compression coefficient of the rock decreased by about 16% and 24%, respectively, when the temperature ranged from −4 °C to 6 °C [8–10]. Different kinds of granites were selected for the FTC and ultrasonic test study [11], and it was found that the wave velocity of different kinds of granites decreased after each FTC, and the decreasing trend was related to lithology, freeze–thaw time and cycle times, with the maximum decrease of about 15%. According to the above analysis, these changes are mainly caused by the expansion of pores and cracks in rocks [12].

However, in many engineering practices, as mentioned above, rock causes a dynamic stress. There has been limited research conducted on the dynamic mechanical properties of rocks following FTC [13]. For example, through the comparison of a quasi-static uniaxial experiment and a SHPB impact experiment, it was found that the compressive strength of granite and limestone under high strain rate is greater than that under static load [14]. Generally speaking, the increase in the number of FTCs or a decrease in the loading rate will lead to a decrease in the fracture strength [15]. The failure process in rock under the action of the FTC is a change process of microstructural damage [4,16,17]: (1) the expansion force generated when the water contained in the rock voids is converted into ice, (2) thermal stress caused by the temperature difference between the surface and the interior of the rock, (3) the water pressure inside the rock caused by the freezing of the surface. In addition, there is also an expansive force phenomenon when ice crystals are formed [18,19].

Looking at the current research status in China and abroad, there are many research projects on freeze–thaw rock, but few on damage caused by the mechanical properties of freeze–thaw rock. Most are confined to the unidirectional stress state, without considering the influence of confining pressure. The parameters of the model are mostly based on the fitting of test data, lack clear physical meaning, and can not reflect the general change law of the selected physical quantity.

Drawing inspiration from the traditional statistical stress–strain model of rock, this paper postulates that the micro-strength of rock subjected to FTC adheres to the Weibull distribution. By incorporating a cumulative damage variable based on the number of FTCs, we have developed a constitutive model for cumulative damage in rock shear strength decay that aligns with the Mohr–Coulomb criterion. The validity and practicality of this proposed damage constitutive model are confirmed through a comparison of its predictions with observed changes in rock shear strength parameters under cyclic freeze–thaw loads.

2. Experimental Setup

Typical rock samples of amphibolite in the study area were selected, and all the kinds of rock sample were made into standard cylinders according to different sample sizes and hardness by using a diamond machine or stone grinder and other processing equipment. The rocks are all standard cylinders with a diameter of 5 cm and a height of 10 cm, and the initial deviation is controlled within 5%. A total of 5 groups were set up, 6 samples were prepared in each group, and 0, 10, 20, 30 and 40 different FTC tests were conducted respectively. According to the test requirements, a total of 30 standard samples were made. Some of the field test samples are shown in Figure 1.

Figure 1. Photos of some rock samples before the freeze–thaw experiment.

3. Preparation of Samples and Their Associated Physical Characteristics

The FTC test process is taken in accordance with the Standard of Test Methods for Engineering Rock Mass [20]. The temperature range is set as −20–20 °C. The temperature–time parameters of the FTC and the main process of the FTC are as follows:

a. Cooling stage: cooling freezing, freeze–thaw test box temperature gradually decreased from room temperature and stabilized at −20 °C, cooling process time is about 100 min.
b. Freezing stage: constant temperature freezing, freeze–thaw test box to maintain −20 °C ambient temperature, constant temperature for 4 h.
c. heating stage: heating and melting, freeze–thaw box placed automatically into the test box, slightly higher than 20 °C warm water to heat the frozen sample, the box temperature gradually stabilized at 20 °C, the heating process time about 100 min.

d. Melting stage: constant temperature water bath, freeze–thaw box environment temperature maintained at 20 °C constant temperature, the sample immersed in 20 °C warm water bath, heating for 4 h.
e. The warm water in the freeze–thaw test box is discharged to complete a full FTC test cycle and enter the next FTC stage until the preset number of FTCs.

In this study, a total of five groups of FTC damaged rock samples were set, which were 0, 10, 20, 30, and 40 FTC, respectively. To ensure that the test rock samples are in a state of full water without water loss, the samples that have completed the FTC continue to be immersed in the constant temperature water bath until the mechanical test is carried out.

3.1. Rock Porosity and Saturation Density

Concurrently, both the quality and wave velocity of the saturated rock samples were measured following each freeze–thaw condition; the values in the figure are the average values of each group of rocks under the same number of FTCs, as shown in Figure 2.

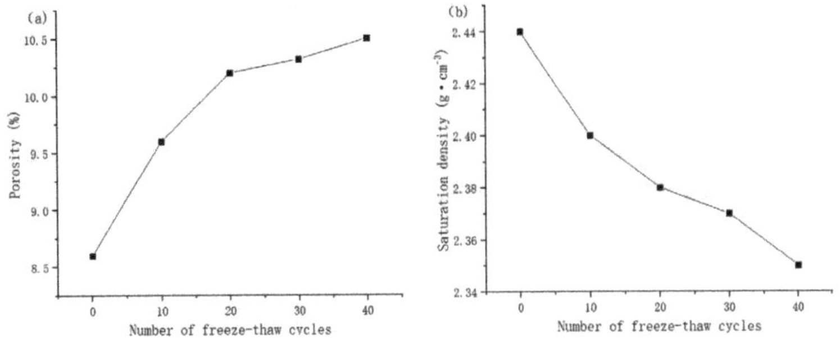

Figure 2. Relation of saturation density and porosity of rocks with the number of FTCs after freeze–thaw test: (**a**) Porosity; (**b**) Saturation density.

As depicted in Figure 2, the physical and hydraulic properties of rock samples exhibit variations with the number of FTCs. However, these differences are relatively minor, and the overall trend remains consistent. Initially, at 0 FTC, the porosity of the rock samples stands at 8.78%, suggesting the presence of a significant number of pore structures under natural saturation. As the samples undergo FTC, the water within experiences a phase transition from water to ice, leading to the expansion of existing pores and cracks, as well as the formation of new pores. Consequently, the porosity increases to 10.80% after 40 FTCs, representing a 23.01% increase. Concurrently, the saturated density of the rock samples decreases with the increasing number of FTCs. This decrease is primarily attributed to the loss of mineral particles on the sample surface, resulting in a reduction in the saturated mass. Additionally, the expansion of the rock volume due to the development of cracks caused by the frost heave force contributes to this decrease. Notably, within the first 20 FTCs, the saturation density of the samples decreases significantly, from 2.44 g·cm^{-3} to 2.38 g·cm^{-3}. The porosity of the rock samples continues to increase with the increasing number of FTCs, and the growth rate of porosity during the first 20 cycles is significantly higher than that in subsequent cycles. This observation suggests that FTC not only increases the number of pores in the rock but also significantly enhances the openness and connectivity of these pores.

3.2. Rock Mass Loss Rate and P-Wave Velocity

Ultrasonic wave speed, as an effective means of non-destructive rock testing, can effectively explore the development of the interior of the rock, according to the ultrasonic wave speed difference, in order to judge the integrity of the interior of the rock, as shown in Equation (1).

$$K_m = \frac{m_{s1} - m_{s2}}{m_{s1}} \times 100\% \tag{1}$$

Figure 3 shows the relationship between rock mass loss rate and longitudinal wave velocity with the number of FTCs; the values in the figure are the average values of each group of rocks under the same number of FTCs. By observing Figure 3, we can find that, with the reduction of FTC frequency, the loss of rock sample quality is also gradually reduced. However, in the first 10 FTCs, the variation range of rock mass loss rate is significantly greater than that in the subsequent FTCs (20, 30, 40). This is because the sample itself is in a certain breezy state when the damage frequency of the FTC is small, the rock sample is mainly concentrated in the surface layer, and the material spalling on the rock is the main reason for the substantial reduction of its mass. With the increase in FTC frequency, the damage was transferred to the inside of the specimen. The damage of the internal material does not leave the specimen, while the damage of the dry rock sample is 2813 m/s. The reason is that the pores of the saturated water sample are filled with water, and the wave speed in water is much larger than that in air. Second, the wave impedance of water is closer to the solid skeleton of rock than that of air. According to the elastic wave propagation theory, this will weaken the refraction and reflection of stress waves to a certain extent. The propagation path of the stress wave in the rock sample is shortened because of the reduction, which results in the velocity of the P-wave slowing down with time. The decrease in P-wave velocity is not only affected by the freeze–thaw cycle. It is also closely related to its internal structure and mineral composition. In the early FTC stage, the intercrystalline cohesion in rock is weaker than the frost heave force, and a large number of argillaceous cements in rock will significantly soften and argillify in the early FTC stage, which also has a significant impact on the P-wave velocity. The cementation between quartz particles will be weakened and the wave velocity will decrease.

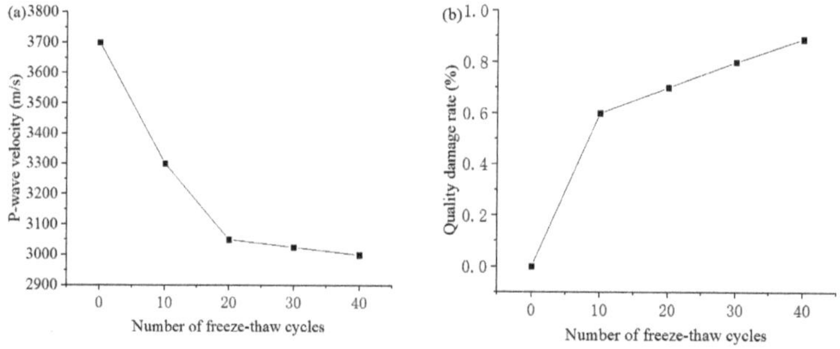

Figure 3. The relationship between rock mass loss rate and p-wave velocity with the number of FTCs: (**a**) P-wave velocity; (**b**) Quality damage rate.

3.3. Analysis of Strength Loss Rate

Table 1 shows the average value of saturated uniaxial compressive strength and loss rate of samples with different FTCs. The experimental data show that with the increase in FTC number, the saturated compressive strength of specimens decreases obviously, and the same FTC number has a similar weakening effect on the saturated compressive strength of different types of rocks.

Table 1. Mean saturated strength and strength loss rate of samples after different FTCs.

Mean Compressive Strength at Unfrozen and Thawing Saturation (MPa)	0 Times of FTC		10 Times of FTC		20 Times of FTC		30 Times of FTC		40 Times of FTC	
	Saturated Water Compressive Strength (MPa)	Loss Rate of Freeze–Thaw Strength (%)	Saturated Water Compressive Strength (MPa)	Loss Rate of Freeze–Thaw Strength (%)	Compressive Strength in Saturated Water (MPa)	Loss Rate of Freeze–Thaw Strength (%)	Compressive Strength in Saturated Water (MPa)	Loss Rate of Freeze–Thaw Strength (%)	Saturated Water Compressive Strength (MPa)	Loss Rate of Freeze–Thaw Strength (%)
125.7	119.5	0	116.9	7.00	113.2	9.94	111.7	11.14	105.6	15.99

With the increase in the number of FTCs, the mechanical properties of rock blocks basically show a downward trend, but there are also abnormal conditions, e.g., the compressive strength of the granulated rock group after 20 FTCs is greater than that after 10 FTCs, which shows the discrete data and non-uniformity of rock samples during the experiment.

3.4. Static Compression Test of Rock Damaged by FTC

The Φ50 mm × 100 mm cylinder rock samples obtained had undergone 0 (no freeze–thaw damage, i.e., the water-saturated sample), 10, 20, 30, and 40 FTC damage, respectively. The samples under different freeze–thaw test conditions were all in a water-saturated state. The changes in static elastic modulus, Poisson's ratio, and freeze–thaw coefficient of the rock sample along with the number of FTCs are shown in Figure 4; the values in the figure are the average values of each group of rocks under the same number of FTCs.

$$K_f = \frac{\overline{R_f}}{\overline{R_s}} \quad (2)$$

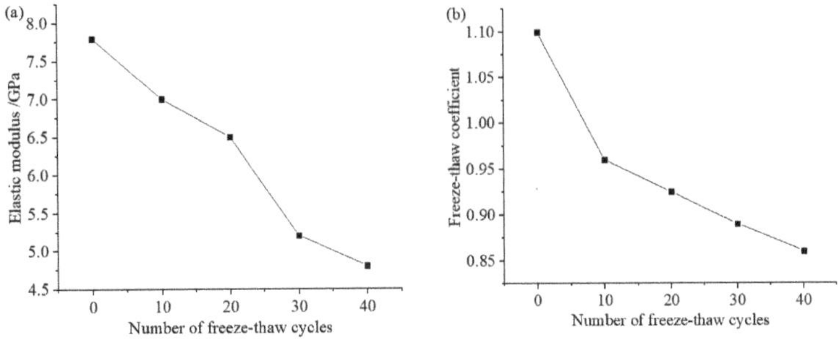

Figure 4. Changes in static elastic modulus and freeze–thaw coefficient of rocks under different cyclic damage states of freeze–thaw: (**a**) Elastic modulus; (**b**) Freeze–thaw coefficient.

The experimental results show that. with the increase in FTC times, the Poisson's ratio of rock samples increases gradually, and the freeze–thaw coefficient increases but then decreases. The quality loss rate of the samples also increased. This indicates that FTC damage gradually develops from the surface to the interior, which is consistent with the law of rock mass affected by temperature under a natural environment. It can be seen that FTC damage is a process that develops from the surface to the interior, which is also consistent with the effect of temperature on in situ rock mass under a natural environment.

4. Constitutive Relationship and Damage Variables

4.1. Establishment of the Constitutive Model

Damage theory is widely used in damage models, among which a large number of damage models are established based on the assumption of Lemaitre strain equivalence, as shown in Equation (3) below. This hypothesis shows that the attenuation of the shear strength of the rock is caused by the destruction of the internal mechanical structure of the rock mass. When the rock is subjected to shear failure, the material bearing the shear strength inside can be divided into two parts: the undamaged material and the cavity. The cavity is the part that does not bear any shear strength. As shown in Figure 5, the blank part is the undamaged part of the material, and the shaded part is the damaged part of the material.

$$\tau_1 = \tau_1'(1 - D) \quad (3)$$

where τ_1 is the nominal shear stress, τ_1' is the effective shear stress, and D is the damage percentage. According to Equation (4), when the rock is destroyed, i.e., when D = 1, the nominal shear stress of the rock and soil must be 0, i.e., the residual strength must be 0, which is contrary to the actual situation, as shown in Equation (4) below.

$$\tau_1 = \tau_1'(1 - D) + \tau_1^r D \tag{4}$$

where τ_1'' is the shear stress of the damaged part of the material, and the meaning of other physical quantities is the same as that of the model shown in Equation (4). According to Equation (4), when the rock and soil material is in a state of destruction, i.e., when D = 1, and $\tau_1 = \tau_1'(1 - D) = 0$, so $\tau_1 = \tau_1''$. The stress state and rock strength parameters of some undamaged rocks meet the molar Coulomb failure criterion, as shown in Equations (5) and (6) and Figure 6.

$$\tau' = \frac{1}{2}(\sigma_1' - \sigma_3') \sin 2\theta_r \tag{5}$$

$$\sigma' = \frac{1}{2}(\sigma_1' + \sigma_3') + \frac{1}{2}(\sigma_1' - \sigma_3') \cos 2\theta_r \tag{6}$$

θ_r is the residual friction Angle of the fracture zone, C_r is the residual cohesion force, respectively, and σ_1^r is the first principal stress. Equations (7) and (8) can be obtained from Figure 5.

$$\tau_1 A = \tau_1' A_1 + \tau_1^r A_2 \tag{7}$$

$$A = A_1 + A_2 \tag{8}$$

Equation (4) is a new damage model for rock materials. If $\sigma' = \sigma$, Equation (9) can be obtained by combining Equations (6) and (10).

$$\sigma = \frac{1}{2}(\sigma_1' + \sigma_3') + \frac{1}{2}(\sigma_1' - \sigma_3') \cos 2\theta_r \tag{9}$$

In addition, since it is the horizontal residual shear strength of the rock, it also conforms to the Mohr–Coulomb criterion, i.e., τ_1^r

$$\sigma_1^r = \sigma_3^r \tan^2(45° + \frac{\theta_r}{2}) + 2C_r \tan(45° + \frac{\theta_r}{2}) \tag{10}$$

$$\tau_1^r = \frac{1}{2}(\sigma_1^r - \sigma_3^r) \sin(90° + \theta_r) \tag{11}$$

where σ_3^r is the third principal stress determined according to the residual strength of the rock.

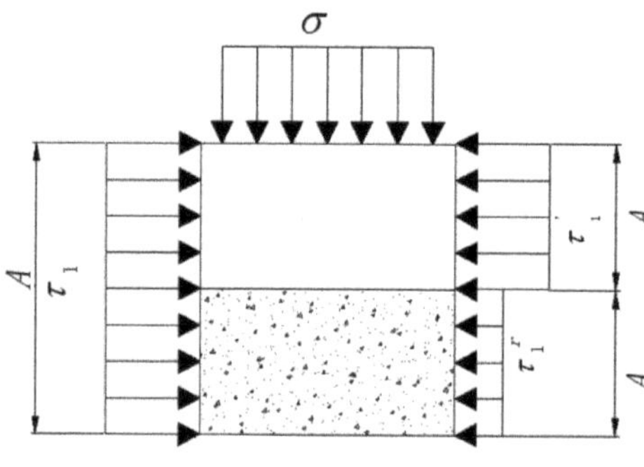

Figure 5. Shear stress diagram of rock microelements.

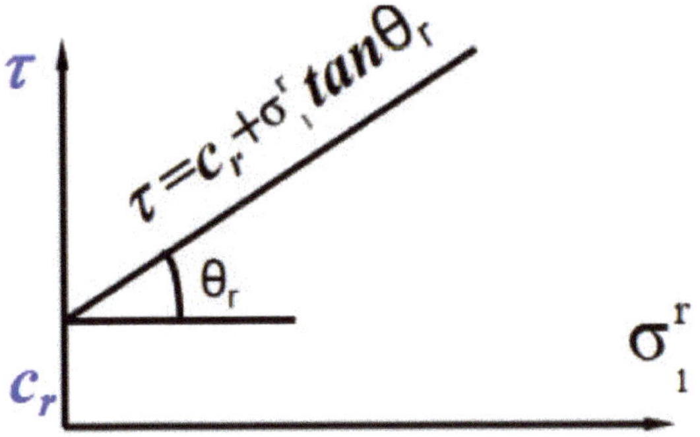

Figure 6. Diagram of the Mohr–Coulomb criterion.

Therefore, by combining Equation (11) with Equation (10), Equation (12) can be obtained:

$$\tau_1^r = \frac{1}{2}\frac{2\sin(45^\circ + \frac{\theta_r}{2})\sigma_3^r + 2C_r\cos(45^\circ + \frac{\theta_r}{2})}{1 - \sin(45^\circ + \frac{\theta_r}{2})}\sin(45^\circ + \frac{\theta_r}{2}) \quad (12)$$

The shear stress of the undamaged part of the rock also conforms to the Mohr–Coulomb criterion under different normal stress conditions, as shown in Equation (13) below.

$$\tau_1' = \sigma\tan\theta + C \quad (13)$$

Thus, we can obtain Equations (14) and (15) by substituting Equations (12) and (13) into Equation (7).

$$\tau_1 = (\sigma\tan\theta + C)(1 - D) + B_1 D \quad (14)$$

$$B_1 = \frac{1}{2}\frac{2\sin(45^\circ + \frac{\theta_r}{2})\sigma_3^r + 2C_r\cos(45^\circ + \frac{\theta_r}{2})}{1 - \sin(45^\circ + \frac{\theta_r}{2})}\sin(45^\circ + \frac{\theta_r}{2}) \quad (15)$$

$$\gamma = \frac{2C\cos\theta}{\sin\theta - 1} \quad (16)$$

$$F = \frac{2\sigma - \sigma_3'(1 - \cos 2\theta)}{1 + \cos 2\theta} + \beta\sigma_3' + \gamma \quad (17)$$

$$D = \begin{cases} 1 - \exp[-(F/F_0)^m] & F > 0 \\ 0 & F \le 0 \end{cases} \quad (18)$$

In Equation (18), m, and F_0 are Weibull distribution parameters of rock microunit strength F.

4.2. Determination of Parameters of Rock Damage Constitutive Model

Combine Equation (18) with Equation (14), as shown in Equation (19). It is also assumed that, under different normal stress conditions, the peak shear stress of the rock is $\sigma_{SC}.\tau_{sc}$

$$\left.\frac{\partial\tau}{\partial\sigma}\right|_{\tau=\tau_{SC},\sigma=\sigma_{SC}} = 0 \quad (19)$$

From the partial decomposition of Equation (12), Equation (20) is obtained:

$$\frac{\partial\tau}{\partial\sigma} = \tan\theta_r - D\tan\theta_r + (B_1 - C - \sigma\tan\theta_r)\frac{\partial D}{\partial\sigma} \quad (20)$$

Therefore, the combination of Equations (19) and (20) can be derived from Equation (21):

$$\left.\frac{\partial D}{\partial\sigma}\right|_{\tau=\tau_{SC},\sigma=\sigma_{SC}} = \frac{(D_{SC} - 1)\tan\theta_r}{(B_1 - C - \sigma_{SC}\tan\theta_r)} \quad (21)$$

Equations (20) and (22) can be obtained by partial decomposition of Equation (18):

$$\frac{\partial D}{\partial \sigma} = \exp[-(\frac{F}{F_0})^m][m(\frac{F_{sc}}{F_0})^{m-1}]\frac{1}{F_0}\frac{\partial F}{\partial \sigma} \tag{22}$$

Equation (23) can be solved by partial differential of Equation (17):

$$\frac{\partial F}{\partial \sigma} = \frac{2}{1+\cos 2\theta} \tag{23}$$

Equations (24) and (25) can be obtained by combining Equations (18) and (23).

$$\exp[-(\frac{F}{F_0})^m] = 1 - D \tag{24}$$

$$(\frac{F}{F_0})^{m-1} = -\frac{F}{F_0}\ln(1-D) \tag{25}$$

Combine Equation (21) with Equation (25) to obtain Equation (26).

$$m = \frac{F_{sc}\tan\theta(1+\cos 2\theta_r)}{2(B_1 - C - \sigma_{SC}\tan\theta_r)\ln(1-D_{SC})} \tag{26}$$

where, when $\tau = \tau_{SC}$, $\sigma = \sigma_{SC}$, the micro-element strength F of the rock can be determined by Equation (17), i.e.,

$$F_{sc} = \frac{2\sigma_{SC} - \sigma_3'(1-\cos 2\theta_r)}{1+\cos 2\theta_r} + \beta\sigma_3' + \gamma \tag{27}$$

Combine Equations (24) and (27) to obtain Equation (28).

$$F_0 = F_{sc}[-\ln(1-D_{SC})]^{\frac{-1}{m}} \tag{28}$$

Combine Equations (12)–(28) to obtain Equation (29). $\sigma = \sigma_{SC}$ $\tau = \tau_{SC}$

$$D_{SC} = \frac{\tau_{SC} - (\sigma_{SC}\tan\theta_r + C)}{B_1 - (\sigma_{SC}\tan\theta_r + C)} \tag{29}$$

$$D_n = 1 - \frac{E_n}{E_0} \tag{30}$$

where E_0 and E_n are elastic modulus of rock before and after receiving FTC for n times, respectively. The damage of rock compressive strength caused by FTC was coupled and analyzed, i.e., the total damage variable expression of FTC was obtained, as shown in Equation (31).

$$D_m = D + D_n - DD_n \tag{31}$$

By combining Equations (24), (30) and (31), the total evolution equation of rock damage variables after coupling of FTC loads of different times is obtained, as shown in Equation (32):

$$D_m = 1 - \frac{E_n}{E_0}\exp[-(\frac{F}{F_0})^m] \tag{32}$$

When $\gamma_1 = \gamma_n$, the corresponding value of F is F_C, which is obtained by combining Equations (24), (25) and (32), as shown in Equation (33).

$$(\frac{F_c}{F_0})^m = -\ln\left\{\left[1 - \frac{\tau_{\max} - E_n\gamma_n}{\tau_r - E_n\gamma_n}\right]\cdot\frac{E_0}{E_n}\right\} = B \tag{33}$$

Equations (36) and (37) are obtained from Equations (27) and (34).

$$m = \frac{E_n\gamma_n}{(E_n\gamma_n - \tau_r)B} \tag{34}$$

$$F_0 = F_c\left[\left(1 - \frac{\tau_r}{E_n\gamma_n}\right)\cdot m\right]^{\frac{1}{m}} \tag{35}$$

Equations (23)–(25) are the relationship between model parameters and rock mechanics characteristic parameters. The macroscopic mechanical properties of rock contained in the parameter expressions can be easily obtained from the conventional mechanical properties tests. Especially when it is difficult to obtain the whole stress–strain curve of rock due to the limitation of test conditions, the constitutive model reflecting the whole process of rock failure can still be obtained. According to Equations (32), (34) and (35), the total evolution equation of rock damage variables after different numbers of FTCs can be obtained, as shown in Equation (36).

$$D_m = 1 - \frac{E_n}{E_0}\exp\left\{\left[\left(\frac{E_n\gamma_1}{\tau_r - E_n\gamma_n}\right)\left[\frac{(E_n\gamma_n - \tau_r)\left\{-\ln\left[\left(1 - \frac{\tau_{max} - E_n\gamma_n}{\tau_r - E_n\gamma_n}\right)\cdot\frac{E_0}{E_n}\right]\right\}}{E_n\gamma_n}\right]\right]\left(\frac{E_n\gamma_1 - \tau_r}{E_n'\gamma_n - \tau_r}\right)^{\frac{E_n\gamma_n}{(E_n\gamma_n - \tau_r)\{-\ln\{[1 - \frac{\tau_{max} - E_n\gamma_n}{\tau_r - E_n\gamma_n}]\cdot\frac{E_0}{E_n}\}\}}}\right\} \tag{36}$$

The damage evolution rate of frequency FTC under the action of different frequency FTC can be derived by Equation (36), as shown in Equation (37).

$$\dot{D_m} = (1 - D_m) * \frac{{E_n}^2\gamma_n(E_n\gamma_1 - \tau_r)^{m-1}}{(E_n\gamma_n - \tau_r)^{m+1}} \tag{37}$$

5. Discussion

5.1. Comparison of Theoretical Models

In order to verify and illustrate the rationality and superiority of the damage constitutive model proposed in this paper to simulate the whole process of rock strain softening deformation, the stress–strain curves of the rock under the conditions of confining pressure of 2 MPa and number of FTCs of 10 were analyzed, respectively, by using the method described in the literature [21], as shown in Figure 7.

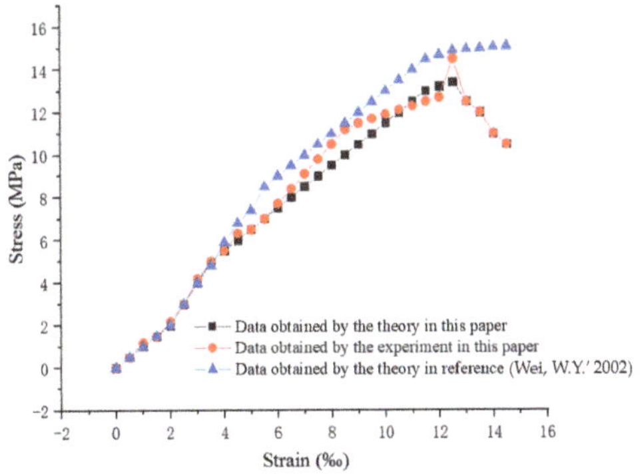

Figure 7. Comparison between test and theoretical curves [21].

(1) Figure 7 shows the theoretical curves of the rock damage constitutive model under various confining pressure and freeze–thaw cycle conditions calculated by using the mechanical test results of the freeze–thaw cycle shown in Tables 1 and 2 and compared with the experimental curves. Although the stress–strain curve obtained by the model in

reference [21] can better reflect the strain softening characteristics of rock, it cannot reflect the residual strength characteristics of rock after failure.

Table 2. Parameter values of cumulative damage model of rock shear strength under different normal stresses and FTCs.

Number of FTCs	Confining Pressure/MPa	Extreme Stress /MPa	Modulus of Elasticity /GPa
10	0	4.23	0.867
10	2	14.5	1.295
10	5	19.6	1.452
10	10	24.3	1.565
20	0	3.8	0.761
20	2	12.7	1.156
20	5	18.9	1.289
20	10	23.5	1 325
30	0	3.74	0.529
30	2	11.3	0.890
30	5	18.1	1.066
30	10	22.9	1.240
40	0	3.3	0.515
40	2	10.5	0.710
40	5	17.2	0.917
40	10	21.2	0.932

(2) The proposed model can not only reflect the strain softening characteristics and residual strength characteristics of rock after failure, which is in good agreement with the measured curve, but can also simulate the mechanical behavior of rock under various confining pressures and freeze–thaw cycles in a unified function form and does not contain unconventional mechanical parameters. Therefore, the proposed model is more reasonable and superior to the existing models.

5.2. Damage Percentage Analysis of Rocks under Different FTCs

Based on the verified cumulative damage constitutive model describing the decay of rock shear strength, combined with Equation (36) for calculating rock damage percentage, the changes in rock damage mechanical properties under different freeze–thaw cycles can be obtained, as shown in Figure 8.

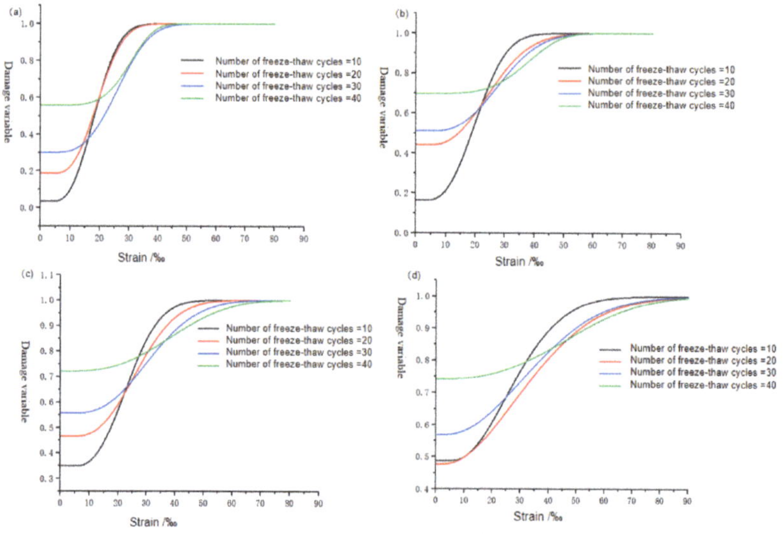

Figure 8. Damage percentage of rocks under different FTCs: (**a**) Confining pressure = 0; (**b**) Confining pressure = 2 MPa; (**c**) Confining pressure = 5 MPa; (**d**) Confining pressure = 10 MPa.

(1) It can be seen from Figure 8 that, under the same number of freeze–thaw cycles, the initial damage percentage of rocks gradually increases with the increase in freeze–thaw cycles. In addition, it can be seen from Figure 8a that with the increase in freeze–thaw cycles, the strain value required for the sliding zone soil to reach the damage variable of 1 under the same normal stress load also increases, i.e., as shown in Figure 8c, when the rock passes 10, 20, 30 and 40 times and the freeze–thaw cycle reaches the failure value 1, its shear strain is 52‰, 68‰, 75‰ and 81‰, respectively. A larger number of freeze–thaw cycles can lead to a larger shear displacement of the rock, and the change is obvious. Under the same number of freeze–thaw cycles, a larger number of freeze–thaw cycles will cause greater damage to the rock material.

(2) It can be seen from Figure 8b–d that, with the increase in the number of freeze–thaw cycles, the change in rock damage value tends to be gentle under different freeze–thaw cycles. At the same time, when the peak value of the curve describing the change of damage value reaches 1, the required final strain value gradually increases, indicating that more cracks appear in the rock with the increase in the number of freeze–thaw cycles. With the increase in the number of freeze–thaw cycles, cracks appear continuously in the rock, and obvious plastic zones appear along with them, making the rock produce plastic hardening characteristics under the accumulation of freeze–thaw cycles. Meanwhile, with the increase in the number of freeze–thaw cycles, the rock is damaged to different degrees. With the increase in freeze–thaw cycles, the greater the damage degree of the rock, the more damaged parts in the rock. The energy required to produce further damage is also greater, so the ductile failure characteristics of the rock are becoming more and more obvious.

5.3. Analysis of Damage Evolution Rate of Rocks under Different FTCs

Combined with Equation (37), we analyzed the evolution law of rock damage change rate under different FTCs. As shown in Figure 9, with the increase in FTCs, the maximum damage rate of the rock decreases because of the time delay in the occurrence of damage. At the same time, the increase in FTCs increases the peak strain value of the rock damage change rate curve, indicating that the ductile failure characteristics of rock are gradually significant.

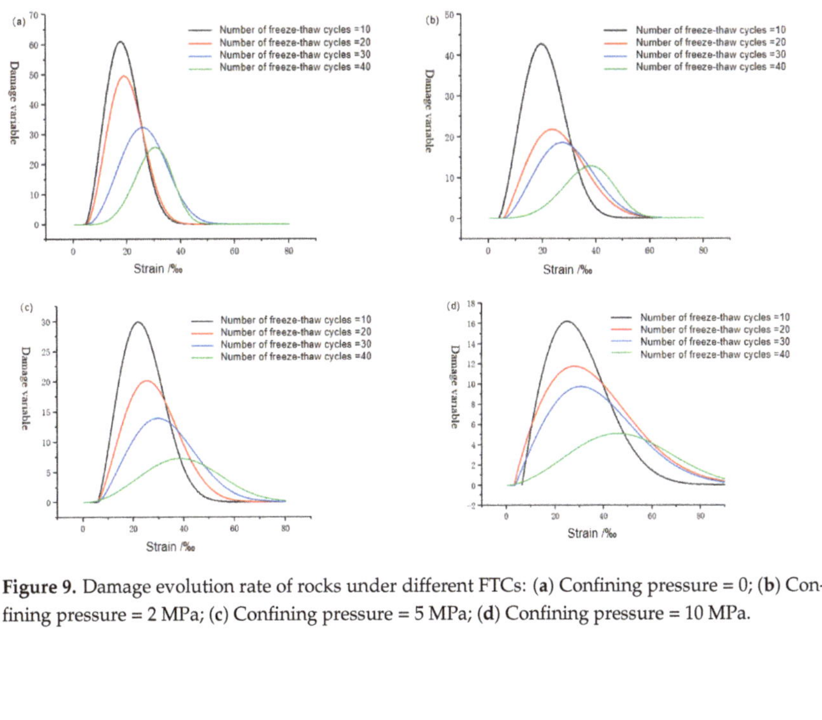

Figure 9. Damage evolution rate of rocks under different FTCs: (**a**) Confining pressure = 0; (**b**) Confining pressure = 2 MPa; (**c**) Confining pressure = 5 MPa; (**d**) Confining pressure = 10 MPa.

6. Conclusions

The stress-strain curves of rocks with various FTC strengths were obtained by dynamic compression tests under different loading rates and confining pressure. The study found:

(1) Through different FTC tests, the physical properties of rock samples change significantly. The FTC has a great influence on rock parameters, including saturation density, porosity and P-wave velocity. With the increase in FTC frequency, the mineral particles on the surface of the sample peel off, and the internal particles soften, resulting in the decrease in saturated mass, the expansion of pore structure, and the increase in mass loss rate. At the same time, the damage accumulation reduces the wave velocity.

(2) With the increase in the number of FTCs, the change in rock damage value gradually flattens, the final strain value increases, and the cracks increase. With the increase in FTCs, the peak strain value of rock damage change rate increases, indicating that the ductile failure characteristics are more and more significant.

(3) Although this study involves four FTC conditions, the results may be limited. In the future, by increasing the number of FTCs and confining pressure conditions, the mechanical properties of rocks under the coupling of freeze–thaw and confining pressure can be more comprehensively understood, and the universality of test results can be improved.

Author Contributions: Conceptualization, Y.L., Z.W., H.C. and T.W.; methodology, Y.L.; software, Y.L.; validation, Z.W.; writing-original draft preparation, T.W. and Z.W.; writing—review and editing, T.W. and H.C.; project administration, H.C. and H.C.; funding acquisition, T.W. and H.C. All authors have read and agreed to the published version of the manuscript.

Funding: This study was sponsored by the National Natural Science Foundation of China (Grant No. 52308393) and the State Key Laboratory of Precision Blasting and Hubei Key Laboratory of Blasting Engineering, Jianghan University (Grant No. PBSKL2023B2).

Data Availability Statement: The raw data supporting the conclusions of this article will be made available by the authors on request.

Conflicts of Interest: The authors declare no conflicts of interest.

References

1. Zeng, Z.; Kong, L.; Wang, M.; Wang, J. Effects of remolding and wetting-drying-freezing-thawing cycles on the pore structures of Yanji mudstones. *Cold Reg. Sci. Technol.* **2020**, *174*, 103037. [CrossRef]
2. Soltani, A. Comments on "effect of wetting–drying–freezing–thawing cycles on the swelling behaviour of the Yanji mudstone"[Environ Earth Sci (2019) 78: 435]. *Environ. Earth Sci.* **2019**, *78*, 512. [CrossRef]
3. Wang, J.; Song, C.; Hou, A.; Miao, Y.; Yang, G.; Zhang, J. Effects of freezing-thawing cycle on peatland active organic carbon fractions and enzyme activities in the Da Xing'anling Mountains, Northeast China. *Environ. Earth Sci.* **2014**, *72*, 1853–1860. [CrossRef]
4. Zeng, Z.; Kong, L. Effect of wetting–drying–freezing-thawing cycles on the swelling behavior of the Yanji mudstone. *Environ. Earth Sci.* **2019**, *78*, 435. [CrossRef]
5. Martinez, J.; Benavente, D.; Gomez-Heras, M.; Marco-Castaño, L.; García-del-Cura, M.Á. Non-linear decay of building stones during freeze-thaw weathering processes. *Constr. Build. Mater.* **2013**, *38*, 443–454. [CrossRef]
6. Yamabe, T.; Neaupane, K.M. Determination of some thermo-mechanical properties of Sirahama rock under subzero temperature condition. *Int. J. Rock Mech. Min. Sci.* **2001**, *38*, 1029–1034. [CrossRef]
7. Park, C.; Synn, J.H.; Shin, H.S.; Cheon, D.S.; Lim, H.D.; Jeon, S.W. Experimental study on the thermal characteristics of rock at low temperatures. *Int. J. Rock Mech. Min. Sci.* **2004**, *41*, 81–86. [CrossRef]
8. Sondergeld, C.H.; Rai, C.S. Velocity and resistivity changes during FTC in Berea rock. *Geophysics* **2007**, *72*, E99–E105. [CrossRef]
9. Del Roa, L.M.; Lopez, F.; Esteban, F.J.; Tejado, J.J.; Mota, M.I.; González, I.; Ramos, A.; San Emeterio, J.L. Ultrasonic study of alteration processes in granites caused by freezing and thawing. In Proceedings of the IEEE Ultrasonics Symposium 2005, Rotterdam, The Netherlands, 18–21 September 2005.
10. Janach, W. The role of bulking in brittle failure of rocks under rapid compression. *Int. J. Rock Mech. Min. Sci. Geomech. Abstr.* **1976**, *13*, 177–186. [CrossRef]
11. Lankford, J. The role of tensile microfracture in the strain rate dependence of compressive strength of fine-grained limestone—Analogy with strong ceramics. *Int. J. Rock Mech. Min. Sci. Geomech. Abstr.* **1981**, *18*, 173–175. [CrossRef]
12. Olsson, W.A. The compressive strength of tuff as a function of strain rate from 10^{-6} to 10^{3}/sec. *Int. J. Rock Mech. Min. Sci. Geomech. Abstr.* **1991**, *28*, 115–118. [CrossRef]
13. Zhou, K.P.; Li, B.; Li, J.L.; Deng, H.W.; Feng, B.I.N. Microscopic damage and dynamic mechanical properties of rock under freeze-thaw environment. *Trans. Nonferrous Met. Soc. China* **2015**, *25*, 1254–1261. [CrossRef]

14. Kawakata, H.; Cho, A.; Yanagidani, T.; Shimada, M. The observations of faulting in westerly granite under triaxial compression by X-ray CT scan. *Int. J. Rock Mech. Min. Ences* **1997**, *34*, 151.e1–151.e12. [CrossRef]
15. Kawakata, H.; Cho, A.; Kiyama, T.; Yanagidani, T.; Kusunose, K.; Shimada, M. Three-dimensional observations of faulting process in Westerly granite under uniaxial and triaxial conditions by X-ray CT scan. *Tectonophysics* **1999**, *313*, 293–305. [CrossRef]
16. Kyoya, T.; Baek, Y.; Onda, C. On deterioration of Oya-tuff in freezing-thawing cycle and its quantitative description. *Doboku Gakkai Ronbunshu* **2000**, *652*, 103–114. [CrossRef] [PubMed]
17. Sun, Y.; Zhai, C.; Qin, L.; Xu, J.; Yu, G. Coal pore characteristics at different freezing temperatures under conditions of freezing-thawing cycles. *Environ. Earth Sci.* **2018**, *77*, 525. [CrossRef]
18. Tiwari, N.; Satyam, N.; Singh, K. Effect of Curing on Micro-Physical Performance of Polypropylene Fiber Reinforced and Silica Fume Stabilized Expansive Soil Under Freezing Thawing Cycles. *Sci. Rep.* **2020**, *10*, 7624. [CrossRef] [PubMed]
19. Sellares, J.; Mas, S.; Melo, E.; Sanchez, F.; Marin, J.; Gea, J.; Barreiro, E. Oxidative stress time course in the rat diaphragm after freezing-thawing cycles. *Respir. Physiol. Neurobiol.* **2006**, *155*, 156–166. [CrossRef] [PubMed]
20. *GBT 50266-2013*; Test Method Standard for Engineering Rock Mass. China Planning Press: Beijing, China, 2013. (In Chinese)
21. Wei, W.Y.; Wei, L.D. Study on statistical damage constitutive model of rock. *Chin. J. Rock Mech. Eng.* **2002**, *21*, 787–791. (In Chinese)

MDPI AG
Grosspeteranlage 5
4052 Basel
Switzerland
Tel.: +41 61 683 77 34

Applied Sciences Editorial Office
E-mail: applsci@mdpi.com
www.mdpi.com/journal/applsci

www.ingramcontent.com/pod-product-compliance
Lightning Source LLC
LaVergne TN
LVHW070451061126
851858LV00011B/214